Introduction to Health and Safety at Work

Introduction to Health and Safety at Work

The Handbook for the NEBOSH National General Certificate

Phil Hughes MSc, FIOSH, RSP, Chairman NEBOSH 1995–2001

Ed Ferrett PhD, BSc, (Hons Eng), CEng, MIMechE, MIEE, MI0SH, Deputy Chairman NEBOSH

ELSEVIER
BUTTERWORTH
HEINEMANN

AMSTERDAM   BOSTON   HEIDELBERG   LONDON   NEW YORK   OXFORD
PARIS   SAN DIEGO   SAN FRANCISCO   SINGAPORE   SYDNEY   TOKYO

Elsevier Butterworth-Heinemann
Linacre House, Jordan Hill, Oxford OX2 8DP
200 Wheeler Road, Burlington MA 01803

First published 2003
Reprinted 2003 (twice)

British Library Cataloguing in Publication Data
Hughes, Philip
    Introduction to health and safety at work: the handbook for the NEBOSH
    National General Certificate
    1. Industrial hygiene – Examinations – Study guides
    2. Industrial safety – Examinations – Study guides
    I. Title II. Ferrett, Ed
    363.1′1

Library of Congress Cataloguing in Publication Data
A catalogue record for this book is available from the Library of Congress

For information on all Butterworth-Heinemann publications
visit our website at www.bh.com

ISBN 0 7506 5730 8

Typeset by Keyword Typesetting Services, Wallington, Surrey
Printed and bound in Italy

Contents

Contents

About the authors

Phil Hughes is a well-known UK safety professional with over thirty years worldwide experience as Head of Environment, Health and Safety at two large multinationals, Courtaulds and Fisons. Phil started in health and safety in the Factory Inspectorate at the Derby District in 1969 and moved to Courtaulds in 1974. He joined IOSH in that year and became Chairman of the Midland Branch, National Treasurer and then President in 1990–1. Phil has been very active on the NEBOSH Board for over ten years and served as Chairman from 1995–2001. He is also a Professional Member of the American Society of Safety Engineers and has lectured widely throughout the world. Phil received the RoSPA Distinguished service award in May 2001 and became a director in 2002.

Ed Ferrett is an experienced health and safety consultant who has practised for over twenty years. He spent thirty years in Higher and Further Education retiring as the Head of the Faculty of Technology of Cornwall College in 1993. Since then he has been an independent consultant to several public and private sector organizations, the Regional Health and Safety Adviser for the Government Office (West Midlands) and Vice Chair of NEBOSH. Ed is currently Chair of West of Cornwall Primary Care NHS Trust. He has delivered many health and safety courses and is a manager of NEBOSH courses at the Cornwall Business School. Ed is a Chartered Engineer and a Member of IOSH.

Preface

The legal health and safety requirements for places of work are numerous and complex; it is the intention of the authors to offer an introduction to the subject for all those who have the maintenance of good health and safety standards as part of their employment duties or those who are considering the possibility of a career as a health and safety professional. Health and safety is well recognized as an important component of the activities of any organization, not only because of the importance of protecting people from harm, but also because of the growth in the direct and indirect costs of accidents which have exceeded retail price inflation by a considerable amount in the last few years with the number of civil claims and awards increasing each year. It is very important that the basic health and safety legal requirements are clearly understood by all organizations, whether public or private, large or small.

A good health and safety performance is normally only achieved when health and safety is effectively managed so that significant risks are identified and reduced by adopting appropriate high quality control measures.

This *Introduction to Health and Safety at Work* is based on the QCA (Qualification and Curriculum Authority) accredited NEBOSH General Certificate syllabus as revised in 2002. It has been developed specifically for students who are studying for the NEBOSH National General Certificate in Occupational Safety and Health. It was felt appropriate to produce a textbook that mirrored the General Certificate syllabus in its revised unitized form and in a single volume to the required breadth and depth. The syllabus, which follows the general pattern for health and safety management set by the Health and Safety Executive in their guidance HSG 65, is risk and management based so it does not start from the assumption that health and safety is best managed by looking first at the causes of failures. Fortunately, failures such as accidents and ill-health are relatively rare and random events in most workplaces. A full copy of the syllabus and guide can be obtained from NEBOSH direct.

The book is also intended as a useful reference guide for managers and directors with health and safety responsibilities and for safety representatives. The final chapter, which summarizes all the most commonly used Acts and Regulations, was written to provide an easily accessible reference source and a basic understanding for students during and after the course and many others in industry and commerce such as managers, supervisors and safety representatives.

Acknowledgements

Throughout the book, definitions used by the relevant legislation, the Health and Safety Commission, the Health and Safety Executive and advice published in Approved Codes of Practice or various Health and Safety Commission/Executive publications have been utilized. Most of the references produced at the end of each Act or Regulation summary in Chapter 17, are drawn from HSE Books range of publications.

At the end of each chapter, there are some examination questions taken from recent NEBOSH National General Certificate papers. Some of the questions may include topics which are covered in more than one chapter. The answers to these questions are to be found within the preceding chapter of the book. NEBOSH publishes a very full examiners' report after each public examination which gives further information on each question. Most accredited NEBOSH training centres will have copies of these reports and further copies may be purchased directly from NEBOSH. The authors would like to thank NEBOSH for giving them permission to use these questions.

The authors' grateful thanks go to Liz Hughes for her proofreading and patience, and to Jill Ferrett for encouraging Ed Ferrett to keep at the daunting task of text preparation. We would also like to thank Stephen Vickers, Chief Executive of NEBOSH, for his unwavering support for the project and various HSE staff for their generous help and advice.

List of principal abbreviations

Most abbreviations are defined within the text. Abbreviations are not always used of it is not appropriate within the particular context of the sentence. The most commonly used ones are as follows:

ACOP	Approved Code of Practice
CBI	Confederation of British Industry
CDM	Construction (Design and Management) Regulations
CHIP	Chemicals (Hazard Information and Packaging) Regulations
COSHH	Control of Substances Hazardous to Health Regulations
DSE	Display screen equipment
EPA	Environmental Protection Act 1990
EMAS	Employment Medical Advisory Service
EU	European Union
FPA	Fire Precautions Act 1971
HSC	Health and Safety Commission
HSE	Health and Safety Executive
HSW Act	Health and Safety at Work etc. Act 1974
IOSH	Institution of Occupational Safety and Health
LOLER	Lifting Operations and Lifting Equipment Regulations
MEL	Maximum exposure limit
MHSW	Management of Health and Safety at Work Regulations
MHOR	Manual Handling Operations Regulations
NEBOSH	National Examination Board in Occupational Safety and Health
OEL	Occupational exposure limit
OES	Occupational exposure standard
PPE	Personal protective equipment
PUWER	Provision and Use of Work Equipment Regulations
RIDDOR	Reporting of Injuries, Diseases and Dangerous Occurrences Regulations
ROES	Representative(s) of employee safety
STEL	Short-term exposure limit
TUC	Trades Union Congress
WRULD	Work-related upper limb disorder

Health and safety foundations

1.1 Introduction

Occupational health and safety is relevant to all branches of industry, business and commerce and includes traditional industries, information technology companies, the National Health Service, care homes, schools, universities, leisure facilities and offices.

The purpose of this unit is to introduce the foundations on which appropriate health and safety systems may be built. Occupational health and safety affects all aspects of work and may simply require a trained competent manager in a low hazard organization. In a high hazard manufacturing plant, many different specialists, such as engineers (electrical, mechanical and civil), lawyers, medical doctors and nurses, trainers, work planners and supervisors, may be required to assist the professional health and safety practitioner in ensuring that there are satisfactory health and safety standards within the organization.

There are many obstacles to the achievement of good standards. The pressure of production or performance targets, financial constraints and the complexity of the organization are typical examples of such obstacles. However, there are some powerful incentives for organizations to strive for high health and safety standards. These incentives are moral, legal and economic.

1.2 Some basic definitions

Before a detailed discussion of health and safety issues can take place, some basic occupational health and safety definitions are required:

Health – the protection of the bodies and minds of people from illness resulting from the materials, processes or procedures used in the workplace.
Safety – the protection of people from physical injury. The borderline between health and safety is ill-defined and the two words are normally used together to indicate concern for the physical and mental well-being of the individual at the place of work.
Welfare – the provision of facilities to maintain the health and well-being of individuals at the workplace. Welfare facilities include washing and sanitation arrangements, the provision of drinking water, heating, lighting, accommodation for clothing, seating (when

Figure 1.1 At work.

required by the work activity), eating and rest rooms. First aid arrangements are also considered as welfare facilities.

Occupational or work-related ill-health – is concerned with those illnesses or physical and mental disorders that are either caused or triggered by workplace activities. Such conditions may be induced by the particular work activity of the individual or by activities of others in the workplace. The time interval between exposure and the onset of the illness may be short (e.g. asthma attacks) or long (e.g. deafness or cancer).

Environmental protection – arrangements to cover those activities in the workplace which affect the environment (in the form of flora, fauna, water, air and soil) and, possibly, the health and safety of employees and others. Such activities include waste and effluent disposal and atmospheric pollution.

Accident – defined by the Health and Safety Executive as 'any unplanned event that results in injury or ill health of people, or damage or loss to property, plant, materials or the environment or a loss of a business opportunity'. Other authorities define an accident more narrowly by excluding events that do not involve injury or ill-health. This book will always use the Health and Safety Executive definition.

Near miss – is any incident that could have resulted in an accident. Knowledge of near misses is very important since research has shown that, approximately, for every 10 'near miss' events at a particular location in the workplace, a minor accident will occur.

Dangerous occurrence – is a 'near miss' which could have led to serious injury or loss of life. Dangerous occurrences are defined in the Reporting of Injuries, Diseases and Dangerous Occurrences Regulations 1995 (often known as RIDDOR) and are always reportable to the Enforcement Authorities. Examples include the collapse of a scaffold or a crane or the failure of any passenger carrying equipment.

Hazard and risk – a hazard is the *potential* of a substance, activity or process to cause harm. Hazards take many forms including, for example, chemicals, electricity and working from a ladder. A hazard can be ranked relative to other hazards or to a possible level of danger.

A risk is the *likelihood* of a substance, activity or process to cause harm. A risk can be reduced and the hazard controlled by good management.

It is very important to distinguish between a *hazard* and a *risk* – the two terms are often confused and activities such as construction work are called *high risk* when they are *high hazard*. Although the hazard will continue to be high, the risks will be reduced as controls are implemented. The level of risk remaining when controls have been adopted is known as the *residual risk*. There should only be *high residual risk* where there is poor health and safety management and inadequate control measures.

1.3 The legal framework for health and safety

1.3.1 Sub-divisions of law

There are two sub-divisions of the law that apply to health and safety issues: criminal law and civil law.

Criminal law

Criminal law consists of rules of behaviour laid down by the Government or the State and, normally, enacted by Parliament through Acts of Parliament. These rules or Acts are imposed on the people for the protection of the people. Criminal law is enforced by several different Government Agencies who may prosecute individuals for contravening criminal laws. It is important to note that, except for very rare cases, only these Agencies are able to decide whether to prosecute an individual or not.

An individual who breaks criminal law is deemed to have committed an offence or crime and, if he is prosecuted, the court will determine whether he is guilty or not. If he is found guilty, the court could sentence him to a fine or imprisonment. Due to this possible loss of liberty, the level of proof required by a criminal court is very high and is known as proof 'beyond reasonable doubt', which is as near certainty as possible. While the prime object of a criminal court is the allocation of punishment, the court can award compensation to the victim or injured party. One example of criminal law is the Road Traffic Acts which are enforced by the police. However, the police are not the only criminal law enforcement agency. The Health and Safety at Work Act is another example of criminal law and this is enforced either by the Health and Safety Executive or Local Authority Environmental Health Officers. Other agencies which enforce criminal law include the Fire Authority, the Environmental Agency, Trading Standards and Customs and Excise.

There is one important difference between procedures for criminal cases in general and criminal cases involving health and safety. The prosecution in a criminal case has to prove the guilt of the accused beyond reasonable doubt. While this obligation is not totally removed in health and safety cases, section 40 of the Health and Safety at Work Act 1974 transferred, where there is a duty to do something 'so far as is reasonably practicable' or 'so far as is practicable' or 'use the best practicable means', the onus of proof to the accused to show that there was no better way to discharge his duty under the Act. However, when this burden of proof is placed on the accused, they need only satisfy the court on the balance of probabilities that what they are trying to prove has been done.

Civil law

Civil law concerns disputes between individuals or individuals and companies. An individual sues another individual or company to address a civil wrong or tort (or delict in Scotland). The individual who brings the complaint to court is known as the plaintiff (pursuer in Scotland) and the individual or company who is being sued is known as the defendant (defender in Scotland).

The civil court is concerned with liability and the extent of that liability rather than guilt or non-guilt. Therefore, the level of proof required is based on the 'balance of probability', which is a lower level of certainty than that of 'beyond reasonable doubt' required by the criminal court. If a defendant is found to be liable, the court would normally order him to pay compensation and possibly costs to the plaintiff. However, the lower the balance of probability, the lower the level of compensation awarded. In extreme cases, where the balance of probability is just over 50%, the plaintiff may 'win' his case but lose financially because costs may not be awarded and the level of compensation is low. The level of compensation may also be reduced through the defence of **contributory negligence,** which is discussed later under 'Common Torts and Duties'. For cases involving health and safety, civil disputes usually follow accidents or illnesses and concern negligence or a breach of statutory duty. The vast majority of cases are settled 'out of court'. While actions are often between individuals, where the defendant is an employee who was acting in the course of his employment during the alleged incident, the defence of the action is transferred to his employer – this is known as **vicarious liability**. The civil action then becomes one between the individual and a company.

1.4 The legal system in England and Wales

The description which follows applies to England and Wales (and, with a few minor differences to Northern Ireland). Only the court functions concerning health and safety are mentioned. Figure 1.2 shows the court hierarchy in schematic form.

1.4.1 Criminal law

Magistrates Courts

Most criminal cases begin and end in the Magistrates Courts. Health and safety cases are brought before the court by enforcement officers (Health and Safety Executive or Local Authority Environmental Health Officers) and they are tried by a bench of three lay magistrates (known as Justices of the Peace) or a single district judge. The lay

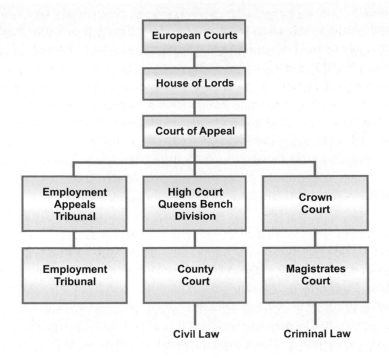

Figure 1.2 The court system in England and Wales relevant to health and safety.

magistrates are members of the public, usually with little previous experience of the law, whereas a district judge is legally qualified.

The Magistrates Court has limited powers with a maximum fine of £5000 (for employees) to £20 000 for employers or for those who ignore prohibition notices. Magistrates are also able to imprison for up to six months for breaches of enforcement notices. The vast majority of health and safety criminal cases are dealt with in the Magistrates Court.

Crown Court

The Crown Court hears the more serious cases, which are passed to them from the Magistrates Court – normally because the sentences available to the magistrates are felt to be too lenient. Cases are heard by a judge and jury, although some cases are heard by a judge alone. The penalties available to the Crown Court are an unlimited fine and up to two years imprisonment for breaches of enforcement notices. The Crown Court also hears appeals from the Magistrates Court.

Appeals from the Crown Court are made to the Court of Appeal (Criminal Division) who may then give leave to appeal to the most senior court in the country – the House of Lords. The most senior judge at the Court of Appeal is the Lord Chief Justice.

1.4.2 Civil law

County Court

The lowest court in civil law is the County Court which only deals with minor cases (for compensation claims of up to £50 000 if the High Court agrees). Cases are normally heard by a judge sitting alone. For personal injury claims of less than £5000, a small claims court is also available.

High Court

Most health and safety civil cases are heard in the High Court (Queens Bench Division) before a judge only. It deals with compensation claims in excess of £50 000 and acts as an appeal court for the County Court.

Appeals from the High Court are made to the Court of Appeal (Civil Division). The House of Lords receives appeals from the Court of Appeal or, on matters of law interpretation, directly from the High Court. The most senior judge at the Court of Appeal is the Master of the Rolls.

The judges in the House of Lords are called Law Lords and are sometimes called upon to make judgements on points of law, which are then binding on lower courts. Such judgements form the basis of Common Law, which is covered later.

Other courts – Employment Tribunals

These were established in 1964 and primarily deal with employment and conditions of service issues, such as unfair dismissal. However, they also deal with appeals over health and safety enforcement notices, disputes between recognized safety representatives and their employers and cases of unfair dismissal involving health and safety issues. There are usually three members who sit on a Tribunal. These members are appointed and are often not legally qualified. Appeals from the Tribunal may be made to the Employment Appeal Tribunal or, in the case of enforcement notices, to the High Court. Appeals from Tribunals can only deal with the clarification of points of law.

1.5 The legal system in Scotland

Scotland has both criminal and civil courts but prosecutions are initiated by the procurator-fiscal rather than the Health and Safety Executive. The lowest criminal court is called the District Court and deals with minor offences. The Sheriff Court has a similar role to that of the Magistrates Court (for criminal cases) and the County Court (for civil cases), although it can deal with more serious cases involving a sheriff and jury.

The High Court of Judiciary, in which a judge and jury sit, has a similar role to the Crown Court and appeals are made to the Court of Criminal Appeal. The Outer House of the Court of Session deals with civil cases in a similar way to the English High Court. The Inner house of the Court of Session is the Appeal Court for civil cases.

For both appeal courts, the House of Lords is the final court of appeal. There are Industrial Tribunals in Scotland with the same role as those in England.

1.6 European Courts

There are two European Courts – the European Court of Justice and the European Court of Human Rights.

The European Court of Justice, based in Luxembourg, is the highest court in the European Union (EU). It deals primarily with community law and its interpretation. It is normally concerned with breaches of community law by Member States and cases may be brought by other Member States or institutions. Its decisions are binding on all Member States. There is currently no right of appeal.

The European Court of Human Rights, based in Strasbourg, is not directly related to the EU – it covers most of the countries in Europe including the 15 EU member states. As its title suggests, it deals with human rights and fundamental freedoms. With the introduction of the Human Rights Act 1998 in October 2000, many of the human rights cases will be heard in the UK.

1.7 Sources of law (England and Wales)

There are two sources of law – common law and statute law.

1.7.1 Common Law

Common law dates from the eleventh century when William I set up Royal Courts to apply a uniform (common) system of law across the whole of England. Prior to that time, there was a variation in law, or the interpretation of the same law, from one town or community to another. Common law is based on judgements made by courts (or strictly judges in courts). In general, courts are bound by earlier judgements on any particular point of law – this is known as 'precedent'. Lower courts must follow the judgements of higher courts. Hence judgements made by the Law Lords in the House of Lords form the basis of most of the common law currently in use.

In health and safety, the legal definition of negligence, duty of care and terms such as 'practicable' and 'as far as is reasonably practicable' are all based on legal judgements and form part of the common law. Common law also provides the foundation for most civil claims made on health and safety issues.

1.7.2 Statute law

Statute law is law which has been laid down by Parliament as Acts of Parliament. In health and safety, an Act of Parliament, the Health and Safety at Work Act 1974, lays down a general legal framework. Specific health and safety duties are, however, covered by Regulations or Statutory Instruments – these are also examples of statute law. If there is a conflict between statute and common law, statute law takes precedence. However, as with common law, judges interpret statute law usually when it is new or ambiguous. Although for health and safety, statute law is primarily the basis of criminal law, there is a tort of breach of statutory duty which can be used when a person is seeking damages following an accident or illness. Breaches of the Health and Safety at Work Act 1974 cannot be used for civil action but breaches of most of the Regulations produced by the Act may give rise to civil actions.

1.7.3 The relationship between the sub-divisions and sources of law

The two sub-divisions of law may use either of the two sources of law. For example, murder is a common law crime. In terms of health and safety, however, criminal law is only based on statute law, whereas civil law may be based on either common law or statute law. This relationship is shown in Figure 1.3.

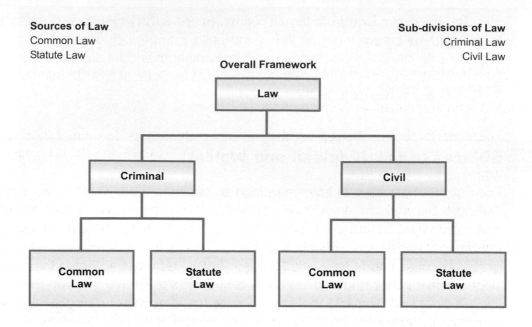

Figure 1.3 Sub-divisions and sources of law.

In summary, criminal law seeks to protect everyone in society whereas civil law seeks to recompense the individual citizen.

1.8 Common law torts and duties

1.8.1 Negligence

The only tort (civil wrong) of real significance in health and safety is negligence. Negligence is the lack of reasonable care or conduct which results in the injury (or financial loss) of or to another. Whether the act or omission was reasonable is usually decided as a result of a court action.

There have been two important judgements that have defined the legal meaning of negligence. In 1856, negligence was judged to involve *actions or omissions* and the need for *reasonable and prudent* behaviour. In 1932, Lord Atkin said,

> *You must take reasonable care to avoid acts or omissions which you reasonably* foreseе *would be likely to injure your* neighbour. *Who then, in law is my neighbour? The answer seems to be persons who are so closely and directly affected by my act that I ought reasonably to have them in contemplation as being so affected when I am directing my mind to the acts or omissions which are called in question.*

It can be seen, therefore, that for negligence to be established, it must be reasonable and foreseeable that the injury could result from the act or omission. In practice, the Court may need to decide whether the injured party is the neighbour of the perpetuator. A collapsing scaffold could easily injure a member of the public who could be considered a neighbour to the scaffold erector.

An employee who is suing his employer for negligence, needs to establish the following three criteria:

➤ a duty was owed to him by his employer
➤ there was a breach of that duty
➤ the breach resulted in the injury, disease, damage and/or loss.

These tests should also be used by anyone affected by the employer's undertaking (such as contractors and members of the public) who is suing the employer for negligence.

If the employer is unable to defend against the three criteria, two further partial defences are available. It could be argued that the employee was fully aware of the risks that were taken by not complying with safety instructions, (known as *volenti non fit injuria* or 'the risk was willingly accepted'). This defence is unlikely to be totally successful because courts have ruled that employees have not accepted the risk voluntarily since economic necessity forces them to work.

The second possible defence is that of 'contributory negligence' where the employee is deemed to have contributed to the negligent act. This defence, if successful, can significantly reduce the level of compensation (up to 80% in some cases).

1.8.2 Duties of care

Several judgements have established that employers owe a duty of care to each of their employees. This duty cannot be assigned to others, even if a consultant is employed to advise on health and safety matters or if the employees are sub-contracted to work with another employer. These duties may be sub-divided into four groups. Employers must:

➤ provide a safe place of work
➤ provide safe plant and equipment
➤ provide a safe system of work
➤ provide safe and competent fellow employees.

Employer duties under common law are often mirrored in statute law. This, in effect, makes them both common law and statutory duties.

The requirements of a safe workplace, including the maintenance of floors and the provision of walkways and safe stairways, for example, are also contained in the Workplace (Health, Safety and Welfare) Regulations 1992.

The requirement to provide competent fellow employees includes the provision of adequate supervision, instruction and training. As mentioned earlier, employers are responsible for the actions of their employees (**vicarious liability**) provided that the action in question took place during the normal course of his employment.

1.9 Levels of statutory liability

There are three levels of statutory liability which form a hierarchy of duties. These levels are used extensively in health and safety statutory (criminal) law but have been defined by judges under common law. The three levels of duty are absolute, practicable and reasonably practicable.

1.9.1 Absolute liability

This is the highest level of liability and, normally, occurs when the risk of injury is so high that injury is inevitable unless safety precautions are taken. It is a rare requirement regarding physical safeguards, although it was more common before 1992 when certain sections of the Factories Act 1961 were still in force. No assessment of risk is required but the duty is absolute and the employer has no choice but to undertake the duty. The verbs used in the Regulations are 'must' and 'shall'.

An example of this is Regulation 11(1) of the Provision and Use of Work Equipment Regulations 1998 concerning contact with a rotating stock bar which projects beyond a headstock of a lathe. Although this liability is absolute, it may still be defended using, for example, the argument that 'all reasonable precautions and all due diligence' were taken. This particular defence is limited to certain health and safety regulations such as The Electricity at Work Regulations 1989 and The Control of Substances Hazardous to Health 2002.

Many of the health and safety management requirements contained in health and safety law place an absolute duty on the employer. The need for written safety policies and risk assessments when employee numbers rise above a basic threshold are examples of this.

1.9.2 Practicable

This level of duty is more often used than the absolute duty as far as the provision of safeguards is concerned and, in many ways, has the same effect.

A duty that 'the employer ensure, so far as is practicable, that any control measure is maintained in an efficient state' means that if the duty is technically possible or feasible then it must be done irrespective of any difficulty, inconvenience or cost. Examples of this duty may be found in the Provision and Use of Work Equipment Regulations 1998 (Regulation 11(2) (a and b)) and the Control of Lead at Work Regulations 2002 where Regulation 8 states, 'Every employer who provides any control measure . . . shall ensure, so far as is practicable, that it is maintained in an efficient state . . . in good repair . . .'.

1.9.3 Reasonably practicable

This is the most common level of duty in health and safety law and was defined by Judge Asquith in Edwards vs the National Coal Board (1949) as follows:

> 'Reasonably practicable' is a narrower term than 'physically possible', and seems to me to imply that computation must be made by the owner in which the quantum of risk is placed on one scale and the sacrifice involved in the measures necessary for averting the risk (whether in time, money or trouble) is placed in the other, and that, if it be shown that there is a gross disproportion between them − the risk being insignificant in relation to the sacrifice − the defendants discharge the onus on them.

In other words, if the risk of injury is very small compared to the cost, time and effort required to reduce the risk, then no action is necessary. It is important to note that

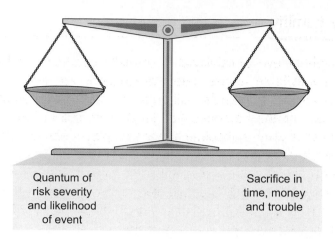

Figure 1.4 Diagrammatic view of 'reasonably practicable'.

money, time and trouble must 'grossly outweigh' not balance the risk. This duty requires judgement on the part of the employer (or his adviser) and clearly needs a risk assessment to be undertaken with conclusions noted. Continual monitoring is also required to ensure that risks do not increase. There are numerous examples of this level of duty, including the Manual Handling Operations Regulations 1992 and The Control of Substances Hazardous to Health 2002. The term 'suitable and sufficient' is used to define the scope and extent required for health and safety risk assessment and may be interpreted in a similar way to reasonably practicable. More information is given on this definition in Chapters 5 and 12.

1.10 The influence of the European Union (EU) on health and safety

As Britain is part of the European Union, much of the health and safety law originates in Europe. Proposals from the European Commission may be agreed by member states. The member states are then responsible for making them part of their domestic law.

In Britain itself and in much of Europe, health and safety law is based on the principle of risk assessment described above. The main role of the EU in health and safety is to harmonize workplace and legal standards and remove barriers to trade across member states. In the future, all health and safety statute law in the UK will be derived from European Union Directives. A directive from the EU is legally binding on each member state and must be incorporated into the national law of each member state. Directives set out specific minimum aims which must be covered within the national law. Some states incorporate Directives more speedily than others.

Directives are proposed by the European Commission, comprising 21 commissioners, who are citizens of each of the member states. The proposed Directives are sent to the European Parliament which is directly elected from the member states. The European Parliament may accept, amend or reject the proposed Directives. The proposed Directives are then passed to the Council of Ministers who may accept the proposals on a qualified majority vote, unless the European Parliament has rejected them, in which case, they can only be accepted on a unanimous vote of the Council. The Council of Ministers consists of one senior Government minister from each of the member states.

The powers of the EU in health and safety law are derived from the Treaty of Rome 1957 and the Single European Act 1986. For health and safety, the Single European Act added two additional Articles to the Treaty – Article 100A and Article 118A. Article 100A is concerned with health and safety standards of equipment and plant and its Directives are implemented in the UK by the Department of Trade and Industry.

Article 118A is concerned with minimum standards of health and safety in employment and its Directives are implemented by the Health and Safety Commission/Executive.

The objective of the Single European Act 1986 is to produce a 'level playing field' for all member states so that goods and services can move freely around the EU without any one state having an unfair advantage over another. The harmonization of health and safety requirements across the EU is one example of the 'level playing field'.

Figure 1.5 Health and Safety Law poster – must be displayed or brochure given to employees. Source HSE.

The first introduction of an EU Directive into UK Health and Safety law occurred on 1 January 1993 when a Framework Directive on Health and Safety management and five daughter directives were introduced using powers contained in the Health and Safety at Work Act 1974 (Figures 1.5 and 1.6). These directives, known as the European Six Pack, covered the following areas:

➤ Management of Health and Safety at Work
➤ Workplace
➤ Provision and Use of Work Equipment

Figure 1.6 Health and Safety at Work Act.

➤ Manual Handling
➤ Personal Protective Equipment
➤ Display Screen Equipment.

Since 1993, several other EU Directives have been introduced into UK law. Summaries of the more common UK regulations are given in Chapter 17.

1.11 The Health and Safety at Work Act 1974

1.11.1 Background to the Act

The Health and Safety at Work Act resulted from the findings of the Robens Report, published in 1972. Earlier legislation had tended to relate to specific industries or workplaces. This resulted in over 5m workers being unprotected by any health and safety legislation. Contractors and members of the public were generally ignored. The law was more concerned with the requirement for plant and equipment to be safe rather than the development of parallel arrangements for raising the health and safety awareness of employees.

A further serious problem was the difficulty that legislation had in keeping pace with developments in technology. For example, following a court ruling in 1955 which, in effect, banned the use of grinding wheels throughout industry, it took 15 years to produce the Abrasive Wheels Regulations 1970 to address the problem raised by the 1955 court judgement (John Summers and Sons Ltd. v. Frost). In summary, health and safety legislation before 1974 tended to be reactive rather than proactive.

Lord Robens was asked, in 1970, to review the provision made for the health and safety of people at work. His report produced conclusions and recommendations upon which the Health and Safety at Work Act 1974 was based. The principal recommendations were as follows:

➤ there should be a single Act that covers all workers and that Act should contain general duties which should 'influence attitudes';
➤ the Act should cover all those affected by the employer's undertaking such as contractors, visitors, students and members of the public;
➤ there should be an emphasis on health and safety management and the development of safe systems of work. This would involve the encouragement of employee participation in accident prevention. (This was developed many years later into the concept of the health and safety culture);
➤ enforcement should be targeted at 'self-regulation' by the employer rather than reliance on prosecution in the courts.

These recommendations led directly to the introduction of the Health and Safety at Work etc. Act in 1974.

1.11.2 An overview of the Act

Health and Safety Commission (HSC)
The Health and Safety at Work Act established the HSC and gave it the responsibility to draft new Regulations and to enforce them either through its executive arm, known as

the Health and Safety Executive (HSE), or through the Local Authority Environmental Health Officers (EHO). The HSC has equal representation from employers, trade unions and special interest groups.

Regulations

The Health and Safety at Work Act is an **Enabling Act** which allows the Secretary of State to make further laws (known as regulations) without the need to pass another Act of Parliament. Regulations are law, approved by Parliament. These are usually made under the Health and Safety at Work Act, following proposals from the HSC. This applies to regulations based on EC Directives as well as 'home-grown' ones.

The Health and Safety at Work Act, and general duties in the Management Regulations, aim to help employers to set goals, but leave them free to decide how to control hazards and risks which they identify. Guidance and Approved Codes of Practice give advice, but employers are free to take other routes to achieving their health and safety goals, so long as they do what is reasonably practicable. But some hazards are so great, or the proper control measures so expensive, that employers cannot be given discretion in deciding what to do about them. Regulations identify these hazards and risks and set out specific action that must be taken. Often these requirements are absolute – employers have no choice but to follow them and there is no qualifying phrase of 'reasonably practicable' included.

Some regulations apply across all organizations – the Manual Handling Regulations would be an example. These apply wherever things are moved by hand or bodily force. Equally, the Display Screen Equipment Regulations apply wherever visual display units are used at work. Other regulations apply to hazards unique to specific industries, such as mining or construction.

Wherever possible, the HSC will set out the regulations as goals, and describe what must be achieved, but not how it must be done.

Sometimes it is necessary to be *prescriptive*, and to spell out what needs to be done in detail, since some standards are absolute. For example, all mines should have two exits; contacts with live electrical conductors should be avoided. Sometimes European law requires prescription.

Some activities or substances are so dangerous that they have to be licensed, for example, explosives and asbestos removal. Large, complex installations or operations require 'safety cases', which are large-scale risk assessments subject to scrutiny by the regulator. An example would be the recently privatized railway companies. They are required to produce safety cases for their operations.

Approved Code of Practice (ACOP)

An ACOP is produced for most sets of regulations by the HSC and attempts to give more details on the requirements of the regulations. It also attempts to give the level of compliance needed to satisfy the regulations. ACOPs have a special legal status (sometimes referred to as quasi-legal). The relationship of an ACOP to a regulation is similar to the relationship of the Highway Code to the Road Traffic Acts. A person is never prosecuted for contravening the Highway Code but can be prosecuted for contravening the Road Traffic Acts. If a company is prosecuted for a breach of health and safety law and it is proved that it has not followed the relevant provisions of the ACOP, a court can find them at fault, unless the company can show that it has complied with the law in some other way.

Since most health and safety prosecutions take place in a Magistrates Court, it is likely that the lay magistrates will consult the relevant ACOP as well as the regulations when dealing with a particular case. Therefore, in practice, an employer must have a good reason for not adhering to an ACOP.

Codes of Practice generally are only directly legally binding if:

➤ the regulations or Act indicates that they are, for example, The Safety Signs and Signals Regulations Schedule 2 specify British Standard Codes of practice for alternative hand signals; or
➤ they are referred to in an Enforcement Notice.

Guidance

Guidance comes in two forms – legal and best practice. The Legal Guidance series of booklets is issued by the HSC and/or the HSE to cover the technical aspects of health and safety regulations. These booklets generally include the Regulations and the ACOP, where one has been produced.

Best practice guidance is normally published in the HSG series of publications by the HSE. Examples of best practice guidance books include *Health and Safety in Construction* HSG 150 and *Lighting at Work* HSG 38.

An example of the relationship between these three forms of requirement/advice can be shown using a common problem found throughout industry and commerce – minimum temperatures in the workplace. Regulation 7 of the Workplace (Health, Safety and Welfare) Regulations 1992 states, '(1) During working hours, the temperature in all workplaces inside buildings shall be reasonable'. The ACOP states 'The temperature in workrooms should normally be at least 16 degrees Celsius unless much of the work involves severe physical effort in which case the temperature should be at least 13 degrees Celsius . . .'

It would, therefore, be expected that employers would not allow their workforce to work at temperatures below those given in the ACOP unless it was **reasonable** for them to do so (if, for example, the workplace was a refrigerated storage unit).

Best practice guidance to cover this example is given in HSG 194 (*Thermal Comfort in the Workplace*) in which possible solutions to the maintenance of employee welfare in low temperature environments is given.

1.11.3 General duties and key sections of the Act

A summary of the Health and Safety at Work Act is given under Chapter 17 (17.4) – only an outline will be given here.

Section 2 Duties of employers to employees
To ensure, so far as is reasonably practicable, the health, safety and welfare of all employees. In particular:

➤ safe plant and systems of work
➤ safe use, handling, transport and storage of substances and articles
➤ provision of information, instruction, training and supervision
➤ safe place of work, access and egress
➤ safe working environment with adequate welfare facilities

➤ a written safety policy together with organizational and other arrangements (if more than four employees)

➤ consultation with safety representatives and formation of safety committees where there are recognized trade unions.

Section 3 Duties of employers to others affected by this undertaking
A duty to safeguard those not in their employment but affected by the undertaking. This includes members of the public, contractors, patients, customers and students.

Section 4 Duties of employers to ensure that premises are safe
Safe access and egress and that any plant or substances are safe and without risk to health.

Section 6 Duties of suppliers
Persons who design, manufacture, import or supply any article or substance for use at work must ensure, so far as is reasonably practicable, that they are safe and without risk to health.

Section 7 Duties of employees
Two main duties:

➤ to take reasonable care for the health and safety of themselves and others affected by their acts or omissions;

➤ to cooperate with the employer and others to enable them to fulfil their legal obligations.

Section 8
No person is to misuse or interfere with safety provisions (sometimes known as the 'horseplay section').

Section 9
Employees cannot be charged for health and safety requirements such as personal protective equipment.

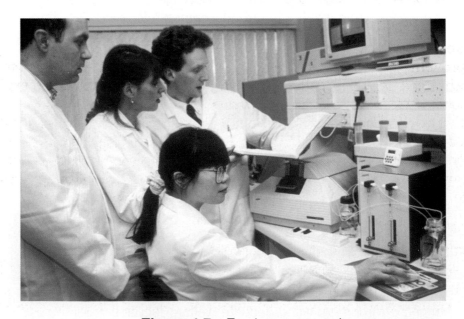

Figure 1.7 Employees at work.

Section 37 Personal liability of directors
Where an offence is committed by a body corporate or can be attributable to any neglect of a director or other senior officer of the body, both that body and the person are liable to prosecution.

1.11.4 Enforcement of the Act

Powers of inspectors
Inspectors under the HSW Act work either for the HSE or the Local Authority. Local Authorities are responsible for retail and service outlets, such as shops (retail and wholesale), restaurants, garages, offices, residential homes, entertainment and hotels. The HSE are responsible for all other work premises including the Local Authorities themselves. Both groups of inspectors have the same powers. The detailed powers of inspectors are given in Chapter 17. In summary an inspector has the right to:

Figure 1.8 The inspector inspects.

> enter premises at any reasonable time, accompanied by a police officer, if necessary;
> examine, investigate and require the premises to be left undisturbed;
> take samples, photographs and, if necessary, dismantle and remove equipment or substances;
> require the production of books or other relevant documents and information;
> seize, destroy or render harmless any substance or article;
> issue enforcement notices and initiate prosecutions.

Enforcement Notices
There are two types of enforcement notices:
Improvement Notice – This identifies a contravention of the law and specifies a date by which the situation is to be remedied. An appeal must be made to the Employment Tribunal within 21 days during which period the notice is suspended.
Prohibition Notice – This is used to halt an activity which the inspector feels could lead to a serious personal injury. The notice will identify which legal requirement is being or is likely to be contravened. The notice takes effect as soon as it is issued. As with the improvement notice, an appeal may be made to the Employment Tribunal but, in this case, the notice remains in place during the appeal process.

Penalties

Magistrates Court (Summary Offences)
For health and safety offences, employers may be fined up to £20 000 and employees (or individuals) up to £5000.
For failure to comply with an enforcement notice or a court order, anybody may be imprisoned for up to 6 months.

Crown Court (Indictable Offences)
Fines are unlimited in the Crown Court and imprisonment for up to 2 years for failure to comply with an enforcement notice or a court order.

Summary of the actions available to an inspector
Following a visit by an inspector to a premises, the following actions are available to an inspector:

➤ take no action
➤ give verbal advice
➤ give written advice
➤ serve an improvement notice
➤ serve a prohibition notice
➤ prosecute.

In any particular situation, more than one of these actions may be taken.

1.12 The Management of Health and Safety at Work Regulations 1999

As mentioned earlier, on 1 January 1993, following an EC Directive, the Management of Health and Safety at Work Regulations became law in the UK. These regulations were updated in 1999 and are described in detail in Chapter 17. In many ways the regulations were not introducing concepts or replacing the 1974 Act – they simply reinforced or amended the requirements of the Health and Safety at Work Act. Some of the duties of employers and employees were re-defined.

1.12.1 Employers' duties

Employers must:

➤ undertake suitable and sufficient written risk assessments when there are more than four employees;
➤ put in place effective arrangements for the planning, organization, control, monitoring and review of health and safety measures in the workplace (including health surveillance). Such arrangements should be recorded if there are more than four employees;
➤ employ (to be preferred) or contract competent persons to help them comply with health and safety duties;
➤ develop suitable emergency procedures. Ensure that employees and others are aware of these procedures and can apply them;
➤ provide employees and others with health and safety information, in particular information resulting from risk assessment or emergency procedures;
➤ cooperate in health and safety matters with other employers who share the same workplace;
➤ provide non-employees working on the work site with relevant health and safety information;
➤ provide employees with adequate and relevant health and safety training;
➤ provide temporary workers with appropriate health and safety information.

1.12.2 Employees' duties

Employees must:

➤ use any equipment or substance in accordance with any training or instruction given by the employer;
➤ report to the employer any serious or imminent danger;
➤ report any shortcomings in the employer's protective health and safety arrangements.

1.13 Role and function of external agencies

The Health and Safety Commission, Health and Safety Executive and the Local Authorities (a term used to cover County, District and Unitary Councils) are all external agencies that have a direct role in the monitoring and enforcement of health and safety standards. There are, however, three other external agencies that have a regulatory influence on health and safety standards in the workplace.

Figure 1.9   The main external agencies that impact on the workplace.

1.13.1 Fire Authority

The Fire Authority is situated within a local authority and is normally associated with fire fighting and giving general advice. It has also been given powers to regulate fire precautions within places of work under fire precautions law. The powers of the Fire Authority are very similar to those of the HSE on health and safety matters. The Fire Authority grants fire certificates to workplaces and conducts routine and random fire inspections. It must be consulted during the planning stage of proposed building alterations when such alterations may affect the fire safety of the building (e.g. means of escape). It also needs to be informed when inflammables and/or explosives are to be used or stored.

The Fire Authority can issue both improvement and prohibition notices, although the appeal procedure is different from that for health and safety enforcement notices. The Authority can prosecute for offences against fire precaution law.

1.13.2 The Environment Agency (Scottish Environmental Protection Agency)

The Environment Agency was established in 1995 and was given the duty to protect and improve the environment. It is the regulatory body for environmental matters and has an influence on health and safety issues. It is responsible for authorizing and regulating emissions from industry – this includes enforcement of section 5 of the Health and Safety at Work Act 1974.

Other duties and functions of the agency include:

➤ ensuring effective controls of the most polluting industries;
➤ monitoring radioactive releases from nuclear sites;
➤ ensuring that discharges to controlled waters are at acceptable levels;
➤ setting standards and issuing permits for the collection, transporting, processing and disposal of waste (including radioactive waste);
➤ enforcement of the Producer Responsibility Obligations (Packaging Waste) Regulations 1997. These resulted from an EC Directive which seeks to minimize packaging and recycle at least 50% of it.

The Agency may prosecute in the criminal courts for the infringement of environmental law – in one case a fine of £4m was imposed.

1.13.3 Insurance companies

Insurance companies play an important role in the improvement of health and safety standards. Since 1969, it has been a legal requirement for employers to insure against liability for injury or disease to their employees arising out of their employment. This is called employers' liability insurance. Certain public sector organizations are exempted from this requirement because any compensation is paid from public funds. Other forms of insurance include fire insurance and public liability insurance (to protect members of the public).

Premiums for all these types of insurance are related to levels of risk which is related to standards of health and safety. In recent years, there has been a considerable increase in the number and size of compensation claims and this has placed further pressure on insurance companies.

Insurance companies are becoming effective health and safety regulators by weighting the premium offered to an organization according to its safety and/or fire precaution record.

1.14 | Moral, legal and financial arguments for health and safety management

Figure 1.10 Good standards prevent harm and save money.

1.14.1 Moral arguments

The moral arguments are reflected by the occupational accident and disease rates.

Accident rates

Accidents at work can lead to serious injury and even death. Although accident rates are discussed in greater detail in later chapters, some trends are shown in Tables 1.1–1.4. A major accident is a serious accident typically involving a fracture of a limb or a 24-hour stay in a hospital. An 'over three-day accident' is an accident that leads to more than 3 days off work. Statistics are collected on all people who are injured at places of work not just employees.

It is important to note that since 1995 suicides and trespassers on the railways have been included in the HSE figures – this has led to a significant increase in the figures. Table 1.2 shows, for the year 1998/99, the breakdown in accidents between employees, self-employed and members of the public.

Table 1.1 Accidents involving all people at a place of work

Injury	1996/97	1997/98	1998/99	1999/00
Death	654	667	622	656
Major	65 014	58 615	52 853	52 300
Over 3-day	129 568	135 773	133 144	129 599

Table 1.3 shows the figures for employees only.

Table 1.3 shows that while there has been a decline in fatalities, major accidents have hardly changed. Table 1.4 gives an indication of accidents in different employment sectors.

These figures indicate that there is a need for health and safety awareness even in occupations which many would consider very low hazard, such as schools and hotels. In

Table 1.2 Accidents for different groups of people for 1998/99

	Deaths	Major	Over 3 day
Total	622	52 853	133 144
Employees	188 (30%)	28 368	132 295
Self-employed	65 (10%)	685	849
Members of the public	369 (60%)	23 800	n/a

Table 1.3 Accidents involving employees at the place of work

Injury	1996/97	1997/98	1998/99	1999/00
Death	207	212	188	161
Major	27 964	29 187	28 368	27 563
Over 3-day	127 286	134 789	132 295	128 889

Table 1.4 Accidents to all people in various employment sectors (1999/2000)

Sector	Deaths	Majors
Education	5 (1%)	9687 (18%)
Hotel & catering	6 (1%)	1955 (4%)
Service industries	474 (72%)	37 710 (72%)
Manufacturing	44 (7%)	8081 (15%)
Construction	85 (13%)	5040 (10%)
Agriculture	55 (7%)	920 (2%)
Total	656	52 300

fact, over 70% of all deaths occur in the service sector and manufacturing is considerably safer than construction and agriculture.

Although there has been a decrease in fatalities over recent years, there is still a very strong moral case for improvement in health and safety performance.

Disease rates

Work-related ill-health and occupational disease can lead to absence from work and, in some cases, to death. Such occurrences may also lead to costs to the state (the Industrial Injuries Scheme) and to individual employers (sick pay and, possibly, compensation payments).

In 1995, a major survey was undertaken into self-reported work-related illness. In that year there were 2.25m reported work-related illnesses leading to 18m working days lost. The top four illnesses were:

> musculoskeletal disorders 1.2m
> stress related 0.5m
> lower respiratory disease 202 000
> deafness, tinnitus and other ear problems 170 000

9.9 working days were lost due to musculoskeletal disorders causing each sufferer to have 13 days off work on average.

Tables 1.5 and 1.6 show the trends for both lung and non-lung diseases over a three-year period from 1996.

Table 1.5   Lung disease statistics

	1996/97	1997/98	1998/99
Asbestos related	897	906	1027
Occupational asthma	298	222	196
Chronic bronchitis and/or emphysema	3030	3423	1451
Total	4662	5383	3437

Table 1.6   Non-lung disease statistics

	1996/97	1997/98	1998/99
Musculoskeletal	764	600	465
Occupational deafness	413	258	316
Vibration white finger	3288	3033	3155
Dermatitis and allergic rhinitis	688	470	355
Total	5535	4844	4846

1.14.2 Legal arguments

The legal arguments, concerning the employer's duty of care in criminal and civil law, have been covered earlier. Some statistics on legal enforcement indicate the legal consequences resulting from breaches in health and safety law. There have been some very high compensation awards for health and safety cases in the civil courts and fines in excess of £100 000 in the criminal courts. Table 1.7 shows the number of enforcement notices served over a three-year period. Most notices are served in the manufacturing sector followed by construction and agriculture. Local Authorities serve 40% of the improvement notices and 20% of the prohibition notices.

Table 1.8 shows the number of prosecutions over the same three-year period. HSE present 80% of the prosecutions and the remainder are presented by Local Authority Environmental Health Officers. Most of these prosecutions are for infringements of the Construction Regulations and the Provision and Use of Work Equipment Regulations.

There are some clear legal reasons for sound health and safety management systems.

Table 1.7 Number of enforcement notices issued over a three-year period

Year	Improvement notice	Prohibition notice
1996/97	7650	4964
1997/98	7731	5680
1998/99	11 493	5877

Table 1.8 Number of prosecutions over a three-year period

Year	Informations	Convictions	Average fine £
1996/97	1854	1518	4463
1997/98	2133	1724	5020
1998/99	2183	1849	4722

1.14.3 Financial arguments

Costs of accidents
Any accident or incidence of ill-health will cause both direct and indirect costs and incur an insured and an uninsured cost. It is important that all of these costs are taken into account when the full cost of an accident is calculated. In a study undertaken by the HSE, it was shown that indirect costs or hidden costs could be 36 times greater than

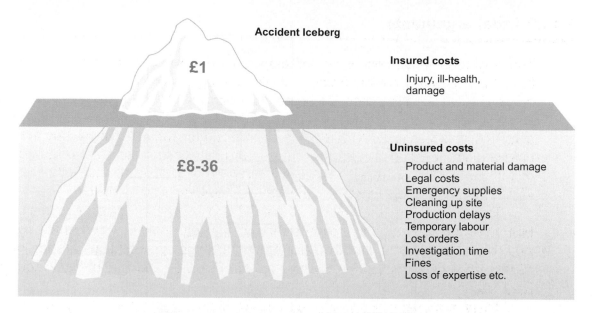

Figure 1.11 Insured and uninsured costs.

direct costs of an accident. In other words, the direct costs of an accident or disease represent the tip of the iceberg when compared to the overall costs (Figure 11.1).

Direct costs
These are costs that are directly related to the accident. They may be insured (claims on employers' and public liability insurance, damage to buildings, equipment or vehicles) or uninsured (fines, sick pay, damage to product, equipment or process).

Indirect costs
Again these may be insured (business loss, product or process liability) or uninsured (loss of goodwill, extra overtime payments, accident investigation time, production delays).

Therefore, insurance policies can never cover all the costs of an accident or disease, either because some items are not covered by the policy or the insurance excess is greater than the particular item cost.

1.15 The framework for health and safety management

Most of the key elements required for effective health and safety management are very similar to those required for good quality, finance and general business management. Commercially successful organizations usually have good health and safety management systems in place. The principles of good and effective management provide a sound basis for the improvement of health and safety performance.

HSE, in HSG 65, have identified five key elements involved in a successful health and safety management system. The following chapters will describe and discuss this framework in detail. The five elements are:

1   A clear health and safety policy – Evidence shows that a sound, well thought out policy contributes to business efficiency and continuous improvement throughout the opera-

25

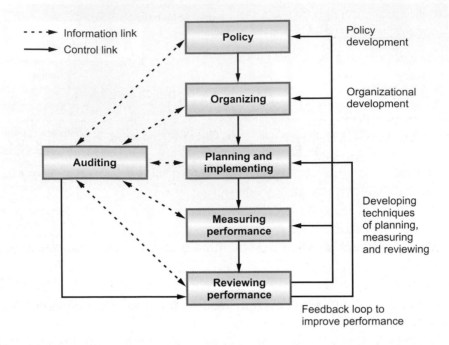

Figure 1.12 Key elements of successful health and safety management. Source HSE. Crown copyright material is reproduced with the permission of the Controller of HMSO and the Queen's Printer for Scotland.

tion. The demonstration of senior management involvement is evidence to all stake-holders that responsibilities to people and the environment are taken seriously.

2   A well-defined health and safety organization – The shared understanding of the organization's values and beliefs, at all levels of the company or concern is an essential component of a positive health and safety culture. An effective organization will be noted for good staff involvement and participation; high quality communications; the promotion of competency; and the empowerment of all employees to make informed contributions.

3   A clear health and safety plan – This involves the setting and implementation of performance standards and procedures through an effective health and safety management system. The plan is based on risk assessment methods to decide on priorities and set objectives for controlling or eliminating hazards and reducing risks. Measuring success requires the establishing of performance standards against which achievements can be identified.

4   The measurement of health and safety performance – This includes both active and reactive monitoring to see how effectively the health and safety management system is working. Active monitoring involves looking at the premises, plant and substances plus the people, procedures and systems. Reactive monitoring discovers through investigation of accidents and incidents why controls have failed. It is also important to measure the organization against its own long-term goals and objectives.

5   The audit and review of health and safety performance – The results of monitoring and independent audits should be systematically reviewed to see if the management system is achieving the right results. This is not only required by the HSW Act but is part of any company's commitment to continuous improvement. Comparisons should be made with

internal performance indicators and the external performance of organizations with exemplary practices and high standards.

Including health and safety performance in meaningful annual reports is considered best practice.

1.16 Practice NEBOSH questions for Chapter 1

1. (a) Explain, giving an example in each case, the circumstances under which a health and safety inspector may serve:
   (i) an improvement notice
   (ii) a prohibition notice
   (b) State the effect on each type of enforcement notice of appealing against it. (March 2001)

2. Outline the differences between civil law and criminal law. (March 2001)

3. Outline, with an example of each, the differences between health and safety regulations and HSC approved codes of practice. (March 2001)

4. (a) Outline four powers available to an inspector when investigating a workplace accident.
   (b) Identify the two types of enforcement notice that may be served by an inspector, stating the conditions that must be satisfied before each type of notice is served. (December 2001)

5. Outline the general duties placed on employees by:
   (i) the Health and Safety at Work etc. Act 1974
   (ii) the Management of Health and Safety at Work Regulations 1999. (June 2001)

6. (a) Explain three possible defences to a civil law claim of negligence.
   (b) State the circumstance in which an employer may be held vicariously liable for the negligence of an employee. (December 2000)

7. (a) Explain the meaning of the phrase 'so far as is reasonable practicable'.
   (b) State the general and specific duties of employers under section 2 of the Health and Safety at Work etc. Act 1974. (December 2000)

8. (a) Define the term 'negligence'.
   (b) Outline the three standard conditions that must be met for an employee to prove a case of negligence against an employer. (March 2000)

9. List the powers given to inspectors appointed under the Health and Safety at Work etc. Act 1974. (March 2000)

10. (a) Outline the main functions of:
    (i) criminal law
    (ii) civil law.
    (b) Explain the principle differences between common law and statute law. (June 1999)

Policy

2.1 Introduction

Every organization should have a clear policy for the management of health and safety so that everybody associated with the organization is aware of its health and safety aims and objectives. For a policy to be effective, it must be honoured in the spirit as well as the letter. A good health and safety policy will also enhance the performance of the organization in areas other than health and safety, help with the personal development of the workforce and reduce financial losses.

Figure 2.1 Well presented policy documents.

2.2 Legal requirements

Section 2(3) of the Health and Safety at Work Act 1974 requires employers, with more than four employees, to prepare and revise on a regular basis, a written health and safety policy together with the necessary organization and arrangements to carry it out and to bring the statement and any revision of it to the notice of their employees. This does not mean that organizations with four or less employees do not need to have a safety policy – it simply means that it does not have to be written down. The number of employees is the maximum number at any one time, whether they are full time, part time or seasonal.

This obligation on employers was introduced for the first time by the HSW Act and is related to the reliance in the Act on self-regulation by employers to improve health and safety standards rather than on enforcement alone. A good health and safety policy involves the development, monitoring and review of the standards needed to address and reduce the risks to health and safety produced by the organization.

The law requires that the written health and safety policy should entail:

> a health and safety policy statement which includes the health and safety aims and objectives of the organization

> a health and safety organizational structure detailing the people with health and safety responsibilities and their duties

> the health and safety arrangements in place in terms of systems and procedures.

The Management of Health and Safety at Work Regulations 1999 also requires the employer to 'make and give effect to such arrangements as are appropriate, having regard to the nature of his activities and the size of his undertaking, for the effective planning, organization, control, monitoring and review of the preventative and protective measures.' It further requires that these arrangements must be recorded when there are more than four employees.

When an inspector visits a premises, it is very likely that they will wish to see the health and safety policy as an initial indication of the management attitude to health and safety. There have been instances of prosecutions being made due to the absence of a written health and safety policy. (Such cases are, however, usually brought before the courts because of additional concerns.)

2.3 Key elements of a health and safety policy

2.3.1 Policy statement

The health and safety policy statement should contain the aims (which are not measurable) and objectives (which are measurable) of the organization or company. Aims will probably remain unchanged during policy revisions, whereas objectives will be reviewed and modified or changed each year. The statement should be written in clear and simple language so that it is easily understandable. It should also be fairly brief and broken down into a series of smaller statements or bullet points. The statement should be signed and dated by the most senior person in the organization. This will indicate the frequency with which the policy statement is reviewed. The policy statement should be written by the organization and not by external consultants since it needs to address the specific health and safety issues and hazards within the organization. In large organizations, it may be necessary to have health and safety policies for each department and/or site with an overarching general policy incorporating the individual policies. Such an approach is often used by local authorities and multinational companies.

The following points should be included or considered when a health and safety policy statement is being drafted:

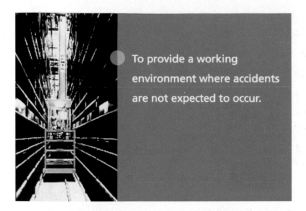

To provide a working environment where accidents are not expected to occur.

Figure 2.2 Part of a policy commitment.

➤ the aims should cover health and safety, welfare and relevant environmental issues
➤ the position of the senior person in the organization or company who is responsible for health and safety (normally the chief executive)
➤ the names of the Health and Safety Adviser and any safety representatives
➤ a commitment to the basic requirements of the Health and Safety at Work Act (access, egress, risk assessments, safe plant and systems of work, use, handling, transport and handling of articles and substances, information, training and supervision)
➤ a commitment to the additional requirements of the Management of Health and Safety at Work Regulations (risk assessment, emergency procedures, health surveillance and employment of competent persons)
➤ duties towards the wider general public and others (contractors, customers, students etc.)
➤ the principal hazards in the organization
➤ specific policies of the organization (e.g. smoking policy, violence to staff etc.)
➤ a commitment to employee consultation possible using a safety committee or plant council
➤ duties of employees (particularly those defined in the Management of Health and Safety at Work Regulations 1999)
➤ specific targets for the immediate and long-term future.

The policy statement should be posted on prominent notice boards throughout the workplace and brought to the attention of all employees at induction.

2.3.2 Organization of health and safety

This section of the policy defines the names, positions and duties of those within the organization or company who have a responsibility for health and safety. This will include:

➤ managers (e.g. directors, works managers, human resource managers and supervisors)
➤ specialists (e.g. health and safety adviser, occupational nurse, first aiders, fire officer, chemical analyst and electrician). For smaller companies, some of these specialists may well be employed on a consultancy basis
➤ employee representatives.

For the health and safety organization to work successfully, it must be supported from the top (preferably at Board level) and some financial resource made available. It is also important that certain key functions are included in the organization structure. These include:

> accident investigation and reporting
> health and safety training and information
> health and safety monitoring and audit
> health surveillance
> monitoring of plant and equipment and its maintenance
> liaison with external agencies
> management and/or employee safety committees – the management committee will monitor day-to-day problems and any concerns of the employee health and safety committee.

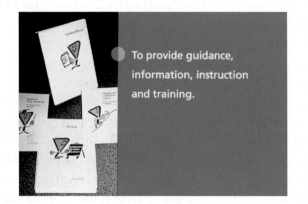
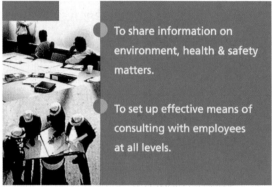

To provide guidance, information, instruction and training.

To share information on environment, health & safety matters.

To set up effective means of consulting with employees at all levels.

Figure 2.3 Good information, training and working with employees is essential.

The role of the health and safety adviser is to provide specialist information to managers in the organization and to monitor the effectiveness of health and safety procedures. The adviser is not 'responsible' for health and safety or its implementation, that is the role of the line managers.

Finally, the job descriptions, which define the duties of each person in the health and safety organizational structure, must not contain responsibility overlaps or blur chains of command. Each individual must be clear about their responsibilities and the limits of those responsibilities.

2.3.3 Arrangements for health and safety

The arrangements section of the health and safety policy comprises details of the means used to carry out the policy statement. This will include health and safety rules and procedures and the provision of facilities, such as a first aid room and wash rooms. It is common for risk assessments (including COSHH, manual handling and PPE assessments) to be included in the arrangements section, particularly for those hazards referred to in the policy statement. It is important that arrangements for fire and other emergencies and for information, instruction, training and supervision are also covered. Local codes of practice (e.g. for fork lift drivers) should be included.

The following list covers the more common items normally included in the arrangements section of the health and safety policy:

➤ employee health and safety code of practice
➤ accident and illness reporting and investigation procedure
➤ emergency procedures, first aid
➤ electrical equipment (maintenance and testing)
➤ control of hazardous substances, manual handling, PPE
➤ machinery safety (including safe systems of work), lifting and pressure equipment
➤ permits to work procedures
➤ health and safety inspection and audit procedures
➤ procedures for contractors and visitors
➤ catering and food hygiene procedures
➤ terms of reference and constitution of the safety committee.

The three sections of the health and safety policy are usually kept together in a health and safety manual and copies distributed around the organization.

2.4 Review of health and safety policy

It is important that the health and safety policy is monitored and reviewed on a regular basis. For this to be successful, a series of benchmarks needs to be established. Such benchmarks, or examples of good practice, are defined by comparison with the health and safety performance of other parts of the organization or the national performance of the occupational group of the organization. The Health and Safety Executive publish an annual report, statistics and bulletins, all of which may be used for this purpose. Typical benchmarks include accident rates per employee and accident or disease causation.

A positive promotion of health and safety performance will achieve far more than simply prevent accidents and ill-health. It will:

➤ support the overall development of personnel
➤ improve communication and consultation throughout the organization
➤ minimize financial losses due to accidents and ill-health and other incidents
➤ directly involve senior managers in all levels of the organization
➤ improve supervision, particularly for young persons and those on occupational training courses
➤ improve production processes
➤ improve the public image of the organization or company.

It is apparent, however, that some health and safety policies appear to be less than successful. There will be many reasons for this. The most common are:

➤ the statements in the policy and the health and safety priorities are not understood by the workforce
➤ minimal resources are made available for the implementation of the policy
➤ too much emphasis on rules for employees and too little on management policy
➤ lack of parity with other activities of the organization (such as finance and quality control) due to mistaken concerns about the costs of health and safety and the effect of those costs on overall performance
➤ lack of senior management involvement in health and safety, particularly at board level

➤ employee concerns that their health and safety issues are not being addressed or that they are not receiving adequate health and safety information.

In summary, a successful health and safety policy is likely to lead to a successful organization or company. A checklist for assessing any safety policy has been produced by the HSE and is reproduced in Appendix 2.1.

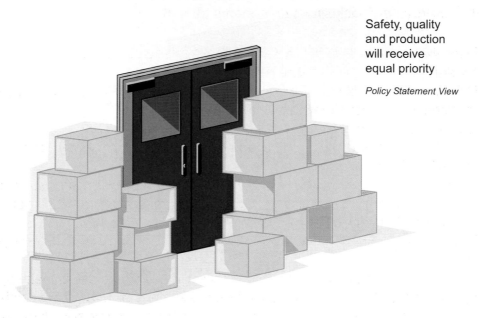

Safety, quality
and production
will receive
equal priority

Policy Statement View

Figure 2.4   Sound policy but not put into practice − blocked fire exit.

2.5 Practice NEBOSH questions for Chapter 2

1. (a) Outline the circumstances under which a written health and safety policy is legally required.
   (b) Identify the purposes of each of the following sections of a health and safety policy document:
      (i) 'statement of intent'
      (ii) 'organization'
      (iii) 'arrangements' (June 2001)
2. (a) Outline the legal requirements whereby an employer must prepare a written health and safety policy.
   (b) Identify the three main sections of a health and safety policy document and explain the purpose and general content of each section. (March 2000)
3. Outline the issues that are typically included in the arrangements section of a health and safety document. (March 1998)
4. Outline the circumstances that may give rise to a need for a health and safety policy to be revised. (March 1997)

Appendix 2.1 – Health and Safety Policy checklist

The following checklist is intended as an aid to the writing and review of a safety policy. It is derived from the booklet *Writing a safety policy statement* published by the HSE in booklet HSC 6.

General policy and organization

➤ Does the statement express a commitment to health and safety and are your obligations towards your employees made clear?

➤ Does it say which senior manager is responsible for seeing that it is implemented and for keeping it under review, and how this will be done?

➤ Is it signed and dated by you or a partner or senior director?

➤ Have the views of managers and supervisors, safety representatives and of the safety committee been taken into account?

➤ Were the duties set out in the statement discussed with the people concerned in advance, and accepted by them, and do they understand how their performance is to be assessed and what resources they have at their disposal?

➤ Does the statement make clear that cooperation on the part of all employees is vital to the success of your health and safety policy?

➤ Does it say how employees are to be involved in health and safety matters, for example, by being consulted, by taking part in inspections, and by sitting on a safety committee?

➤ Does it show clearly how the duties for health and safety are allocated and are the responsibilities at different levels described?

➤ Does it say who is responsible for the following matters (including deputies where appropriate)?

> ➤ reporting investigations and recording accidents

> ➤ fire precautions, fire drill, evacuation procedures

> ➤ first aid

> ➤ safety inspections

> ➤ the training programme

> ➤ ensuring that legal requirements are met, for example regular testing of lifts and notifying accidents to the health and safety inspector.

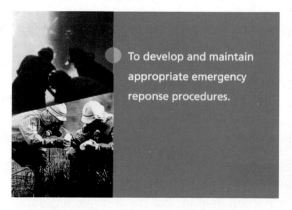

To develop and maintain appropriate emergency reponse procedures.

Figure 2.5 Emergency procedures.

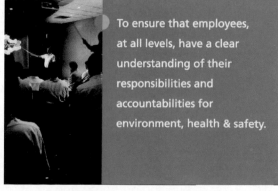

To ensure that employees, at all levels, have a clear understanding of their responsibilities and accountabilities for environment, health & safety.

Figure 2.6 Responsibilities.

Arrangements that need to be considered

➤ Keeping the workplace, including staircases, floors, ways in and out, washrooms etc. in a safe and clean condition by cleaning, maintenance and repair.

Plant and substances

➤ Maintenance of equipment such as tools, ladders etc. Are they in a safe condition?
➤ Maintenance and proper use of safety equipment such as helmets, boots, goggles, respirators etc.
➤ Maintenance and proper use of plant, machinery and guards.
➤ Regular testing and maintenance of lifts, hoists, cranes, pressure systems, boilers and other dangerous machinery, emergency repair work, and safe methods of doing it.
➤ Maintenance of electrical installations and equipment.
➤ Safe storage, handling and, where applicable, packaging, labelling and transport of dangerous substances.
➤ Controls of work involving harmful substances such as lead and asbestos.
➤ The introduction of new plant, equipment or substances into the workplace by examination, testing and consultation with the workforce.

Other hazards

➤ Noise problems – wearing of hearing protection, and control of noise at source.
➤ Preventing unnecessary or unauthorized entry into hazardous areas.
➤ Lifting of heavy or awkward loads.
➤ Protecting the safety of employees against assault when handling or transporting the employer's money or valuables.
➤ Special hazards to employees when working on unfamiliar sites, including discussion with site manager where necessary.
➤ Control of works transport, e.g. fork lift trucks, by restricting use to experienced and authorized operators or operators under instruction (which should deal fully with safety aspects).

Figure 2.7 Fork lift truck.

Emergencies

➤ Ensuring that fire exits are marked, unlocked and free from obstruction.
➤ Maintenance and testing of fire-fighting equipment, fire drills and evacuation procedures.
➤ First aid, including name and location of person responsible for first aid and deputy, and location of first aid box.

Communication

> Giving your employees information about the general duties under the Health and Safety at Work Act and specific legal requirements relating to their work.

> Giving employees necessary information about substances, plant, machinery, and equipment with which they come into contact. Discussing with contractors, before they come on site, how they can plan to do their job, whether they need equipment of yours to help them, whether they can operate in a segregated area or when part of the plant is shut down and, if not, what hazards they may create for your employees and vice versa.

Training

> Training employees, supervisors and managers to enable them to work safely and to carry out their health and safety responsibilities efficiently.

Supervising

> Supervising employees so far as necessary for their safety – especially young workers, new employees and employees carrying out unfamiliar tasks.

Keeping check

> Regular inspections and checks of the workplace, machinery appliances and working methods.

Organizing for health and safety

3.1 Introduction

This chapter is about managers in businesses, or other organizations, setting out clear responsibilities and lines of communications for everyone in the enterprise. The unit also covers the legal responsibilities that exist between people who control premises and those who use them; between contractors and those who hire them; and the duties of suppliers, manufacturers and designers of articles and substances for use at work. Chapter 2 concerned policy, which is an essential first step. The policy will only remain as words on paper, however good the intentions, until there is an effective organization set up to implement and monitor its requirements.

The policy sets the direction for health and safety within the enterprise and forms the written intentions of the principals or directors of the business. The organization needs to be clearly communicated and people need to know what they are responsible for in the day-to-day operations. A vague statement that 'everyone is responsible for

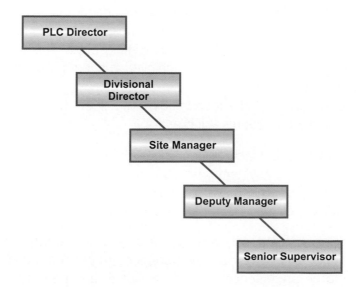

Figure 3.1 Everyone from senior managers down has health and safety responsibilities.

health and safety' is misleading and fudges the real issues. Everyone is responsible but management in particular. There is no equality of responsibility under law between those who provide direction and create policy and those who are employed to follow. Principals, or employers in terms of the HSW Act, have substantially more responsibility than employees.

Some policies are written so that most of the wording concerns strict requirements laid on employees and only a few vague words cover managers' responsibilities. Generally, such policies do not meet the HSW Act or the Management of Health and Safety at Work Regulations 1999, which require an effective policy with a robust organization and arrangements to be set up.

3.2 Control

Like all management functions, establishing control and maintaining it day in day out is crucial to effective health and safety management. Managers, particularly at senior levels, must take proactive responsibility for controlling issues that could lead to ill-health, injury or loss. A nominated senior manager at the top of the organization needs to oversee policy implementation and monitoring. The nominated person will need to report regularly to the most senior management team and will be a director or principal of the organization.

Health and safety responsibilities will need to be assigned to line managers and expertise must be available, either inside or outside the enterprise, to help them achieve the requirements of the HSW Act and the regulations made under the Act. The purpose of the health and safety organization is to harness the collective enthusiasm, skills and effort of the entire workforce with managers taking key responsibility and providing clear direction. The prevention of accidents and ill-health through management systems of control becomes the focus rather than looking for individuals to blame after the incident occurs.

The control arrangements should be part of the written health and safety policy. Performance standards will need to be agreed and objectives set which link the outputs required to specific tasks and activities for which individuals are responsible. For example, the objective could be to carry out a workplace inspection once a week to an agreed checklist and rectify faults within three working days. The periodic, say annual, audit would check to see if this was being achieved and if not the reasons for non-compliance with the objective.

People should be held accountable for achieving the agreed objectives through existing or normal procedures such as:

➤ Job descriptions, which include health and safety responsibilities;
➤ Performance appraisal systems, which look at individual contributions;
➤ Arrangements for dealing with poor performance;
➤ Where justified, the use of disciplinary procedures.

Such arrangements are only effective if health and safety issues achieve the same degree of importance as other key management concerns and a good performance is considered to be an essential part of the career and personal development.

3.3 Employers' responsibilities

Employers have duties under both criminal and civil law. The civil law duties are covered in Chapter 1. The general duties of employers' under HSW Act relate to:

> The health, safety and welfare at work of employees and other workers, whether part-time, casual, temporary, homeworkers, on work experience, Government training schemes or on site as contractors – i.e. anyone working under their control or direction
> The health and safety of anyone who visits or uses the workplace
> The health and safety of anyone who is allowed to use the organization's equipment
> The health and safety of those affected by the work activity, for example neighbours, and the general public.

Other duties on employers are covered in Chapter 1 and the summary of the HSW Act in Chapter 17.

3.4 Employees' responsibilities

Employees have specific responsibilities under the HSW Act, which are:

> To take reasonable care for the health and safety of themselves and of other persons who may be affected by their acts or omissions at work. This involves the same wide group that the employer has to cover, not just the people on the next desk or bench;
> To cooperate with employers in assisting them to fulfil their statutory duties;
> Not to interfere with deliberately or misuse anything provided, in accordance with health and safety legislation, to further health and safety at work.

3.5 Organizational health and safety responsibilities

In addition to the legal responsibilities on management, there are many specific responsibilities imposed by each organization's health and safety policy. Appendix 3.1 to this chapter shows a typical summary of the health and safety responsibilities and accountability of each level of the line organization to provide an understanding of how health and safety responsibilities and accountability are integrated within the total organization. The responsibilities cover Directors, senior managers, site managers, department managers, supervisors and employees. Many organizations will not fit this exact structure but most will have those who direct, those who manage or supervise and those who have no line responsibility but have responsibilities to themselves and fellow workers.

Because of the special role and importance of directors, these are covered here in detail.

3.5.1 Directors' responsibilities

The Chairman of the Health and Safety Commission said at the launch of the guidance on Directors' responsibilities:

Health and safety is a boardroom issue. Good health and safety reflects strong leadership from the top and that is what we want to see. The company whose chairperson or chief executive is the champion of health and safety sends the kind of message which delivers good performance on the ground.

Those who are at the top have a key role to play, which is why boards are being asked to nominate one of their members to be a 'health and safety' director. But appointing a health and safety director or department does not absolve the Board from its collective responsibility to lead and oversee health and safety management.

The guidance from the HSC published in 2001, sets out the following action points for Directors:

➤ the Board needs to accept formally and publicly its collective role in providing health and safety leadership in its organization;

➤ each member of the Board needs to accept their individual role in providing health and safety leadership for their organization;

➤ the Board needs to ensure that all board decisions reflect its health and safety intentions, as articulated in the health and safety policy statement. It is important for boards to remember that, although health and safety functions can (and should) be delegated, legal responsibility for health and safety rests with the employer;

➤ the Board needs to recognize its role in engaging the active participation of workers in improving health and safety;

➤ the Board needs to ensure that it is kept informed of, and alert to, relevant health and safety risk management issues. The Health and Safety Commission recommends that boards appoint one of their number to be the 'Health and Safety Director'.

Directors need to ensure that the Board's health and safety responsibilities are properly discharged. The Board will need to:

➤ carry out an annual review of health and safety performance;

➤ keep the health and safety policy statement up to date with current board priorities and review the policy at least every year;

➤ ensure that there are effective management systems for monitoring and reporting on the organization's health and safety performance;

➤ ensure that any significant health and safety failures and their investigation are communicated to board members;

➤ ensure that when decisions are made the health and safety implications are fully considered;

➤ ensure that regular audits are carried out to check that effective health and safety risk management systems are in place.

By appointing a 'Health and Safety Director' there will be a board member who can ensure that these health and safety risk management issues are properly addressed, both by the Board and more widely throughout the organization.

The Chairman and/or Chief Executive have a critical role to play in ensuring risks are properly managed and that the Health and Safety Director has the necessary competence, resources and support of other board members to carry out their functions. Indeed, some boards may prefer to see all the health and safety functions assigned to their Chairman and/or Chief Executive. As long as there is clarity about the health and safety responsibilities and functions, and the Board properly addresses the issues, this is acceptable.

The health and safety responsibilities of all board members should be clearly articulated in the organization's statement of health and safety policy and arrangements. It is important that the role of the Health and Safety Director should not detract either from the responsibilities of other directors for specific areas of health and safety risk management or from the health and safety responsibilities of the Board as a whole.

3.6 Role and functions of health and safety and other advisers

Figure 3.2 Safety practitioner at the front line.

3.6.1 Competent person

One or more competent persons must be appointed to help managers comply with their duties under health and safety law. The essential point is that managers should have access to expertise to help them fulfil the legal requirements. However, they will always remain as advisers and do not assume responsibility in law for health and safety matters. This responsibility always remains with line managers and cannot be delegated to an adviser whether inside or outside the organization. The appointee could be:

➤ an employer themselves if they are sure they know enough about what to do. This may be appropriate in a small low hazard business;
➤ one or more employees, as long as they have sufficient time and other resources to do the task properly;
➤ someone from outside the organization who has sufficient expertise to help.

The HSE have produced two free leaflets entitled *Need help on health and safety? Guidance for employers on when and how to get advice on health and safety* and *Selecting a Health & Safety Consultancy*.

If an employer decides to seek outside help they need to be sure that no employees are competent to assist. Many health and safety issues can be tackled by people with an understanding of current best practice and an ability to judge and solve problems. Some help is needed long term, others for a one-off short period. There are a wide range of experts available for different types of health and safety problem. For example:

➤ engineers for specialist ventilation or chemical processes;

➤ occupational hygienists for assessment and practical advice on exposure to chemical (dust, gases, fumes, etc.), biological (viruses, fungi, etc.) and physical (noise, vibration, etc.) agents;

➤ occupational health professionals for medical examinations and diagnosis of work-related disease, pre-employment and sickness advice, health education, etc.;

➤ ergonomists for advice on suitability of equipment, comfort, physical work environment, work organization, etc.;

➤ physiotherapists for treatment and prevention of musculoskeletal disorders, etc.;

➤ radiation protection advisers for advice on compliance with the Ionizing Radiation Regulations 1999;

➤ health and safety practitioner for general advice on implementation of legislation, health and safety management, risk assessment, control measures and monitoring performance, etc.

3.6.2 Health and safety adviser

Status and **competence** are essential to the role of health and safety and other advisers. They must be able to advise management and employees or their representatives with authority and independence. They need to be able to advise on:

➤ creating and developing health and safety policies. These will be for existing activities plus new acquisitions or processes;

➤ the promotion of a positive health and safety culture. This includes helping managers to ensure that an effective health and safety policy is implemented;

➤ health and safety planning. This will include goal-setting, deciding priorities and establishing adequate systems and performance standards. Short- and long-term objectives need to be realistic;

➤ day-to-day implementation and monitoring of policy and plans. This will include accident and incident investigation, reporting and analysis;

➤ performance reviews and audit of the whole health and safety management system.

3.6.3 To do this properly, health and safety advisers need to

➤ have proper training and be suitably qualified – e.g. NEBOSH Diploma parts 1 & 2, relevant degree and, where appropriate, a Registered Safety Practitioner; NEBOSH Certificate in small to medium-sized low hazard premises, like offices and retail stores;

➤ keep up-to-date information systems on such topics as civil and criminal law, health and safety management and technical advances;

➤ know how to interpret the law as it applies to their own organization;

➤ actively participate in the establishment of organizational arrangements, systems and risk control standards relating to hardware and human performance. Health and safety advisers will need to work with management on matters such as legal and technical standards;

➤ undertake the development and maintenance of procedures for reporting, investigating, recording and analysing accidents and incidents;

> develop and maintain procedures to ensure that senior managers get a true picture of how well health and safety is being managed (where a benchmarking role may be especially valuable). This will include monitoring, review and auditing;
> be able to present their advice independently and effectively.

3.6.4 Relationships within the organization

Health and safety advisers:

> support the provision of authoritative and independent advice;
> report directly to directors on matters of policy and have the authority to stop work if it contravenes agreed standards and puts people at risk of injury;
> are responsible for professional standards and systems. They may also have line management responsibility for other health and safety professionals, in a large group of companies or on a large and/or high-hazard site.

3.6.5 Relationships outside the organization

Health and safety advisers also have a function outside their own organization. They provide the point of liaison with a number of other agencies including the following:

> environmental health officers and licensing officials
> architects and consultants
> HSE and the Fire Authorities
> the police
> HM Coroner or the Procurator Fiscal
> local authorities
> insurance companies
> contractors
> clients and customers
> the public
> equipment suppliers
> the media
> general practitioners
> IOSH and occupational health specialists and services.

3.7 Persons in control of premises

Section 4 of the HSW Act requires that 'Persons in control of **non-domestic** premises' take such steps as are reasonable in their position to ensure that there are no risks to the health and safety of people who are not employees but use the premises. This duty extends to:

> people entering the premises to work
> people entering the premises to use machinery or equipment, for example, a launderette
> access and exit from the premises

Figure 3.3   NEBOSH are in control here.

➤   corridors, stairs, lifts and storage areas.

Those in control of premises are required to take a range of steps depending on the likely use of the premises and the extent of their control and knowledge of the actual use of the premises.

Persons in control of premises must also prevent harmful emissions into the atmosphere under section 5 of the HSW Act, but this has been repealed (Subject to Commencement Order) by the Environmental Protection Act 1990. In the interim period any action for emissions is likely to be under the 1990 Act and not the HSW Act.

3.8   Self-employed

The duties of the self-employed under the HSW Act are fairly limited. The Revitalising Health and Safety Strategy document expressed concern about whether the HSW Act sufficiently covers the area, in view of the huge growth in the use of contractors and sub-contractors throughout the UK. At December 2001 they are still being reviewed to consider how adjustments should be made. Under the HSW Act the self-employed are:

➤   responsible for their own health and safety;
➤   responsible to ensure that others who may be affected are not exposed to risks to their health and safety.

These responsibilities are extended by the Management of Health and Safety at Work Regulations 1999, which requires self-employed people to:

➤   carry out risk assessment;
➤   cooperate with other people who work in the premises and, where necessary, in the appointment of a health and safety coordinator;
➤   provide comprehensible information to other peoples' employees working in their undertaking.

3.9 The supply chain

3.9.1 Supply chain management

Market leaders in every industry are increasing their grip on the chain of supply. They do so by monitoring rather than managing, and also by working more closely with suppliers. The result of this may be that suppliers or contractors are absorbed into the culture of the dominant firm, while avoiding the costs and liabilities of actual management. Powerful procurement departments emerge, to define and impose the necessary quality standards and guard the lists of preferred suppliers.

In the process, the freedom of local operating managers to pick and choose suppliers is reduced. Even though the responsibility to do so is often retained, it is strongly qualified by centrally imposed rules and lists, and assistance or oversight.

In these conditions, suppliers and contractors looking for business with major firms need greater flexibility and wider competence than previously. This often implies increased size and perhaps mergers, though in principle bids could be and perhaps are made by loose partnerships of smaller firms organized to secure such business. Firms applying to tender may not always understand that large firms will usually make inquiries of other firms with experience of their health and safety performance. Many large firms regard health and safety performance as a good indicator of the general competence of a smaller firm, precisely because it is an aspect that is often neglected. An adverse report will, at the very least, mean that a bidder must do even better on other aspects to succeed and, in these circumstances, companies generally make special arrangements to ensure that their safety standards are met by the contractors.

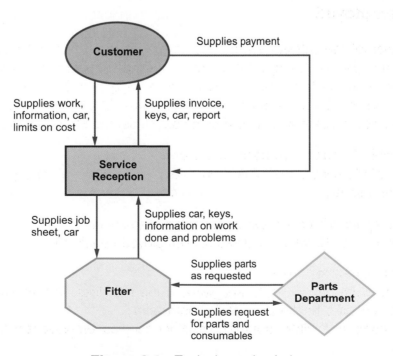

Figure 3.4 Typical supply chain.

3.9.2 Legislation

The HSW Act Section 6 places a duty on everyone in the supply chain, from the designer to the final installer, of articles of plant or equipment for use at work or any article of fairground equipment to:

➤ ensure that the article will be safe and without risks to health at all times when it is being set, used, cleaned or maintained;
➤ carry out any necessary testing and examination to ensure that it will be safe, and;
➤ provide adequate information about its safe setting, use, cleaning, maintenance, dismantling and disposal.

There is an obligation on designers or manufacturers to do any research necessary to prove safety in use. Erectors or installers have special responsibilities to make sure when handed over that the plant or equipment is safe to use.

Similar duties are placed on manufacturers and suppliers of substances for use at work to ensure that the substance is safe when properly used, handled, processed, stored or transported, to provide adequate information and do any necessary research, testing or examining.

Where articles or substances are imported, the suppliers' obligations outlined above attach to the importer, whether a separate importing business or the user himself.

Often items are obtained through hire purchase, leasing or other financing arrangements with the ownership of the items being vested with the financing organization. Where the financing organization's only function is to provide the money to pay for the goods, the supplier's obligations do not attach to them.

3.9.3 Information for customers

The quality movement has drawn attention to the need to ensure that there are processes in place which ensure quality, rather than just inspecting and removing defects when it is too late. In much the same way, organizations need to manage health and safety rather than acting when it is too late.

Customers need information and specifications from the manufacturer or supplier – especially where there is a potential risk involved for them. When deciding what the supplier needs to pass on, careful thought is required about the health and safety factors associated with any product or service.

This means focusing on four key questions and then framing the information supplied so that it deals with each one. The questions are:

➤ Are there any inherent dangers in the product or service being passed on – what could go wrong?
➤ What can the manufacturer or supplier do while working on the product or service to reduce the chance of anything going wrong later?
➤ What can be done at the point of handover to limit the chances of anything going wrong?
➤ What steps should customers take to reduce the chances of something going wrong? What precisely would they need to know?

Depending on what is being provided to customers, the customer information may need to comply with the following legislation:

➤ Supply of Machinery (Safety) Regulations 1992 and amendment 1994
➤ Provision and Use of Work Equipment Regulations 1998
➤ Control of Substances Hazardous to Health Regulations 2002
➤ Chemicals (Hazard Information and Packaging for Supply) Regulations 2002.

This list is not exhaustive.

3.9.4 Buying problems

Examples of problems that may arise when purchasing include:

➤ second-hand equipment which does not conform to current safety standards
➤ starting to use new substances which do not have safety data sheets
➤ machinery which, while well guarded for operators, may pose risks for a maintenance engineer
➤ office chairs which do not provide adequate back support.

A risk assessment should be done on any new product, taking into account the likely life expectancy (e.g. delivery, installation, use, cleaning, maintenance, disposal, etc.). The supplier should be able to provide the information needed to do this. This will help the purchaser make an informed decision on the total costs because the risks will have been identified as will the precautions needed to control those risks. A risk assessment will still be needed for a CE-marked product. The CE marking signifies the manufacturer's declaration that the product conforms to relevant European Directives. Declarations from reputable manufacturers will normally be reliable. However, purchasers should be alert to fake or inadequate declarations and technical standards which may affect the health and safety of the product despite the CE marking. The risk assessment is still necessary to consider how and where the product will be used, what effect it might have on existing operations and what training will be required.

Figure 3.5 Inadequate chair – take care when buying second-hand.

Employers have some key duties when buying plant and equipment:

➤ they must ensure that work equipment is safe, suitable for its purpose and complies with the relevant legislation. This applies equally to equipment which is adapted to be used in ways for which it was not originally designed;
➤ when selecting work equipment, they must consider existing working conditions and health and safety issues;
➤ they must provide adequate health and safety information, instructions, training and supervision for operators. Manufacturers and suppliers are required by law to provide

information that will enable safe use of the equipment, substances, etc. and without risk to health.

Some of the issues that will need to be considered when buying in product or plant include:

➤ ergonomics – risk of work-related upper limb disorders (WRULD)
➤ manual handling needs
➤ access/egress
➤ storage, e.g. of chemicals
➤ risk to contractors when decommissioning old plant or installing new plant
➤ hazardous materials – provision of extraction equipment or personal protective equipment
➤ waste disposal
➤ safe systems of work
➤ training
➤ machinery guarding
➤ emissions from equipment/plant, such as noise, heat or vibration.

3.10 Contractors

3.10.2 Introduction

The use of contractors is increasing as many companies turn to outside resources to supplement their own staff and expertise. A contractor is anyone who is brought in to work who is not an employee. Contractors are used for maintenance, repairs, installation, construction, demolition, computer work, cleaning, security, health and safety and many other tasks. Sometimes there are several contractors on site at any one time. Clients need to think about how their work may affect each other and how they interact with the normal site occupier.

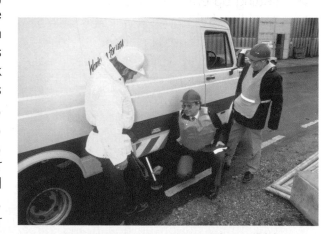

Figure 3.6 Contractors at work.

3.10.2 Legal considerations

The HSW Act applies to all work activities. It requires employers to ensure, so far as is reasonably practicable, the health and safety of:

➤ their employees
➤ other people at work on their site, including contractors
➤ members of the public who may be affected by their work.

All parties to a contract have specific responsibilities under health and safety law, and these cannot be passed on to someone else:

➤ employers are responsible for protecting people from harm caused by work activities. This includes the responsibility not to harm contractors and sub-contractors on site;

➤ employees and contractors have to take care not to endanger themselves, their colleagues or others affected by their work;

➤ contractors also have to comply with the HSW Act and other health and safety legislation. Clearly, when contractors are engaged, the activities of different employers do interact. So cooperation and communication are needed to make sure all parties can meet their obligations;

➤ employees have to cooperate with their employer on health and safety matters, and not do anything that puts them or others at risk;

➤ employees must be trained and clearly instructed in their duties;

➤ self-employed people must not put themselves in danger, or others who may be affected by what they do;

➤ suppliers of chemicals, machinery and equipment have to make sure their products or imports are safe, and provide information on this.

The Management of Health and Safety at Work Regulations apply to everyone at work and encourage employers to take a more systematic approach to dealing with health and safety by:

➤ assessing the risks which affect employees and anyone who might be affected by the site occupier's work, including contractors;

➤ setting up emergency procedures;

➤ providing training;

➤ cooperating with others on health and safety matters, for example, contractors who share the site with an occupier;

➤ providing temporary workers, such as contractors, with health and safety information.

The principles of cooperation, coordination and communication between organizations underpin the Management of Health and Safety at Work Regulations and the CDM Regulations, explained next. See later section 3.11 on joint occupation of premises. For more information on the Management of Health and Safety at Work Regulations, read the summary in Chapter 17.

3.10.3 Construction design and management (CDM) regulations

Businesses often engage contractors at one time or another to build plant, convert or extend premises and demolish buildings. In many cases, where the CDM regulations apply, there is a requirement for the contractor to produce a safety plan containing the following key elements:

➤ information regarding the contractor's Health & Safety policy
➤ the contractor's health and safety organization detailing the responsibilities of individuals
➤ information on the contractor's procedures and standards of safe working
➤ the method statements for the contract in hand
➤ auditing and implementation of the plan.

Smaller contractors may need some guidance to produce a method statement. While it does not need to be lengthy, it should set out those features essential to safe working, for example, access arrangements, personal protective equipment, control of chemical risks, etc.

Copies of relevant risk assessments for the work to be undertaken should be requested. These need not be very detailed but should indicate the risk and the control methods to be used.

The Client, Planning Supervisor and Principal Contractor all have specific roles under CDM Regulations. (For more information see Chapter 14 on Construction activities and the Summary of Legislation in Chapter 17.)

3.10.4 Contractor selection

The selection of the right contractor for a particular job is probably the most important element in ensuring that the risks to the health and safety of everybody involved on the activity and people in the vicinity are reduced as far as possible. Ideally, selection should be made from a list of approved contractors who have demonstrated that they are able to meet the client's requirements.

The selection of a contractor has to be a balanced judgement with a number of factors taken into account. Fortunately, a contractor who works well and meets the client's requirements in terms of the quality and timeliness of the work is likely also to have a better than average health and safety performance. Cost, of course, will have to be part of the judgement but may not provide any indication of which contractor is likely to give the best performance in health and safety terms. In deciding which contractor should be chosen for a task, the following should be considered:

➤ Do they have an adequate health and safety policy?
➤ Can they demonstrate that the person responsible for the work is competent?
➤ Can they demonstrate that competent safety advice will be available?
➤ Do they monitor the level of accidents at their work site?
➤ Do they have a system to assess the hazards of a job and implement appropriate control measures?
➤ Will they produce a method statement, which sets out how they will deal with all significant risks?
➤ Do they have guidance on health and safety arrangements and procedures to be followed?
➤ Do they have effective monitoring arrangements?
➤ Do they use trained and skilled staff who are qualified where appropriate? (Judgement will be required, as many construction workers have had little or no training except training on the job.) Can the company demonstrate that the employees or other workers used for the job have had the appropriate training and are properly experienced and, where appropriate, qualified?
➤ Can they produce good references indicating satisfactory performance?

3.10.5 Contractors authorization

Contractors, their employees, sub-contractors and their employees, should not be allowed to commence work on any client's site without authorization signed by the

Company contact. The authorization should clearly define the range of work that the contractor can carry out and set down any special requirements, for example, protective clothing, fire exits to be left clear, isolation arrangements, etc.

Permits will be required for operations such as hot work, etc. All contractors should keep a copy of their authorization at the place of work. A second copy of the authorization should be kept at the site and be available for inspection.

The Company contact signing the authorization will be responsible for all aspects of the work of the contractor. The contact will need to check as a minimum the following:

➤ that the correct contractor for the work has been selected
➤ that the contractor has made appropriate arrangements for supervision of staff
➤ that the contractor has received and signed for a copy of the contractor's safety rules
➤ that the contractor is clear what is required, the limits of the work and any special precautions that need to be taken
➤ that the contractor's personnel are properly qualified for the work to be undertaken.

The Company contact should check whether sub-contractors will be used. They will also require authorization, if deemed acceptable. It will be the responsibility of the Company contact to ensure that sub-contractors are properly supervised.

Appropriate supervision will depend on a number of factors, including the risk associated with the job, experience of the contractor and the amount of supervision the contractor will provide. The responsibility for ensuring there is proper supervision lies with the person signing the contractor's authorization.

The Company contact will be responsible for ensuring that there is adequate and clear communication between different contractors and Company personnel where this is appropriate.

3.10.6 Safety rules for contractors

In the conditions of contract there should be a stipulation that the contractor and all of their employees adhere to the contractor's safety rules. Contractor's safety rules should contain as a minimum the following points:

➤ **health & safety:** that the contractor operates to at least the minimum legal standard and conforms to accepted industry good practice;
➤ **supervision:** that the contractor provides a good standard of supervision of their own employees;
➤ **sub-contractors:** that they may not use sub-contractors without prior written agreement from the Company;
➤ **authorization:** that each employee must carry an authorization card issued by the Company at all times while on site.

3.10.7 Example of rules for contractors

Contractors engaged by the organization to carry out work on its premises will:

➤ familiarize themselves with so much of the organization's safety policy as affects them and will ensure that appropriate parts of the policy are communicated to their employees, and any sub-contractors and employees of sub-contractors who will do work on the premises;

➤ cooperate with the organization in its fulfilment of its health and safety duties to contractors and take the necessary steps to ensure the like cooperation of their employees;

➤ comply with their legal and moral health, safety and food hygiene duties;

➤ ensure the carrying out of their work on the organization's premises in such a manner as not to put either themselves or any other persons on or about the premises at risk;

➤ where they wish to avail themselves of the organization's first aid arrangements/facilities while on the premises, ensure that written agreement to this effect is obtained prior to first commencement of work on the premises;

➤ where applicable and requested by the organization, supply a copy of its statement of policy, organization and arrangements for health and safety written for the purposes of compliance with The Management of Health and Safety at Work Regulations 1999 and Section 2(3) of the Health and Safety at Work Act 1974;

➤ abide by all relevant provisions of the organization's safety policy, including compliance with health and safety rules;

➤ ensure that on arrival at the premises, they and any other persons who are to do work under the contract report to reception or their designated organization contact.

Without prejudice to the requirements stated above, contractors, sub-contractors and employees of contractors and sub-contractors will, to the extent that such matters are within their control, ensure:

➤ the safe handling, storage and disposal of materials brought onto the premises;

➤ that the organization is informed of any hazardous substances brought onto the premises and that the relevant parts of the Control of Substances Hazardous to Health Regulations in relation thereto are complied with;

➤ that fire prevention and fire precaution measures are taken in the use of equipment which could cause fires;

➤ that steps are taken to minimize noise and vibration produced by their equipment and activities;

➤ that scaffolds, ladders and other such means of access, where required, are erected and used in accordance with statutory requirements and good working practice;

➤ that any welding or burning equipment brought onto the premises is in safe operating condition and used in accordance with all safety requirements;

➤ that any lifting equipment brought onto the premises is adequate for the task and has been properly tested/certified;

➤ that any plant and equipment brought onto the premises is in safe condition and used/ operated by competent persons;

➤ that for vehicles brought onto the premises, any speed, condition or parking restrictions are observed;

➤ that compliance is made with the relevant requirements of the Electricity at Work Regulations 1989;

➤ that connection(s) to the organization's electricity supply is from a point specified by its management and is by proper connectors and cables;

➤ that they are familiar with emergency procedures existing on the premises;

➤ that welfare facilities provided by the organization are treated with care and respect;

➤ that access to restricted parts of the premises is observed and the requirements of food safety legislation are complied with;

➤ that any major or lost-time accident or dangerous occurrence on the organization's premises is reported as soon as possible to their site contact;

➤ that where any doubt exists regarding health and safety requirements, advice is sought from the site contact.

The foregoing requirements *do not* exempt contractors from their statutory duties in relation to health and safety, but are intended to assist them in attaining a high standard of compliance with those duties.

3.11 Joint occupation of premises

The Management of Health and Safety at Work Regulations specifically states that where two or more employers share a workplace – whether on a temporary or a permanent basis – each employer shall:

➤ cooperate with other employers;
➤ take reasonable steps to coordinate between other employers to comply with legal requirements;
➤ take reasonable steps to inform other employers where there are risks to health and safety.

All employers and self-employed people involved should satisfy themselves that the arrangements adopted are adequate. Where a particular employer controls the premises, the other employers should help to assess the shared risks and coordinate any necessary control procedures. Where there is no controlling employer the organizations present should agree joint arrangements to meet regulatory obligations, such as appointing a health and safety coordinator.

3.12 Cooperation with the workforce

It is important to gain the cooperation of all employees if a successful health and safety culture is to become established. This cooperation is best achieved by consultation. The easiest and potentially the most effective method of consultation is the health and safety committee. It will realize its full potential if its recommendations are seen to be implemented and both management and employee concerns are freely discussed. It will not be so successful if it is seen as a talking shop.

The committee should have stated objectives, which mirror the objectives in the organization's health and safety policy statement, and its own terms of reference. Terms of reference could include the following:

Figure 3.7 Working together.

➤ the study of accident and notifiable disease statistics to enable reports to be made of recommended remedial actions
➤ the examination of health and safety audit and statutory inspection reports
➤ the consideration of reports from the external enforcement agency
➤ the review of new legislation, approved codes of practice and guidance and its effect on the organization
➤ the monitoring and review of all health and safety training and instruction activities in the organization
➤ the monitoring and review of health and safety publicity and communication throughout the organization.

There are no fixed rules on the composition of the health and safety committee except that it should be representative of the whole organization. It should have representation from the workforce and the management including at least one senior manager.

There are two pieces of legislation that cover health and safety consultation with employees and both are summarized in Chapter 17.

3.12.1 The Safety Representatives and Safety Committees Regulations 1977

These regulations only apply to those organizations that have recognized trade unions for collective bargaining purposes. The recognized trade union may appoint safety representatives from among the employees and notify the employer in writing.

The safety representative has several functions (not duties) including:

➤ representing employees in consultation with the employer
➤ investigating potential hazards and dangerous occurrences
➤ investigating the causes of accidents
➤ investigating employee complaints relating to health, safety and welfare
➤ making representations to the employer on health, safety and welfare matters
➤ carrying out inspections
➤ representing employees at the workplace in consultation with enforcing inspectors
➤ receiving information
➤ attending safety committee meetings.

Representatives must be allowed time off with pay to fulfil these functions and to undergo health and safety training. They must also be allowed to inspect the workplace at least once a quarter or sooner if there has been a substantial change in the conditions of work.

Finally, if at least two representatives have requested in writing that a safety committee be set up, the employer has three months to comply.

3.12.2 The Health and Safety (Consultation with Employees) Regulations 1996

These regulations were introduced so that employees working in organizations without recognized trade unions would still need to be consulted on health and safety matters. All employees must now be consulted either on an individual basis (e.g. very small companies) or through safety representatives elected by the workforce, known as ROES (Representative(s) of Employee Safety). The guidance to these regulations emphasizes the difference between informing and consulting. Consultation involves listening to the opinion of employees on a particular issue and taking it into account before a decision is made. Informing employees means providing information on health and safety issues such as risks, control systems and safe systems of work.

The functions of these representatives are similar to those under the Safety Representatives and Safety Committees Regulations 1977 as are the rights to health and safety training (i.e. time off and costs covered by the employer).

The employer must consult on the following:

➤ the introduction of any measure or change which may substantially affect employees' health and safety
➤ the arrangements for the appointment of competent persons to assist in following health and safety law
➤ any information resulting from risk assessments or their resultant control measures which could affect the health, safety and welfare of employees
➤ the planning and organization of any health and safety training required by legislation
➤ the health and safety consequences to employees of the introduction of new technologies into the workplace.

However, the employer is not expected to disclose information if:

➤ it violates a legal prohibition
➤ it could endanger national security
➤ it relates specifically to an individual without their consent
➤ it could harm substantially the business of the employer or infringe commercial security
➤ it was obtained in connection with legal proceedings.

Finally, an employer should ensure that the representative receives reasonable training in health and safety at the employer's expense, is allowed time during working hours to perform his duties and provide other facilities and assistance that the representative should reasonably require.

3.13 Practice NEBOSH questions for Chapter 3

1. (a) State the circumstances under which an employer must establish a health and safety committee.
   (b) Give six reasons why a health and safety committee may prove to be ineffective in practice. (March 2001)
2. In relation to the Health and Safety (Consultation with Employees) Regulations 1996, identify:
   (i) the health and safety matters on which employers have a duty to consult their employees
   (ii) four types of information that an employer is not obliged to disclose to an employee representative. (March 2000)
3. With reference to the Management of Health and Safety at Work Regulations 1992. (*Note: now 1999*) Identify:
   (i) the particular matters on which employees should receive health and safety information

    (ii) the specific circumstances when health and safety training should be given to employees. (June 1999)

4. With reference to the Health and Safety (Consultation with Employees) Regulations 1996:

    (i) explain the difference between consulting and informing

    (ii) outline the health and safety matters on which employers must consult their employees. (June 1998)

5. Outline the factors that may determine the effectiveness of a safety committee. (December 1996)

6. Outline the functions of a safety representative as stated in the Safety Representatives and Safety Committee Regulations 1977. (September 1996)

Appendix 3.1 – Typical organizational responsibilities

Senior managers

> Are responsible and held accountable for the relevant organization's health and safety performance.

> Develop strong, positive health and safety attitudes among those employees reporting directly to them.

> Provide guidance to their management team.

> Establish minimum acceptable health and safety standards within the organization.

> Provide necessary staffing and funding for the health and safety effort within the organization.

> Authorize necessary major expenditures.

> Evaluate, approve, and authorize health and safety-related projects developed by members of the organization's health and safety advisers.

> Review and approve policies, procedures, and programmes developed by group staff.

> Maintain a working knowledge of areas of health and safety that are regulated by governmental agencies.

> Keep health and safety in the forefront by including it as a topic in business discussions.

> Review health and safety performance statistics and provide feedback to the management team.

> Review and act upon major recommendations submitted by outside loss-prevention consultants.

> Make health and safety a part of site tours by making specific remarks about observations of acts or conditions that fall short of or exceed health and safety standards.

> Personally investigate fatalities and major property losses.

> Personally chair the organization's main health and safety committee (if applicable).

Site managers

> Are responsible and held accountable for their site's health and safety performance.

> Establish, implement, and maintain a formal, written site health and safety programme, encompassing applicable areas of loss prevention, that is consistent with the organization's health and safety policy.

➤ Establish controls to ensure uniform adherence to health and safety programme elements. These controls should include corrective action and follow-up.

➤ Develop, by action and example, a positive health and safety attitude and a clear understanding of specific responsibilities in each member of management.

➤ Approve and adopt local health and safety policies, rules, and procedures.

➤ Chair the site health and safety committee.

➤ Personally investigate fatalities, serious lost workday case injuries, and major property losses.

➤ Review monthly health and safety activity reports and performance statistics.

➤ Review lost workday cases injury/illness investigation reports.

➤ Review health and safety reports submitted by outside agencies.

➤ Attend site health and safety audits regularly to appraise the programme's effectiveness.

➤ Review annually the health and safety programme's effectiveness and make adjustments where necessary.

➤ Evaluate the functional performance of the health and safety staff and provide guidance or training where necessary.

➤ Personally review, sign and approve corrective action planned for lost-time accidents.

➤ Monitor staff's progress toward achieving their health and safety goals and objectives.

Department managers

➤ Are responsible and held accountable for their department's or area's health and safety performance.

➤ Contact each supervisor frequently (daily) about health and safety.

➤ Hold departmental health and safety meetings for staff at least once a month.

➤ Make daily observations of supervisors' health and safety activities.

➤ Review and approve all job procedures developed as a result of job safety analyses; install approved procedures; require each supervisor to check use of the procedures.

➤ Approve and annually review all departmental health and safety risk assessments, rules and regulations; maintain strict enforcement; and develop plans to ensure employee instruction and re-instruction.

➤ Establish acceptable housekeeping standards, defining specific areas of responsibility, assign areas to supervisors; make a daily spot check of some area; hold formal inspection with supervisors at least monthly, submit written reports with assignments and deadlines for corrections.

➤ Authorize purchases of tools and equipment necessary to attain compliance with site specifications and statutory regulations.

➤ Develop a training plan that includes specific job instructions for new or transferred employees; follow up on training by supervisors.

➤ Review the health and safety performance of their area of responsibility.

➤ Personally investigate all lost workday cases and significant losses and report to the site manager; follow up on corrective action.

➤ Adopt standards for assigning personal protective equipment to employees; insist on strict enforcement; and make spot field checks to determine compliance.

➤ Evaluate supervisors' health and safety performances.

➤ Develop in each member of management strong health and safety attitudes and clear-cut understanding of specific duties and responsibilities.

➤ Instil, by action, example, and training, a sincere health and safety attitude.

➤ Instruct supervisors in site procedures for the care and treatment of sick or injured employees.

Supervisors

➤ Are responsible and held accountable for their group's health and safety performance.

➤ Conduct health and safety meetings for employees at least monthly.

➤ Enforce safe job procedures, such as those developed by job safety analyses.

➤ Report to manager and act upon any weaknesses in safe job procedures, as revealed by health and safety risk assessments and observations.

➤ Report jobs not covered by safe job procedures.

➤ Review unsafe acts and conditions and direct daily health and safety activities to correct the causes.

➤ Instruct employees in health and safety rules and regulations; make records of instruction; and enforce all health and safety rules and regulations.

➤ Make daily inspections of assigned work areas and take immediate steps to correct unsafe or unsatisfactory conditions; report to the manager those conditions that cannot be immediately corrected; instruct employees on housekeeping standards.

➤ Instruct employees that tools/equipment are to be inspected before each use; make spot checks of tools'/equipment's condition.

➤ Instruct each new employee personally on job health and safety requirements in assigned work areas.

➤ Enforce the organization's/site's medical recommendations with respect to an employee's physical limitations. Report on employee's apparent physical limitations to their manager; and request physical examination of the employee.

➤ Enforce personal protective equipment requirements; make spot checks to determine that protective apparel is being used; and periodically appraise condition of equipment.

➤ See that, in a case of serious injury, the injured employee receives prompt medical attention; isolate the area or shut down equipment, as necessary; and immediately report to the manager the facts regarding the employee's accident or illness and any action taken. In serious incident cases, the supervisor determines the cause, takes immediate steps to correct an unsafe condition, and isolates area and/or shuts down the equipment, as necessary. They immediately report facts and action taken to the manager.

➤ Make thorough investigation of all accidents, serious incidents, and cases of ill-health involving employees in assigned work areas; immediately after the accident, prepare a complete accident investigation report and include recommendations for preventing recurrence.

➤ Check changes in operating practices, procedures, and conditions at the start of each shift/day and before relieving the 'on-duty' supervisor (if applicable), noting health and safety-related incidents that have occurred since their last working period.

➤ Make, at the start of each shift/day, an immediate check to determine absentees. If site injury is claimed, an immediate investigation is instituted and the department manager is notified.

➤ Make daily spot checks and take necessary corrective action regarding housekeeping, unsafe acts or practices, unsafe conditions, job procedures and adherence to health and safety rules.

➤ Attend all scheduled and assigned health and safety training meetings.
➤ Instruct personally or provide on-the-job instruction on safe and efficient performance of assigned jobs.
➤ Act on all employee health and safety complaints and suggestions.
➤ Maintain, in their assigned area, health and safety signs and notice boards in a clean and legible condition.

Employees

➤ Are responsible for their own safety and health.
➤ Ensure that their actions will not jeopardize the safety or health of other employees.
➤ Are alert to observe and correct, or report, unsafe practices and conditions.
➤ Maintain a healthy and safe place to work and cooperate with managers in the implementation of health and safety matters.
➤ Make suggestions to improve any aspect of health and safety.
➤ Maintain an active interest in health and safety.
➤ Learn and follow the operating procedures and health and safety rules and procedures for safe performance of the job.
➤ Follow the established procedures if accidents occur.

Promoting a positive health and safety culture

4.1 Introduction

In 1972, the Robens report recognized that the introduction of health and safety management systems was essential if the ideal of self-regulation of health and safety by industry was to be realized. It further recognized that a more active involvement of the workforce in such systems was essential if self-regulation was to work. Self-regulation and the implicit need for health and safety management systems and employee involvement was incorporated into the Health and Safety at Work Act 1974.

Since the introduction of the HSW Act, health and safety standards have improved considerably but there have been some catastrophic failures. One of the worst was the fire on the off-shore oil platform, Piper Alpha, in 1988 when 167 people died. At the subsequent inquiry, the concept of a safety culture was defined by the Director General of the HSE, J R Rimington. This definition has remained as one of the main checklists for a successful health and safety management system.

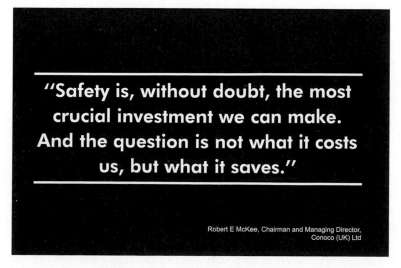

Figure 4.1   Safety investment.

4.2 Definition of a health and safety culture

The health and safety culture of an organization may be described as the development stage of the organization in health and safety management at a particular time. HSG (65) gives the following definition of a health and safety culture:

> *The safety culture of an organization is the product of individual and group values, attitudes, perceptions, competencies and patterns of behaviour that determine the commitment to, and the style and proficiency of, an organization's health and safety management.*
>
> *Organizations with a positive safety culture are characterized by communications founded on mutual trust, by shared perceptions of the importance of safety and by confidence in the efficacy of preventive measures.*

There is concern among some health and safety professionals that many health and safety cultures are developed and driven by senior managers with very little input from the workforce. Others argue that this arrangement is sensible because the legal duties are placed on the employer. A positive health and safety culture needs the involvement of the whole workforce just as a successful quality system does. There must be a joint commitment in terms of attitudes and values. The workforce must believe that the safety measures put in place will be effective and followed even when financial and performance targets may be affected.

4.3 Safety culture and safety performance

4.3.1 The relationship between health and safety culture and health and safety performance

The following elements are the important components of a positive health and safety culture:

- leadership and commitment to health and safety throughout and at all levels of the organization;
- acceptance that high standards of health and safety are achievable as part of a long-term strategy formulated by the organization;
- a detailed assessment of health and safety risks in the organization and the development of appropriate control and monitoring systems;
- a health and safety policy statement outlining short- and long-term health and safety objectives. Such a policy should also include codes of practice and required health and safety standards;
- relevant employee training programmes and communication and consultation procedures;
- systems for monitoring equipment, processes and procedures and the prompt rectification of any defects;
- the prompt investigation of all incidents and accidents and reports made detailing any necessary remedial actions.

If the organization adheres to these elements, then a basis for a good performance in health and safety will have been established. However, to achieve this level of performance, sufficient financial and human resources must be made available for the health and safety function at all levels of the organization.

All managers, supervisors and members of the governing body (e.g. directors) should receive training in health and safety and be made familiar during training sessions with the health and safety targets of the organization. The depth of training undertaken will depend on the level of competence required of the particular manager. Managers should be accountable for health and safety within their departments and be rewarded for significant improvements in health and safety performance. They should also be expected to discipline employees within their departments who infringe health and safety policies or procedures.

4.3.2 Important indicators of a health and safety culture

There are several outputs or indicators of the state of the health and safety culture of an organization. The most important are the numbers of accidents, near misses and occupational ill-health cases occurring within the organization.

While the number of accidents may give a general indication of the health and safety culture, a more detailed examination of accidents and accident statistics is normally required. A calculation of the rate of accidents enables health and safety performance to be compared between years and organizations.

The simplest measure of accident rate is called the incident rate and is defined as:

$$\frac{\text{Total number of accidents}}{\text{Number of persons employed}} \times 1000$$

or the total number of accidents per 1000 employees.

A similar measure (per 100 000) is used by the HSC in its annual report on national accident statistics and enables comparisons to be made within an organization between time periods when employee numbers may change. It also allows comparisons to be made with the national occupational or industrial group relevant to the organization.

There are four main problems with this measure which must be borne in mind when it is used. These are:

➤ there may be a considerable variation over a time period in the ratio of part-time to full-time employees
➤ the measure does not differentiate between major and minor accidents and takes no account of other incidents, such as those involving damage but no injury (although it is possible to calculate an incidence rate for a particular type or cause of accident)
➤ there may be significant variations in work activity during the periods being compared
➤ under-reporting of accidents will affect the accuracy of the data.

Subject to the above limitations, an organization with a high accident incidence rate is likely to have a negative or poor health and safety culture.

There are other indications of a poor health and safety culture or climate. These include:

➤ a high sickness, ill-health and absentee rate amongst the workforce
➤ no resources (in terms of budget, people or facilities) made available for the effective management of health and safety
➤ a lack of compliance with relevant health and safety law and the safety rules and procedures of the organization

> poor selection procedures and management of contractors
> poor levels of communication, cooperation and control
> a weak health and safety management structure
> either a lack or poor levels of health and safety competence
> high insurance premiums.

In summary, a poor health and safety performance within an organization is an indication of a negative health and safety culture.

4.3.3 Factors affecting a health and safety culture

The most important factor affecting the culture is the commitment to health and safety from the top of an organization. This commitment may be shown in many different ways. It needs to have a formal aspect in terms of an organizational structure, job descriptions and a health and safety policy, but it also needs to be apparent during crises or other stressful times. The health and safety procedures may be circumvented or simply forgotten when production or other performance targets are threatened.

Structural reorganization or changes in market conditions will produce feelings of uncertainty among the workforce which, in turn, will affect the health and safety culture.

Poor levels of supervision, health and safety information and training are very significant factors in reducing health and safety awareness and, therefore, the culture.

Finally, the degree of consultation and involvement with the workforce in health and safety matters is crucial for a positive health and safety culture. Most of these factors may be summed up as human factors.

4.4 Human factors and their influence on safety performance

4.4.1 Human factors

Over the years, there have been several studies undertaken to examine the link between various accident types, graded in terms of their severity, and near misses. One of the most interesting was conducted in the USA by H W Heinrich in 1950. He looked at over 300 accidents/incidents and produced the ratio illustrated in Figure 4.2.

This study indicated that for every 10 near misses, there will be an accident. While the accuracy of this study may be debated and other studies have produced different ratios, it is clear that if near misses are continually ignored, an accident will result. Further, the HSE Accident Prevention Unit has suggested that 90% of all accidents are due to human error and 70% of all accidents could have been avoided by ear-

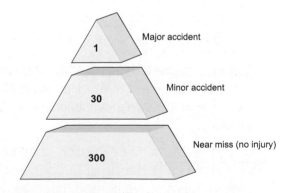

Figure 4.2 Heinrich's accidents/incidents ratio.

lier (proactive) action by management. It is clear from many research projects that the major factors in most accidents are human factors.

The HSE has defined human factors as, 'environmental, organizational and job factors, and human and individual characteristics which influence behaviour at work in a way which can affect health and safety'.

In simple terms, the health and safety of people at work are influenced by:

➤ the organization
➤ the job
➤ personal factors.

These are known as human factors since they each have a human involvement. The personal factors which differentiate one person from another are only one part of those factors – and not always the most important.

Each of these elements will be considered in turn.

4.4.2 The organization

The organization is the company or corporate body and has the major influence on health and safety. It must have its own positive health and safety culture and produce an environment in which it:

➤ manages health and safety throughout the organization, including the setting and publication of a health and safety policy and the establishment of a health and safety organizational structure;
➤ measures the health and safety performance of the organization at all levels and in all departments. The performance of individuals should also be measured. There should be clear health and safety targets and standards and an effective reporting procedure for accidents and other incidents so that remedial actions may be taken;
➤ motivates managers within the organization to improve health and safety performance in the workplace in a proactive rather than reactive manner.

The HSE has recommended that an organization needs to provide the following elements within its management system:

➤ a clear and evident commitment from the most senior manager downwards, which provides a climate for safety in which management's objectives and the need for appropriate standards are communicated and in which constructive exchange of information at all levels is positively encouraged;
➤ an analytical and imaginative approach identifying possible routes to human factor failure. This may well require access to specialist advice;
➤ procedures and standards for all aspects of critical work and mechanisms for reviewing them;
➤ effective monitoring systems to check the implementation of the procedures and standards;
➤ incident investigation and the effective use of information drawn from such investigations;
➤ adequate and effective supervision with the power to remedy deficiencies when found.

It is important to recognize there are often reasons for these elements not being present in many organizations and the management of health and safety being weak. The most common reason is that individuals within the management organization do not understand their roles – or their roles have never been properly explained to them. The higher that a person is within the structure the less likely it is that he has received any health and safety training. Such training at board level is rare.

Objectives and priorities may vary across and between different levels in the structure leading to disputes which affect attitudes to health and safety. For example, a warehouse manager may be pressured to block walkways so that a large order can be stored prior to dispatch.

Motivations can also vary across the organization which may cause health and safety to be compromised. The production controller will require that components of a product are produced as near simultaneously as possible so that final assembly of them is performed as quickly as possible. However, the health and safety adviser will not want to see safe systems of work compromised.

In an attempt to address some of these problems, the HSC produced guidance in 2001 on the safety duties of company directors. Each director and the Board, acting collectively, will be expected to provide health and safety leadership in the organization. The Board will need to ensure that all its decisions reflect its health and safety intentions and that it engages the workforce actively in the improvement of health and safety. The Board will also be expected to keep itself informed of changes in health and safety risks. (See Chapter 3 for more details on directors' responsibilities.)

The following simple checklist may be used to check any organizational health and safety management structure.

Does the structure have:

➤ an effective health and safety management system?
➤ a positive health and safety culture?
➤ arrangements for the setting and monitoring of standards?
➤ adequate supervision?
➤ effective incident reporting and analysis?
➤ learning from experience?
➤ clearly visible health and safety leadership?
➤ suitable team structures?
➤ efficient communication systems and practices?
➤ adequate staffing levels?
➤ suitable work patterns?

HSG(48) gives the following causes for failures in organizational and management structures:

➤ poor work planning, leading to high work pressure
➤ lack of safety systems and barriers
➤ inadequate responses to previous incidents
➤ management based on one-way communications
➤ deficient coordination and responsibilities
➤ poor management of health and safety
➤ poor health and safety culture.

Organizational factors play a significant role in the health and safety of the workplace. However, this role is often forgotten when health and safety is being reviewed after an accident or when a new process or piece of equipment is introduced.

4.4.3 The job

Jobs may be highly dangerous or present only negligible risk of injury. Health and safety is an important element during the design stage of the job and any equipment, machinery or procedures associated with the job. Method study helps to design the job in the most cost effective way and ergonomics helps to design the job with health and safety in mind. Ergonomics is the science of matching equipment and machines to man rather than the other way round. An ergonomically designed machine will ensure that control levers, dials, meters and switches are sited in a convenient and comfortable position for the machine operator. Similarly, an ergonomically designed workstation will be designed for the comfort and health of the operator. Chairs, for example, will be designed to support the back properly throughout the working day.

Physically matching the job and any associated equipment to the person will ensure that the possibility of human error is minimized. It is also important to ensure that there is mental matching of the person's information and decision-making requirements. A person must be capable, either through past experience or through specific training, to perform the job with the minimum potential for human error.

The major considerations in the design of the job, which would be undertaken by a specialist, have been listed by the HSE as follows:

➤ the identification and detailed analysis of the critical tasks expected of individuals and the appraisal of any likely errors associated with those tasks;

➤ evaluation of the required operator decision making and the optimum (best) balance between the human and automatic contributions to safety actions (with the emphasis on automatic whenever possible);

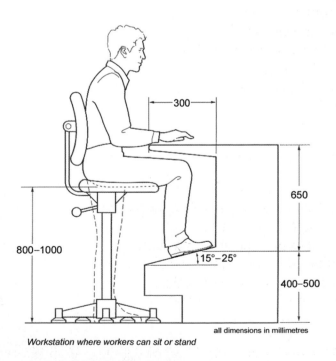

Workstation where workers can sit or stand

Figure 4.3 Well-designed workstation for sitting or standing. Source HSE. Crown copyright material is reproduced with the permission of the Controller of HMSO and the Queen's Printer for Scotland.

> application of ergonomic principles to the design of man–machine interfaces, including displays of plant and process information, control devices and panel layout;
> design and presentation of procedures and operating instructions in the simplest terms possible;
> organization and control of the working environment, including the workspace, access for maintenance, lighting, noise and heating conditions;
> provision of the correct tools and equipment;
> scheduling of work patterns, including shift organization, control of fatigue and stress and arrangements for emergency operations;
> efficient communications, both immediate and over a period of time.

For some jobs, particularly those with a high risk of injury, a job safety analysis should be undertaken to check that all necessary safeguards are in place. All jobs should carry a job description and a safe system of work for the particular job. The operator should have sight of the job description and be trained in the safe system of work before commencing the job. More information on both these latter items is given in Chapter 6.

The following simple checklist may be used to check that the principal health and safety considerations of the job have been taken into account:

> have the critical parts of the job been identified and analysed?
> have the employee's decision-making needs been evaluated?
> has the best balance between human and automatic systems been evaluated?
> have ergonomic principles been applied to the design of equipment displays, including displays of plant and process information, control information and panel layouts?
> has the design and presentation of procedures and instructions been considered?
> has the guidance available for the design and control of the working environment, including the workspace, access for maintenance, lighting, noise and heating conditions, been considered?
> have the correct tools and equipment been provided?
> have the work patterns and shift organization been scheduled to minimize their impact on health and safety?
> has consideration been given to the achievement of efficient communications and shift handover?

HSG(48) gives the following causes for failures in job health and safety:

> illogical design of equipment and instruments
> constant disturbances and interruptions
> missing or unclear instructions
> poorly maintained equipment
> high workload
> noisy and unpleasant working conditions.

It is important that health and safety monitoring of the job is a continuous process. Some problems do not become apparent until the job is started. Other

Figure 4.4 Poor working conditions.

problems do not surface until there is a change of operator or a change in some aspect of the job.

It is very important to gain feedback from the operator on any difficulties experienced because there could be a health and safety issue requiring further investigation.

4.4.4 Personal factors

Personal factors, which affect health and safety, may be defined as any condition or characteristic of an individual which could cause or influence him to act in an unsafe manner. They may be physical, mental or psychological in nature. Personal factors, therefore, include issues such as attitude, motivation, training and human error and their interaction with the physical, mental and perceptual capability of the individual.

These factors have a significant affect on health and safety. Some of them, normally involving the personality of the individual, are unchangeable but others, involving skills, attitude, perception and motivation can be changed, modified or improved by suitable training or other measures. In summary, the person needs to be matched to the job.

Studies have shown that the most common personal factors which contribute to accidents are low skill and competence levels, tiredness, boredom, low morale and individual medical problems.

It is difficult to separate all the physical, mental or psychological factors because they are inter-linked. However, the three most common factors are psychological factors – attitude, motivation and perception.

Attitude is the tendency to behave in a particular way in a certain situation. Attitudes are influenced by the prevailing health and safety culture within the organization, the commitment of the management, the experience of the individual and the influence of the peer group. Peer group pressure is a particularly important factor among young people and health and safety training must be designed with this in mind by using examples or case studies that are relevant to them. Behaviour may be changed by training, the formulation and enforcement of safety rules and meaningful consultation – attitude change often follows.

Motivation is the driving force behind the way a person acts or the way in which people are stimulated to act. Involvement in the decision-making process in a meaningful way will improve motivation as will the use of incentive schemes. However, there are other important influences on motivation such as recognition and promotion opportunities, job security and job satisfaction. Self-interest, in all its forms, is a significant motivator and personal factor.

Perception is the way in which people interpret the environment or the way in which a person believes or understands a situation. In health and safety, the perception of hazards is an important concern. Many accidents occur because people do not perceive that there is a risk. There are many common examples of this, including the use of personal protective equipment (such as hard hats) and guards on drilling machines and the washing of hands before meals. It is important to understand that when perception leads to an increased health and safety risk, it is not always caused by a conscious decision of the individual concerned. The stroboscopic effect caused by the rotation of a drill at certain speeds under fluorescent lighting will make the drill appear stationary. It is a well-known phenomenon, especially among illusionists, that people will often see

71

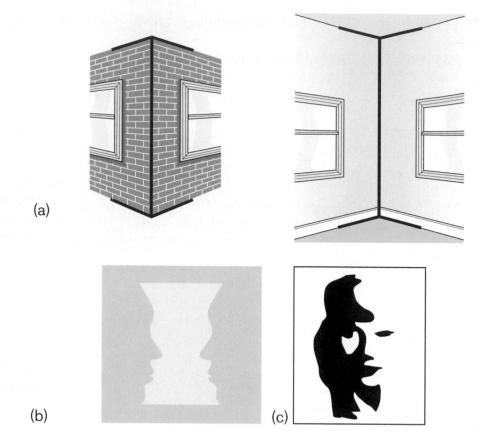

Figure 4.5 Visual perception (a) Are the lines the same length? (b) Face or vase? (c) Face or saxophone player?

what they expect to see rather than reality. Routine or repetitive tasks will reduce attention levels leading to the possibility of accidents.

Other personal factors which can affect health and safety include physical stature, age, experience, health, hearing, intelligence, language, skills, level of competence and qualifications.

Finally, memory is an important personal factor since it is influenced by training and experience. The efficiency of memory varies considerably between people and during the lifetime of an individual. The overall health of a person can affect memory as can personal crises. Due to these possible problems with memory, important safety instructions should be available in written as well as verbal form.

The following checklist given in HSG(48) may be used to check that the relevant personal factors have been covered:

➤ has the job specification been drawn up and included age, physique, skill, qualifications, experience, aptitude, knowledge, intelligence and personality?
➤ have the skills and aptitudes been matched to the job requirements?

Setting goals leads to vigorous activity if:

The goal is realistic.

A serious commitment is made, especially if it is made publicly.

Feedback is received.

Figure 4.6 Motivation and activity.

> have the personnel selection policies and procedures been set up to select appropriate individuals?
> has an effective training system been implemented?
> have the needs of special groups of employees been considered?
> have the monitoring procedures been developed for the personal safety performance of safety critical staff?
> have fitness for work and health surveillance been provided where it is needed?
> has counselling and support for ill-health and stress been provided?

Personal factors are the attributes that employees bring to their jobs and may be strengths or weaknesses. Negative personal factors cannot always be neutralized by improved job design. It is, therefore, important to ensure that personnel selection procedures should match people to the job. This will reduce the possibility of accidents or other incidents.

4.5 Human errors and violations

Human failures in health and safety are either classified as errors or violations. An error is an unintentional deviation from an accepted standard, while a violation is a deliberate deviation from the standard (see Figure 4.7).

4.5.1 Human errors

Human errors fall into three groups – slips, lapses and mistakes, which can be further sub-divided into rule-based and knowledge-based mistakes.

Slips and lapses
These are very similar in that they are caused by a momentary memory loss often due to lack of attention or loss of concentration. They are not related to levels of training, experience or motivation and they can usually be reduced by re-designing the job or equipment or minimizing distractions.

Slips are failures to carry out the correct actions of a task. Examples include the use of the incorrect switch, reading the wrong dial or selecting the incorrect component for an assembly. A slip also describes an action taken too early or too late within a given working procedure.

Lapses are failures to carry out particular actions which may form part of a working procedure. A fork lift truck driver leaving the keys in the ignition lock of his truck is an example of a lapse as is the failure to replace the petrol cap on a car after filling it with petrol. Lapses may be reduced by re-designing equipment so that, for example, an audible horn indicates the omission of a task. They may also be reduced significantly by the use of detailed checklists.

Mistakes
Mistakes occur when an incorrect action takes place but the person involved believes the action to be correct. A mistake involves an incorrect judgement. There are two types of mistake – rule-based and knowledge-based.

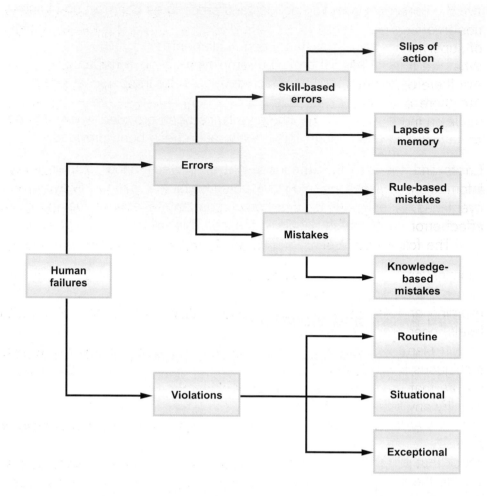

Figure 4.7 Types of human failure. Source HSE. Crown copyright material is reproduced with the permission of the Controller of HMSO and the Queen's Printer for Scotland.

Rule-based mistakes occur when a rule or procedure is remembered or applied incorrectly. These mistakes usually happen when, due to an error, the rule that is normally used no longer applies. For example, a particular job requires the counting of items into groups of ten followed by the adding together of the groups so that the total number of items may be calculated. If one of the groups is miscounted, the final total will be incorrect even though the rule has been followed.

Knowledge-based mistakes occur when well-tried methods or calculation rules are used inappropriately. For example, the depth of the foundations required for a particular building was calculated using a formula. The formula, which assumed a clay soil, was used to calculate the foundation depth in a sandy soil. The resultant building was unsafe.

The HSE have suggested the following points to consider when the potential source of human errors are to be identified:

➤ What human errors can occur with each task? There are formal methods available to help with this task;

> What influences are there on performance? Typical influences include, time pressure, design of controls, displays and procedures, training and experience, fatigue and levels of supervision;
> What are the consequences of the identified errors? What are the significant errors?
> Are there opportunities for detecting each error and recovering it?
> Are there any relationships between the identified errors? Could the same error be made on more than one item of equipment due, for example, to the incorrect calibration of an instrument?

Errors and mistakes can be reduced by the use of instruction, training and relevant information. However, communication can also be a problem, particularly at shift hand-over times. Environmental and organizational factors, involving workplace stress will also affect error levels.

The following steps are suggested to reduce the likelihood of human error:

> examine and reduce the workplace stressors (e.g. noise, poor lighting, etc.) which increase the frequency of errors;
> examine and reduce any social or organizational stressors (e.g. insufficient staffing levels, peer pressure, etc.);
> design plant and equipment to reduce error possibilities – poorly designed displays, ambiguous instructions;
> ensure that there are effective training arrangements;
> simplify any complicated or complex procedures;
> ensure that there is adequate supervision, particularly for inexperienced or young trainees;
> check that job procedures, instructions and manuals are kept up to date and are clear;
> include the possibility of human error when undertaking the risk assessment;
> isolate the human error element of any accident or incident and introduce measures to reduce the risk of a repeat;
> monitor the effectiveness of any measures taken to reduce errors.

4.5.2 Violations

There are three categories of violation – routine, situational and exceptional.

Routine violation occurs when the breaking of a safety rule or procedure is the normal way of working. It becomes routine not to use the recommended procedures for tasks. An example of this is the regular speeding of fork lift trucks in a warehouse so that orders can be fulfilled on time.

There are many reasons given for routine violation – the job would not be completed on time, lack of supervision and enforcement or, simply, the lack of knowledge of the procedures. Routine violations can be reduced by regular monitoring, ensuring that the rules are actually necessary or redesigning the job.

The following features are very common in many workplaces and often lead to routine violations:

> poor working posture due to poor ergonomic design of the workstation or equipment
> equipment difficult to use and/or slow in response
> equipment difficult to maintain or pressure on time available for maintenance
> procedures unduly complicated and difficult to understand

> unreliable instrumentation and/or warning systems
> high levels of noise and other poor aspects to the environment (fumes, dusts, humidity)
> associated personal protective equipment either inappropriate, difficult and uncomfortable to wear or ineffective due to lack of maintenance.

Situational violations occur when particular job pressures at particular times make rule compliance difficult. They may happen when the correct equipment is not available or weather conditions are adverse. A common example is the use of a ladder rather than a scaffold for working at height to replace window frames in a building. Situational violations may be reduced by improving job design, the working environment and supervision.

Exceptional violations rarely happen and usually occur when a safety rule is broken to perform a new task. A good example are the violations which can occur during the operations of emergency procedures such as fires or explosions. These violations should be addressed in risk assessments and during training sessions for emergencies (e.g. fire training).

Everybody is capable of making errors. It is one of the objectives of a positive health and safety culture to reduce them and their consequences as much as possible.

4.6 The development of a positive health and safety culture

No single section or department of an organization can develop a positive health and safety culture on its own. There needs to be commitment by the management, the promotion of health and safety standards, effective communication within the organization, cooperation from and with the workforce and an effective and developing training programme. Each of these topics will be examined in turn to show their effect on improving the health and safety culture in the organization.

4.6.1 Commitment of management

As mentioned earlier, there needs to be a commitment from the very top of the organization and this commitment will, in turn, produce higher levels of motivation and commitment throughout the organization. Probably the best indication of this concern for health and safety is shown by the status given to health and safety and the amount of resources (money, time and people) allocated to health and safety. The management of health and safety should form an essential part of a manager's responsibility and they should be held to account for their performance on health and safety issues. Specialist expertise should be made available when required (e.g. for noise assessment), either from within the workforce or by the employment of external contractors or consultants. Health and safety should be discussed on a regular basis at management meetings at all levels of the organization. If the organization employs sufficient people to make direct consultation with all employees difficult, there should be a health and safety committee at which there is employee representation. In addition, there should be recognized routes for anybody within the organization to receive health and safety information or have their health and safety concerns addressed.

The health and safety culture is enhanced considerably when senior managers appear regularly at all levels of an organization whether it be the shop floor, the hospital ward or the general office and are willing to discuss health and safety issues with staff. A visible management is very important for a positive health and safety culture.

Finally, the positive results of a management commitment to health and safety will be the active involvement of all employees in health and safety, the continuing improvement in health and safety standards and the subsequent reduction in accident and occupational ill-health rates. This will lead, ultimately, to a reduction in the number and size of compensation claims.

HSG(48) makes some interesting suggestions to managers on the improvements that may be made to health and safety which will be seen by the workforce as a clear indication of their commitment. The suggestions are:

➤ review the status of the health and safety committees and health and safety practitioners. Ensure that any recommendations are acted upon or implemented;
➤ ensure that senior managers receive regular reports on health and safety performance and act on them;
➤ ensure that any appropriate health and safety actions are taken quickly and are seen to have been taken;
➤ any action plans should be developed in consultation with employees, based on a shared perception of hazards and risks, be workable and continuously reviewed.

4.6.2 The promotion of health and safety standards

For a positive health and safety culture to be developed, everyone within the organization needs to understand the standards of health and safety expected by the organization and the role of the individual in achieving and maintaining those standards. Such standards are required to control and minimize health and safety risks.

Standards should clearly identify the actions required of people to promote health and safety. They should also specify the competencies needed by employees and they should form the basis for measuring health and safety performance.

Health and safety standards cover all aspects of the organization. Typical examples include:

➤ the design and selection of premises
➤ the design and selection of plant and substances (including those used on site by contractors)
➤ the recruitment of employees and contractors
➤ the control of work activities, including issues such as risk assessment
➤ competence, maintenance and supervision
➤ emergency planning and training
➤ the transportation of the product and its subsequent maintenance and servicing.

Having established relevant health and safety standards, it is important that they are actively promoted within the organization by all levels of management. The most effective method of promotion is by leadership and example. There are many ways to do this such as:

➤ the involvement of managers in workplace inspections and accident investigations
➤ the use of personal protective equipment (e.g. goggles and hard hats) by all managers and their visitors in designated areas

> ensuring that employees attend specialist refresher training courses when required (e.g. first aid and fork lift truck driving)
> full cooperation with fire drills and other emergency training exercises
> comprehensive accident reporting and prompt follow up on recommended remedial actions.

The benefits of good standards of health and safety will be shown directly in less lost production, accidents and compensation claims and lower insurance premiums. It may also be shown in higher product quality and better resource allocation.

An important and central necessity for the promotion of high health and safety standards is health and safety competence. What is meant by 'competence'?

4.6.3 Competence

The word 'competence' is often used in health and safety literature. One definition, made during a civil case in 1962, stated that a competent person is:

> *a person with practical and theoretical knowledge as well as sufficient experi-ence of the particular machinery, plant or procedure involved to enable them to identify defects or weaknesses during plant and machinery examinations, and to assess their importance in relation to the strength and function of that plant and machinery.*

This definition concentrates on a manufacturing rather than service industry requirement of a competent person.

Regulation 6 of the Management of Health and Safety at Work Regulations 1999 requires that 'every employer shall employ one or more competent persons to assist him in undertaking the measures he needs to take to comply with the requirements and prohibitions imposed upon him by or under the relevant statutory provisions'. In other words, competent persons are required to assist the employer in meeting his obligations under health and safety law. This may mean a health and safety adviser in addition to, say, an electrical engineer, an occupational nurse and a noise assessment specialist. The number and range of competent persons will depend on the nature of the business of the organization.

It is recommended that competent employees are used for advice on health and safety matters rather than external specialists (consultants). It is recognized, however, that if employees, competent in health and safety, are not available in the organization, then an external service may be enlisted to help. The key is that management and employees need access to health and safety expertise.

The regulations do not define 'competence' but do offer advice to employers when appointing health and safety advisers who should have:

> a knowledge and understanding of the work involved, the principles of risk assessment and prevention and current health and safety applications
> the capacity to apply this to the task required by the employer in the form of problem and solution identification, monitoring and evaluating the effectiveness of solutions and the promotion and communication of health and safety and welfare advances and practices.

Such competence does not necessarily depend on the possession of particular skills or qualifications. It may only be required to understand relevant current best practice, be aware of one's own limitations in terms of experience and knowledge and be willing to supplement existing experience and knowledge. However, in more complex or technical situations membership of a relevant professional body and/or the possession of an appropriate qualification in health and safety may be necessary. It is important that any competent person employed to help with health and safety has evidence of relevant knowledge, skills and experience for the tasks involved. The appointment of a competent person as an adviser to an employer does not absolve the employer from his responsibilities under the Health and Safety at Work Act and other relevant statutory provision.

Finally, it is worth noting that the requirement to employ competent workers is not restricted to those having a health and safety function but covers the whole workforce.

4.7 Effective communication

Many problems in health and safety arise due to poor communication. It is not just a problem between management and workforce – it is often a problem the other way or indeed at the same level within an organization. It arises from ambiguities or, even, accidental distortion of a message.

There are three basic methods of communication in health and safety – verbal, written and graphic.

Verbal communication is the most common. It is communication by speech or word of mouth. Verbal communication should only be used for relatively simple pieces of information or instruction. It is most commonly used in the workplace, during training sessions or at meetings.

There are several potential problems associated with verbal communication. The speaker needs to prepare the communication carefully so that there is no confusion about the message. It is very important that the recipient is encouraged to indicate their understanding of the communication. There have been many cases of accidents occurring because a verbal instruction has not been understood. There are several barriers to this understanding from the point of view of the recipient, including language and dialect, the use of technical language and abbreviations, background noise and distractions, hearing problems, ambiguities in the message, mental weaknesses and learning disabilities, lack of interest and attention.

Having described some of the limitations of verbal communication, it does have some merits. It is less formal, enables an exchange of information to take place quickly and the message to be conveyed as near to the workplace as possible. Training or instructions that are delivered in this way are called toolbox talks and can be very effective.

Written communication takes many forms from the simple memo to the detailed report.

A memo should contain one simple message and be written in straightforward and clear language. The title should accurately describe the contents of the memo. In recent years, e-mails have largely replaced memos since it has become a much quicker method to ensure that the message gets to all concerned (although a recent report has suggested that many people are becoming overwhelmed by the number of e-mails which they receive!). The advantage of memos and e-mails is that there is a record of the message after it has been delivered. The disadvantage is that they can be ambiguous or difficult to understand or, indeed, lost within the system.

Reports are more substantial documents and cover a topic in greater detail. The report should contain a detailed account of the topic and any conclusions or recommendations. The main problem with reports is that they are often not read properly due to the time constraints on managers. It is important that all reports have a summary attached so that the reader can decide whether it needs to be read in detail.

The most common way in which written communication is used in the workplace is the notice board. For a notice board to be effective, it needs to be well positioned within the workplace and there needs to be a regular review of the notices to ensure that they are up to date and relevant.

There are many other examples of written communications in health and safety, such as employee handbooks, company codes of practice, minutes of safety committee meetings and health and safety procedures.

Graphic communication is communication by the use of drawings, photographs or videos. It is used either to impart health and safety information (e.g. fire exits) or health and safety propaganda. The most common forms of health and safety propaganda are the poster and the video. Both can be used very effectively as training aids since they can retain interest and impart a simple message. Their main limitation is that they can become out of date fairly quickly or, in the case of posters, become largely ignored.

There are many sources of health and safety information which may need to be consulted before an accurate communication can be made. These include regulations, judgements, approved codes of practice, guidance, British and European standards, periodicals, case studies and HSE publications.

4.8 Health and safety training

Health and safety training is a very important part of the health and safety culture and it is also a legal requirement, under the Management of Health and Safety at Work Regulations 1999 and other regulations, for an employer to provide such training. Training is required on recruitment, at induction or on being exposed to new or increased risks due to:

> being transferred to another job or given a change in responsibilities
> the introduction of new work equipment or a change of use in existing work equipment
> the introduction of new technology
> the introduction of a new system of work or the revision of an existing system of work
> an increase in the employment of more vulnerable employees (young or disabled persons)
> particular training required by the organization's insurance company (e.g. specific fire and emergency training).

Additional training may well be needed following a single or series of accidents or near misses, the introduction of new legislation, the issuing of an enforcement notice or as a result of a risk assessment or safety audit.

It is important during the development of a training course that the target audience is taken into account. If the audience are young persons, the chosen approach must be capable of retaining their interest and any illustrative examples used must be within their experience. The trainer must also be aware of external influences, such as peer pressures, and use them to advantage. For example, if everybody wears personal protective equipment then it will be seen as the thing to do. Levels of literacy and numeracy are other important factors.

The way in which the training session is presented by the use of videos, power point, case studies, lectures or small discussion groups needs to be related to the material to be covered and the backgrounds of the trainees. Supplementary information in the form of copies of slides and additional background reading is often useful. The environment used for the training sessions is also important in terms of room layout and size, lighting and heating.

Attempts should be made to measure the effectiveness of the training by course evaluation forms issued at the time of the session, by a subsequent refresher session and by checking for improvements in health and safety performance (such as a reduction in specific accidents).

There are several different types of training, these include induction, job specific, supervisory and management, and specialist.

4.8.1 Induction training

Induction training should always be provided to new employees, trainees and possibly contractors. While such training covers items such as pay, conditions and quality, it must also include health and safety. It is useful if the employee signs a record to the effect that training has been received. This record may be required as evidence should there be a subsequent legal claim against the organization.

Most induction training programmes would include the following topics:

➤ the health and safety policy of the organization including a summary of the organization and arrangements including employee consultation
➤ a brief summary of the health and safety management system including the name of the employee's direct supervisor, safety representative and source of health and safety information
➤ the employee responsibility for health and safety including any general health and safety rules (e.g. smoking prohibitions)
➤ the accident reporting procedure of the organization, the location of the accident book and the location of the nearest first aider
➤ the fire and other emergency procedures including the location of the assembly point
➤ a summary of any relevant risk assessments and safe systems of work
➤ the location of welfare, canteen facilities and rest rooms.

Additional items, which are specific to the organization, may need to be included such as:

➤ internal transport routes and pedestrian walkways (e.g. fork lift truck operations)
➤ the correct use of personal protective equipment and maintenance procedures
➤ manual handling techniques and procedures
➤ details of any hazardous substances in use and any procedures relating to them (e.g. health surveillance).

There should be some form of follow-up with each new employee after 3 months to check that the important messages have been retained. This is sometimes called a refresher course, although it is often better done on a one-to-one basis.

It is very important to stress that the content of the induction course should be subject to constant review and updated following an accident investigation, new legislation, changes in the findings of a risk assessment or the introduction of new plant or processes.

4.8.2 Job-specific training

Job-specific training ensures that the employee undertakes his job in a safe manner. Such training, therefore, is a form of skill training and is often best done 'on the job' – sometimes known as 'toolbox training'. Details of the safe system of work or, in more hazardous jobs, a permit to work system should be covered. In addition to normal safety procedures, emergency procedures and the correct use of personal protective equipment also needs to be included. The results of risk assessments are very useful in the development of this type of training. It is important that any common causes of human errors (e.g. discovered as a result of an accident investigation), any standard safety checks or maintenance requirements are addressed.

It is common for this type of training to follow an operational procedure in the form of a checklist which the employee can sign on completion of the training. The new employee will still need close supervision for some time after the training has been completed.

4.8.3 Supervisory and management training

Supervisory and management health and safety training follows similar topics to those contained in an induction training course but will be covered in more depth. There will also be a more detailed treatment of health and safety law. There has been considerable research over the years into the failures of managers that have resulted in accidents and other dangerous incidents. These failures have included:

> lack of health and safety awareness, enforcement and promotion (in some cases there has been an encouragement to circumvent health and safety rules)
> lack of consistent supervision of and communication with employees
> lack of understanding of the extent of the responsibility of the supervisor.

It is important that all levels of management, including the board, receive health and safety training. This will not only keep everybody informed of health and safety legal requirements, accident prevention techniques and changes in the law, but also encourage everybody to monitor health and safety standards during visits or tours of the organization.

4.8.4 Specialist training

Specialist health and safety training is normally needed for activities that are not related to a specific job but more to an activity. Examples include first aid, fire prevention, fork lift truck driving, overhead crane operation, scaffold inspection and statutory health and safety inspections. These training courses are often provided by specialist organizations and successful participants are awarded certificates. Details of two of these courses will be given here by way of illustration.

Fire prevention training courses include the causes of fire and fire spread, fire and smoke alarm systems, emergency lighting, the selection and use of fire extinguishers and sprinkler systems, evacuation procedures, high risk operations and good house-keeping principles.

A fork lift truck drivers' course would include the general use of the controls, loading and unloading procedures, driving up or down an incline, speed limits, pedestrian awareness (particularly in areas where pedestrians and vehicles are not segregated), security of the vehicle when not in use, daily safety checks and defect reporting, refuelling and/or battery charging and emergency procedures.

Details of other types of specialist training appear elsewhere in the text.

Training is a vital part of any health and safety programme and needs to be constantly reviewed and updated. Many health and safety regulations require specific training (e.g. manual handling, personal protective equipment and display screens). Additional training courses may be needed when there is a major reorganization, a series of similar accidents or incidents or a change in equipment or a process. Finally, the methods used to deliver training must be continually monitored to ensure that they are effective.

4.9 Internal influences

There are many influences on health and safety standards, some are positive and others negative. No business, particularly small businesses, are totally divorced from their suppliers, customers and neighbours. This section considers the internal influences on a business, including management commitment, production demands, communication, competence and employee relations.

4.9.1 Management commitment

Managers, particularly senior managers, can give powerful messages to the workforce by what they do for health and safety. Managers can achieve the level of health and safety performance that they demonstrate they want to achieve. Employees soon get the negative message if directors disregard safety rules and ignore written policies to get urgent production to the customer or to avoid personal inconvenience. It is what they do that counts not what they say. In Chapter 3 managers' organizational roles were listed showing the ideal level of involvement for senior managers. Depending on the size and geography of the organization, senior managers should be personally involved in:

> health and safety inspections or audits
> meetings of the central health and safety or joint consultation committees
> involvement in the investigation of accidents, ill-health and incidents. The more serious the incident the more senior the manager who takes an active part in the investigation.

4.9.2 Production/service demands

Managers need to balance the demands placed by customers and action required to protect the health and safety of their employees. How this is achieved has a strong influence on the standards adopted by the organization. The delivery driver operating to near impossible delivery schedules; the manager agreeing to the strident demands of a large customer regardless of the risks to employees involved, are typical dilemmas facing workers and managers alike. The only way to deal with these issues is to be well organized and agree with the client/employees ahead of the crisis how they should be prioritized.

Rules and procedures should be intelligible, sensible and reasonable. They should be designed to be followed under normal production or service delivery conditions. If following the rules involves very long delays or impossible production schedules, they should be revised rather than ignored by workers and managers alike, until it is too late and an accident occurs. Sometimes the safety rules are simply used in a court of law, in an attempt to defend the company concerned, following the accident.

Managers who plan the impossible schedule or ignore a safety rule to achieve production or service demands, are held responsible for the outcomes. It is acceptable to balance the cost of the action against the level of risk being addressed but it is never acceptable to ignore safety rules or standards simply to get the work done. Courts are not impressed by managers who put profit considerations ahead of safety requirements.

4.9.3 Communication

Communications are covered in depth earlier in the chapter but clearly will have significant influence on health and safety issues. This will include:

➤ poorly communicated procedures that will not be understood or followed
➤ poor verbal communications which will be misunderstood and will demonstrate a lack of interest by senior managers
➤ missing or incorrect signs may cause accidents rather than prevent them
➤ managers who are nervous about face-to-face discussions with the workforce on health and safety issues will have a negative effect.

Managers and supervisors should plan to have regular discussions to learn about the problems faced by employees and discuss possible solutions. Some meetings, like the safety committee, are specifically planned for safety matters, but this should be reinforced by discussing health and safety issues at all routine management meetings. Regular one-to-one talks should also take place in the workplace, preferably to a planned theme or safety topic, to get specific messages across and get feedback from employees.

4.9.4 Competence

Competent people, who know what they are doing and have the necessary skills to do the task correctly and safely, will make the organization effective. Competence can be bought in through recruitment or consultancy but is often much more effective if developed from among employees. It demonstrates commitment to health and safety and a sense of security for the workforce. The loyalty that it develops in the workforce, can be a significant benefit to safety standards. Earlier in this chapter there is more detail on how to achieve competence.

4.9.5 Employee representation

Enthusiastic, competent employee safety representatives given the resources and freedom to fulfil their function effectively can make a major contribution to good health and safety standards. They can provide the essential bridge between managers and employees. People are more willing to accept the restrictions that some precautions bring if they are consulted and feel involved, either directly in small workforces or through their safety representatives. See earlier in the chapter for the details.

4.10 External influences

The role of external organizations is set out in Chapter 1. Here the influence they can have is briefly discussed including societal expectations, legislation and enforcement, insurance companies, trade unions, economics, commercial stakeholders.

4.10.1 Societal expectations

Societal expectations are not static and tend to rise over time, particularly in a wealthy nation like the UK. For example, the standards of safety accepted in a motor car fifty years ago would be considered to be totally inadequate at the beginning of the 21st century. We expect safe, quiet, comfortable cars that do not break down and retain their appearance for many thousands of miles. Industry should strive to deliver these same high standards for the health and safety of employees or service providers. The question is whether societal expectations are as great an influence on workplace safety standards as they are on product safety standards. Society can influence standards through:

> people only working for good employers. This is effective in times of low unemployment;
> national and local news media highlighting good and bad employers;
> schools teaching good standards of health and safety;
> the purchase of fashionable and desirable safety equipment, like trendy crash helmets on mountain bikes;
> only buying products from responsible companies. The difficulty of defining what is responsible has been partly overcome through ethical investment criteria but this is possibly not widely enough understood to be a major influence;
> watching TV and other programmes which improve safety knowledge and encourage safe behaviour from an early age. It will be interesting to see if 'Bob the Builder' and all the books, videos and toys produced to enhance sales will have a positive effect on future building and DIY health and safety standards.

4.10.2 Legislation and enforcement

Good comprehensible legislation should have a positive effect on health and safety standards. Taken together legislation and enforcement can affect standards by:

> providing a level to which every employer has to conform;
> insisting on minimum standards which also enhances peoples' ability to operate and perform well;
> providing a tough, visible threat of getting shut down or a heavy fine;
> stifling development by being too prescriptive. For example, woodworking machines have not developed quickly in the 20th century, partly because the regulations have been so detailed in their requirements that new designs were not feasible;
> providing well presented and easily read guidance for specific industries at reasonable cost or free.

On the other hand a weak enforcement regime can have a powerful negative effect on standards.

4.10.3 Insurance companies

Insurance companies can influence health and safety standards mainly through financial incentives. Employers' liability insurance is a legal requirement in the UK and therefore all employers have to obtain this type of insurance cover. Insurance companies can influence standards through:

> discounting premiums to those in the safest sectors or best individual companies;
> insisting on risk reduction improvements to remain insured. This is not very effective where competition for business is very fierce;
> encouraging risk reduction improvements by bundling services into the insurance premium;
> providing guidance on standards at reasonable cost or free.

4.10.4 Trade unions

Trade unions can influence standards by:

> providing training and education for members;
> providing guidance and advice cheaply or free to members;
> influencing Governments to regulate, enhance enforcement activities and provide guidance;
> influencing employers to provide high standards for their members. This is sometimes confused with financial improvements with health and safety getting a lower priority;
> encouraging members to work for safer employers;
> helping members to get proper compensation for injury and ill-health if it is caused through their work.

4.10.5 Economics

Economics can play a major role in influencing health and safety standards. The following ways are the most common:

> lack of orders and/or money can cause employers to try and ignore health and safety requirements;
> if employers were really aware of the actual and potential cost of accidents and fires they would be more concerned about prevention. The HSE believe that the ratio between insured and uninsured costs of accidents is between 1:8 and 1:36 (see *The Costs of Accidents at Work*, HSE Books, HS(G) 96);
> perversely when the economy is booming activity increases and, particularly in the building industry, accidents can sharply increase. The pressures to perform and deliver for customers can be safety averse;
> businesses that are only managed on short-term performance indicators seldom see the advantage of the long-term gains that are possible with a happy safe and fit workforce.

The cost of accidents and ill-health, in both human and financial terms, needs to be visible throughout the organization so that all levels of employee are encouraged to take preventative measures.

4.10.6 Commercial stakeholders

A lot can be done by commercial stakeholders to influence standards. These include:

> insisting on proper arrangements for health and safety management at supplier companies before they tender for work or contracts;
> checking on suppliers to see if the workplace standards are satisfactory;
> encouraging ethical investments;
> considering ethical standards as well as financial when banks provide funding;
> providing high quality information for customers;
> insisting on high standards to obtain detailed planning permission (where this is possible);
> providing low cost guidance and advice.

4.11 Practice NEBOSH questions for Chapter 4

1. Outline four advantages and four disadvantages of using 'propaganda' posters to communicate health and safety information to the workforce. (March 2001)
2. Give reasons why a verbal instruction may not be clearly understood by an employee. (December 2001)
3. Outline the practical means by which a manager could involve employees in the improvement of health and safety in the workplace. (December 2001)
4. (a) Explain the meaning of the term 'perception'.
   (b) Outline the factors relating to the individual that may influence a person's perception of an occupational risk. (December 2001)
5. Outline the factors that may determine the level of supervision an employee should receive during their initial period within a company. (December 2001)

6. Explain why it is important to use a variety of methods to communicate health and safety information in the workplace. (June 2001)
7. Outline the ways in which employers might motivate their employees to comply with health and safety procedures. (June 2001)
8. Most occupational accidents can be attributed in part to human error. Outline ways of reducing the likelihood of human error in the workplace. (December 2000)
9. Describe, using practical examples, four types of human error that can lead to accidents in the workplace. (June 1997)
10. Explain how induction training programmes for new employees can help to reduce the number of accidents in the workplace. (June 1997)

Risk assessment

5.1 Introduction

Risk assessment is an essential part of the planning stage of any health and safety management system. HSE, in the publication HSG(65) *Successful Health and Safety Management*, states that the aim of the planning process is to minimize risks.

> *Risk assessment methods are used to decide on priorities and to set objectives for eliminating hazards and reducing risks. Wherever possible, risks are eliminated through selection and design of facilities, equipment and processes. If risks cannot be eliminated, they are minimized by the use of physical controls or, as a last resort, through systems of work and personal protective equipment.*

5.2 Legal aspects of risk assessment

The general duties of employers to their employees in section 2 of the Health and Safety at Work Act 1974 imply the need for risk assessment. This duty was also extended by section 3 of the Act to anybody else affected by activities of the employer – contractors, visitors, customers or members of the public. However, the Management of Health and Safety at Work Regulations 1999 are much more specific concerning the need for risk assessment. The following requirements are laid down in those regulations:

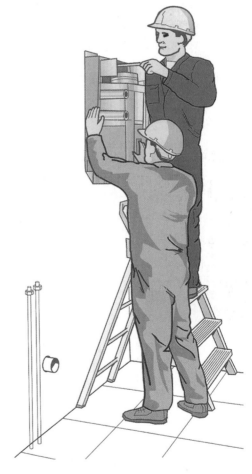

Figure 5.1 Reducing the risk – finding an alternative approach to fitting a wall-mounted boiler. Source HSE. Crown copyright material is reproduced with the permission of the Controller of HMSO and the Queen's Printer for Scotland.

> *the risk assessment shall be 'suitable and sufficient' and cover both employees and non-employees affected by the employer's undertaking (e.g. contractors, members of the public, students, patients, customers etc.);*
>
> *every self-employed person shall make a 'suitable and sufficient' assessment of the risks to which they or those affected by the undertaking may be exposed;*
>
> *any risk assessment shall be reviewed if there is reason to suspect that it is no longer valid or if a significant change has taken place;*
>
> *where there are more than 4 employees, the significant findings of the assessment shall be recorded and any specially at risk group of employees identified. (This does not mean that employers with 4 or less employees need not undertake risk assessments.)*

The term 'suitable and sufficient' is important since it defines the limits to the risk assessment process. A suitable and sufficient risk assessment should:

➤ identify the significant risks and ignore the trivial ones;
➤ identify and prioritize the measures required to comply with any relevant statutory provisions;
➤ remain appropriate to the nature of the work and valid over a reasonable period of time.

Several other regulations require a specific risk assessment to be made. These include:

➤ Ionizing Radiation Regulations 1999
➤ Control of Asbestos at Work Regulations 2002
➤ Noise at Work Regulations 1989
➤ Manual Handling Operations Regulations 1992
➤ Health and Safety (Display Screen Equipment) Regulations 1992
➤ Fire Precautions (Workplace) Regulations 1997
➤ Control of Lead at Work Regulations 2002
➤ Control of Substances Hazardous to Health Regulations 2002.

5.3 Forms of risk assessment

There are two basic forms of risk assessment.

A quantitative risk assessment attempts to measure the risk by relating the probability of the risk occurring to the possible severity of the outcome and then giving the risk a numerical value. This method of risk assessment is used in situations where a malfunction could be very serious (e.g. aircraft design and maintenance or the petrochemical industry).

The more common form of risk assessment is the qualitative assessment, which is based purely on personal judgement and is normally defined as high, medium or low. Qualitative risk assessments are usually satisfactory since the definition (high, medium or low) is normally used to determine the time frame in which further action is to be taken.

The term 'generic' risk assessment is sometimes used and describes a risk assessment which covers similar activities or work equipment in different departments, sites or companies. Such assessments are often produced by specialist bodies, such as trade associations. If used, they must be appropriate to the particular job and they may need to be extended to cover additional hazards or risks.

5.4 Some definitions

Some basic definitions were introduced in Chapter 1 and those relevant to risk assessment are reproduced here.

Electricity is an example of a high hazard since it has the potential to kill a person. The risk associated with electricity – the likelihood of being killed on coming into contact with an electrical device – is, hopefully, low.

5.4.1 Hazard and risk

A hazard is the *potential* of a substance, activity or process to cause harm. Hazards take many forms including, for example, chemicals, electricity and the use of a ladder. A hazard can be ranked relative to other hazards or to a possible level of danger.

A risk is the *likelihood* of a substance, activity or process to cause harm. Risk (or strictly the level of risk) is also linked to the severity of its consequences. A risk can be reduced and the hazard controlled by good management.

It is very important to distinguish between a hazard and a risk – the two terms are often confused and activities often called high risk are in fact high hazard. There should only be high residual risk where there is poor health and safety management and inadequate control measures.

5.4.2 Occupational or work-related ill-health

This is concerned with those illnesses or physical and mental disorders that are either caused or triggered by workplace activities. Such conditions may be induced by the particular work activity of the individual or by activities of others in the workplace. The time interval between exposure and the onset of the illness may be short (e.g. asthma attacks) or long (e.g. deafness or cancer).

5.4.3 Accident

This is defined by the Health and Safety Executive as 'any unplanned event that results in injury or ill-health of people, or damage or loss to property, plant, materials or the environment or a loss of a business opportunity'. Other authorities define an accident more narrowly by excluding events that do not involve injury or ill-health. This book will always use the Health and Safety Executive definition.

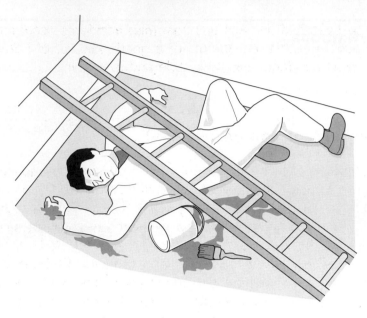

Figure 5.2 Accident at work.

5.4.4 Near miss

This is any incident that could have resulted in an accident. Knowledge of near misses is very important since research has shown that, approximately, for every 10 'near miss' events at a particular location in the workplace, a minor accident will occur.

5.4.5 Dangerous occurrence

This is a 'near miss' which could have led to serious injury or loss of life. Dangerous occurrences are defined in the Reporting of Injuries, Diseases and Dangerous Occurrences Regulations 1995 (often known as RIDDOR) and are always reportable to the Enforcement Authorities. Examples include the collapse of a scaffold or a crane or the failure of any passenger carrying equipment.

In 1969, F E Bird collected a large quantity of accident data and produced a well-known triangle (Figure 5.3).

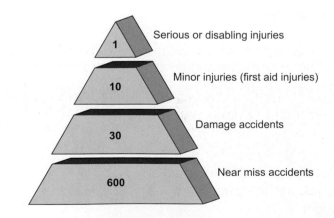

Figure 5.3 F E Bird's well-known accident triangle.

It can be seen that damage and near miss accidents occur much more frequently than injury accidents and are, therefore, a good indicator of hazards. The study also shows that most accidents are predictable and avoidable.

5.5 The objectives of risk assessment

The main objective of risk assessment is to determine the measures required by the organization to comply with relevant health and safety legislation and, thereby, reduce the level of occupational injuries and ill-health. The purpose is to help the employer or self-employed person to determine the measures required to comply with their legal statutory duty under the Health and Safety at Work Act 1974 or its associated Regulations. The risk assessment will need to cover all those who may be at risk, such as customers, contractors and members of the public. In the case of shared workplaces, an overall risk assessment may be needed in partnership with other employers.

In Chapter 1, the moral, legal and financial arguments for health and safety management were discussed in detail. The important distinction between the direct and indirect costs of accidents is reiterated here.

Any accident or incidence of ill-health will cause both direct and indirect costs and incur an insured and an uninsured cost. It is important that all of these costs are taken into account when the full cost of an accident is calculated. In a study undertaken by the HSE, it was shown that indirect costs or hidden costs could be 36 times greater than direct costs of an accident. In other words, the direct costs of an accident or disease represent the tip of the iceberg when compared to the overall costs (*The Cost of Accidents at Work* HS(G) 96).

Direct costs are costs that are directly related to the accident. They may be insured (claims on employers' and public liability insurance, damage to buildings, equipment or vehicles) or uninsured (fines, sick pay, damage to product, equipment or process).

Indirect costs may be insured (business loss, product or process liability) or uninsured (loss of goodwill, extra overtime payments, accident investigation time, production delays).

5.6 Accident categories

There are several categories of accident, all of which will be dealt with in more detail in later chapters. The principal categories are as follows:

> contact with moving machinery or material being machined
> struck by moving, flying or falling object
> hit by a moving vehicle
> struck against something fixed or stationary
> injured while handling, lifting or carrying
> slips, trips and falls on the same level
> falls from a height
> trapped by something collapsing

> drowned or asphyxiated
> exposed to, or in contact, with a harmful substance
> exposed to fire
> exposed to an explosion
> contact with electricity or an electrical discharge
> injured by an animal
> physically assaulted by a person
> other kind of accident.

5.7 Health risks

Risk assessment is not only concerned with injuries in the workplace but also needs to consider the possibility of occupational ill-health. Health risks fall into the following four categories:

> chemical (e.g. paint solvents, exhaust fumes)
> biological (e.g. bacteria, pathogens)
> physical (e.g. noise, vibrations)
> psychological (e.g. occupational stress).

There are two possible health effects of occupational ill-health.

They may be **acute** which means that they occur soon after the exposure and are often of short duration, although in some cases emergency admission to hospital may be required.

They may be **chronic** which means that the health effects develop with time. It may take several years for the associated disease to develop and the effects may be slight (mild asthma) or severe (cancer).

Health risks are discussed in more detail in Chapter 12.

5.8 The management of risk assessment

Risk assessment is part of the planning and implementation stage of the health and safety management system recommended by the HSE in its publication HSG (65). All aspects of the organization, including health and safety management, need to be covered by the risk assessment process. This will involve the assessment of risk in areas such as maintenance procedures, training programmes and supervisory arrangements. A general risk assessment of the organization should reveal the significant hazards present and the general control measures that are in place. Such a risk assessment should be completed first and then followed by more specific risk assessments that examine individual work activities.

HSE has produced a free leaflet entitled *Five steps to risk assessment*. It gives practical advice on assessing risks and recording the findings and is aimed at small and medium-sized companies in the service and manufacturing sectors. The five steps are:

> look for the hazards;
> decide who might be harmed, and how;
> evaluate the risks and decide whether existing precautions are adequate or more should be done;

> record the significant findings;
> review the assessment and revise it if necessary.

Each of these steps will be examined in turn in the next section.

Finally, it is important that the risk assessment team is selected on the basis of its competence to assess risks in the particular areas under examination in the organization. The team leader or manager should have health and safety experience and relevant training in risk assessment. It is sensible to involve the appropriate line manager, who has responsibility for the area or activity being assessed, as a team member. Other members of the team will be selected on the basis of their experience, their technical and/or design knowledge and any relevant standards or regulations relating to the activity or process. At least one team member must have communication and report writing skills. It is likely that team members will require some basic training in risk assessment.

5.9 The risk assessment process

The HSE approach to risk assessment (5 steps) will be used to discuss the process of risk assessment. It is, however, easier to divide the process into six elements:

> hazard identification
> persons at risk
> evaluation of risk level
> risk controls (existing and additional)
> record of risk assessment findings
> monitoring and review.

Each element will be discussed in turn.

5.9.1 Hazard identification

Hazard identification is the crucial first step of risk assessment. Only significant hazards, which could result in serious harm to people, should be identified. Trivial hazards should be ignored.

A tour of the area under consideration by the risk assessment team is an essential part of hazard identification as is consultation with the relevant section of the workforce.

A review of accident, incident and ill-health records will also help with the identification. Other sources of information include safety inspection, survey and audit reports, job or task analysis reports, manufacturers' handbooks or data sheets and approved codes of practice and other forms of guidance.

Hazards will vary from workplace to workplace but the checklist in Appendix 5.1 shows the common hazards that are significant in many workplaces.

It is important that unsafe conditions are not confused with hazards, during hazard identification. Unsafe conditions should be rectified as soon as possible after observation. Examples of unsafe conditions include missing machine guards, faulty warning systems and oil spillage on the workplace floor.

5.9.2 Persons at risk

Employees and contractors who work full time at the workplace are the most obvious groups at risk and it will be necessary check that they are competent to perform their particular tasks. However, there may be other groups who spend time in or around the workplace. These include young workers, trainees, new and expectant mothers, cleaners, contractor and maintenance workers and members of the public. Members of the public will include visitors, patients, students or customers as well as passers by.

The risk assessment must include any additional controls required due to the vulnerability of any of these groups, perhaps caused by inexperience or disability. It must also give an indication of the numbers of people from the different groups who come into contact with the hazard and the frequency of these contacts.

5.9.3 Evaluation of risk level

During most risk assessment it will be noted that some of the risks posed by the hazard have already been addressed or controlled. The purpose of the risk assessment, therefore, is to reduce the remaining risk. This is called the **residual risk**.

The goal of risk assessment is to reduce all residual risks to as low a level as reasonably practicable. In a relatively complex workplace, this will take time so that a system of ranking risk is required – the higher the risk level the sooner it must be addressed and controlled.

For most situations, a **qualitative** risk assessment will be perfectly adequate. (This is certainly the case for NEBOSH Certificate candidates and is suitable for use during the practical assessment.) During the risk assessment, a judgement is made as to whether the risk level is high, medium or low in terms of the risk of somebody being injured. This designation defines a timetable for remedial actions to be taken thereby reducing the risk. High-risk activities should normally be addressed in days, medium risks in weeks and low risks in months or in some cases no action will be required. It will usually be necessary for risk assessors to receive some training in risk level designation.

A **quantitative** risk assessment attempts to quantify the risk level in terms of the likelihood of an incident and its subsequent severity. Clearly the higher the likelihood and severity, the higher the risk will be. The likelihood depends on such factors as the control measures in place, the frequency of exposure to the hazard and the category of person exposed to the hazard. The severity will depend on the magnitude of the hazard (e.g. voltage, toxicity etc.). HSE suggest in HSG(65) a simple 3×3 matrix to determine risk levels.

Likelihood of occurrence	Likelihood level
Harm is certain or near certain to occur	High 3
Harm will often occur	Medium 2
Harm will seldom occur	Low 1

Severity of harm	Severity level
Death or major injury (as defined by RIDDOR)	Major 3
3 day injury or illness (as defined by RIDDOR)	Serious 2
All other injuries or illnesses	Slight 1

Risk = Severity × Likelihood

Likelihood	Severity		
	Slight 1	Serious 2	Major 3
Low 1	Low 1	Low 2	Medium 3
Medium 2	Low 2	Medium 4	High 6
High 3	Medium 3	High 6	High 9

Thus:

    6 – 9 High risk

    3 – 4 Medium risk

    1 – 2 Low risk

It is possible to apply such methods to organizational risk or to the risk that the management system for health and safety will not deliver in the way in which it was expected or required. Such risks will add to the activity or occupational risk level. In simple terms, poor supervision of an activity will increase its overall level of risk. A risk management matrix has been developed which combines these two risk levels as shown below.

		Occupational risk level		
		Low	Medium	High
Organizational risk level	Low	L	L	M
	Medium	L	M	H
	High	M	H	Unsatisfactory

L Low risk   **M** Medium risk   **H** High risk

Whichever type of risk evaluation method is used, the level of risk simply enables a timetable of risk reduction to be formulated. The legal duty requires that all risks should be reduced to as low as is reasonably practicable.

5.10 Risk control measures

The next stage in the risk assessment is the control of the risk. In established workplaces, some control of risk will already be in place. The effectiveness of these controls needs to be assessed so that an estimate of the residual risk may be made. Many hazards have had specific acts, regulations or other recognized standards developed to reduce associated risks. Examples of such hazards are fire, electricity, lead and asbestos. The relevant legislation and any accompanying approved codes of practice or guidance should be consulted first and any recommendations implemented. Advice on control measures may also be available from trade associations, trade unions or employers' organizations.

    Where there are existing preventative measures in place, it is important to check that they are working properly and that everybody affected has a clear understanding of the measures. It may be necessary to strengthen existing procedures, for example, by the introduction of a permit to work system. More details on the principles of control are contained in Chapter 6.

5.11 Hierarchy of risk control

When assessing the adequacy of existing controls or introducing new controls, a hierarchy of risk controls should be considered. The Management of Health and Safety at Work Regulations 1999 Schedule 1 specifies the general principles of prevention which are set out in the European Council Directive. These principles are:

1. Avoiding risks;
2. Evaluating the risks which cannot be avoided;
3. Combating the risks at source;
4. Adapting the work to the individual, especially as regards the design of the workplace, the choice of work equipment and the choice of working and production methods, with a view, in particular, to alleviating monotonous work and work at a predetermined work-rate and to reducing their effects on health;
5. Adapting to technical progress;
6. Replacing the dangerous by the non-dangerous or the less dangerous;
7. Developing a coherent overall prevention policy which covers technology, organization of work, working conditions, social relationships and the influence of factors relating to the working environment;
8. Giving collective protective measures priority over individual protective measures and;
9. Giving appropriate instruction to employees.

These principles are not exactly a hierarchy but must be considered alongside the usual hierarchy of risk control which is as follows:

> elimination
> substitution
> engineering controls (e.g. isolation, insulation and ventilation)
> reduced or limited time exposure
> good housekeeping
> safe systems of work
> training and information
> personal protective equipment
> welfare
> monitoring and supervision
> review.

5.12 Prioritization of risk control

The prioritization of the implementation of risk control measures will depend on the risk rating (high, medium and low) but the time scale in which the measures are introduced will not always follow the ratings. It may be convenient to deal with a low level risk at the same time as a high level risk or before a medium level risk. It may also be that work on a high risk control system is delayed due to a late delivery of an essential component –

this should not halt the overall risk reduction work. It is important to maintain a continuous programme of risk improvement rather than slavishly following a predetermined priority list.

5.13 Record of risk assessment findings

It is very useful to keep a written record of the risk assessment even if there are less than five employees in the organization. For an assessment to be 'suitable and sufficient' only the significant hazards and conclusions need be recorded. The record should also include details of the groups of people affected by the hazards and the existing control measures and their effectiveness. The conclusions should identify any new controls required and a review date. The HSE booklet *Five steps to risk assessment* provides a very useful guide and examples of the detail required for most risk assessments.

There are many other possible layouts which can be used for the risk assessment record. One example is given in Appendix 5.2.

The written record provides excellent evidence to a health and safety inspector of compliance with the law. It is also useful evidence if the organization should become involved in a civil action.

The record should be accessible to employees and a copy kept with the safety manual containing the safety policy and arrangements.

5.14 Monitoring and review

As mentioned earlier, the risk controls should be reviewed periodically. This is equally true for the risk assessment as a whole. Review and revision may be necessary when conditions change as a result of the introduction of new machinery, processes or hazards. There may be new information on hazardous substances or new legislation. There could also be changes in the workforce, for example, the introduction of trainees. The risk assessment only needs to be revised if significant changes have taken place since the last assessment was done. A major accident or incident or a series of minor ones provides a good reason for a review of the risk assessment.

5.15 Special cases

There are several groups of persons who require an additional risk assessment due to their being more 'at risk' than other groups. Three such groups will be considered – young persons, expectant and nursing mothers and disabled workers.

5.15.1 Young persons

A risk assessment involving young people needs to consider the particular vulnerability of young persons in the workplace. Young workers clearly have a lack of experience and awareness of risks in the workplace, a tendency to be subject to peer pressure and a willingness to work hard. Many young workers will be trainees or on unpaid work experience. An amendment to the Health and Safety at Work Act enabled trainees on Government sponsored training schemes to be treated as employees as far as health and safety is concerned. The Management of Health and Safety (At Work) Regulations 1999 defines a young person as anybody under the age of 18 years and stipulates that a special risk assessment must be completed which takes into account their immaturity and inexperience and the assessment must be completed before the young person starts work. If the young person is of school age (16 years or less), the parents or guardian of the child should be notified of the outcome of the risk assessment and details of any safeguards which will be used to protect the health and safety of the child. The following key elements should be covered by the risk assessment:

➤ details of the work activity, including any equipment or hazardous substances
➤ details of any prohibited equipment or processes
➤ details of health and safety training to be provided
➤ details of supervision arrangements.

More detailed guidance is available from HSE Books – HSG 165 *Young people at work* and HSG 199 *Managing health and safety on work experience.*

5.15.2 Expectant and nursing mothers

The Management of Health and Safety at Work Regulations 1999 incorporates the Pregnant Workers Directive from the EU. If any type of work could present a particular risk to expectant or nursing mothers, the risk assessment must include an assessment of such risks. Should these risks be unavoidable, then the woman's working conditions or hours must be altered to avoid the risks. The alternatives are for her to be offered other work or be suspended from work on full pay. The woman must notify the employer in writing that she is pregnant, or has given birth within the previous six months and/or is breastfeeding.

Typical factors which might affect such women are:

➤ manual handling
➤ chemical or biological agents
➤ ionizing radiation
➤ passive smoking
➤ lack of rest room facilities
➤ temperature variations
➤ prolonged standing or sitting
➤ stress and violence to staff.

Detailed guidance is available in *New and expectant mothers at work,* HSG 122, HSE Books.

5.15.3 Workers with a disability

Organizations have been encouraged for many years to employ workers with disabilities and to ensure that their premises provide suitable access for such people. From a health and safety point of view, it is important that workers with a disability are covered by special risk assessments so that appropriate controls are in place to protect them. For example, employees with a hearing problem will need to be warned when the fire alarm sounds or a fork lift truck approaches. Special vibrating signals or flashing lights may be used. Similarly workers in wheelchairs will require a clear, wheelchair friendly, route to a fire exit and onwards to the assembly point. Safe systems of work and welfare facilities need to be suitable for any workers with disabilities.

The Disability Discrimination Act will come into force progressively after 2000 and requires equal opportunities for employment and access to workplaces to be extended to all people with disabilities.

5.15.4 Lone workers

People who work alone, like those in small workshops, remote areas of a large site, social workers, sales personnel or mobile maintenance staff, should not be at more risk than other employees. It is important to consider whether the risks of the job can be properly controlled by one person. Other considerations in the risk assessment include:

➤ whether the particular workplace presents a special risk to someone working alone
➤ is there safe egress and exit from the workplace?
➤ can all the equipment and substances be safely handled by one person?
➤ is violence from others a risk?
➤ would women and young persons be specially at risk?

Figure 5.4 A lone worker – special arrangements are required.

➤ is the worker medically fit and suitable for working alone?
➤ are special training and supervision required?

For details of precautions see Chapter 6 on controls.

5.16 Practice NEBOSH questions for Chapter 5

1. (a) Identify the factors that may place young persons at a greater risk of accidents at work.
   (b) Outline the measures that could be taken to minimize the risks to young persons. (March 2001)
2. Identify the factors to be considered to ensure the health and safety of persons who are required to work on their own away from the workplace. (December 2001)
3. Outline the factors that may increase risks to pregnant employees. (June 2001)
4. A factory manager intends to introduce a new work process for which a risk assessment is required under regulation 3 of the Management of Health and Safety at Work Regulations 1999.

(i) Outline the factors that should be considered when carrying out the risk assessment.

(ii) Explain the criteria that must be met for the assessment to be deemed 'suitable and sufficient'.

(iii) Identify the various circumstances that may require a review of the risk assessment at a later date. (December 2000)

5. (a) Explain the meaning of the term 'risk' as used in occupational safety and health.

(b) Outline the logical steps to take in managing risks at work. (June 1996)

Appendix 5.1 – Hazard checklist

The following checklist may be helpful

1. Equipment/mechanical
Entanglement
Friction/abrasion
Cutting
Shearing
Stabbing/puncturing
Impact
Crushing
Drawing-in
Air or high pressure fluid injection
Ejection of parts
Pressure/vacuum
Display screen equipment
Hand tools

2. Transport
Works vehicles
Mechanical handling
People/vehicle interface

3. Access
Slips, trips and falls
Falling or moving objects
Obstruction or projection
Working at height
Confined spaces
Excavations

4. Handling/lifting
Manual handling
Mechanical handling

5. Electricity
Fixed installation
Portable tools and equipment

6. Chemicals
Dust/fume/gas
Toxic
Irritant
Sensitizing
Corrosive
Carcinogenic
Nuisance

7. Fire and Explosion
Flammable materials/gases/liquids
Explosion
Means of escape/alarms/detection

8. Particles and dust
Inhalation
Ingestion
Abrasion of skin or eye

9. Radiation
Ionizing
Non-ionizing

10. Biological
Bacterial
Viral
Fungal

11. Environmental
Noise
Vibration
Light
Humidity
Ventilation
Temperature
Overcrowding

12. The individual
Individual not suited to work
Long hours
High work rate
Violence to staff
Unsafe behaviour of individual
Stress
Pregnant/nursing women
Young people

13. Other factors to consider
Poor maintenance
Lack of supervision
Lack of training
Lack of information
Inadequate instruction
Unsafe systems

Appendix 5.2 – Example of a risk assessment record

General Health and Safety Risk Assessment	No

Firm/Company	Department
Contact Name	**Nature of Business**
Telephone Number	

Principal Hazards

Risks to employees and members of the public could arise due to the following hazards:

1. hazardous substances
2. electricity
3. fire
4. dangerous occurrences or other emergency incidents
5.
6.

Persons at Risk

Employees, contractors, and members of the public.

Main Legal Requirements

1. Health and Safety at Work Act 1974 – Section 2 and 3
2. Management of Health and Safety at Work Regulations 1999
3. The Noise at Work Regulations 1989
4. Common Law Duty of Care
5. ?

Significant Risks

- Acute and chronic health problems caused by the use or release of hazardous substances.
- Injuries to employees and members of the public due to equipment failure such as electric shock.
- Injuries to employees and members of the public from slips, trips and falls.
- Injuries to employees and members of the public caused by fire.
- ?

Consequences

Fractures, bruising, smoke inhalation and burns, acute and chronic health problems and death.

Existing Control Measures

Possible examples include:

1. All sub-contractors are vetted prior to appointment.
2. All hazardous and harmful materials are identified and the risks to people assessed. COSHH assessments are provided and the appropriate controls are implemented. Health surveillance is provided as necessary.
3. Fire risk assessment has been produced. Fire procedures are in place and all employees are trained to deal with fire emergencies. A carbon dioxide fire extinguisher is available at every work site.
4. A minimum of flammable substances are used on the premises, no more than a half days supply at a time. Kept in fire resistant store.
5. No smoking is allowed on the premises.
6. Manual handling is kept to a minimum. Where there is a risk of injury manual handling assessments are carried out.
7. Method statements are used for complex and/or hazardous jobs and are followed at all times.
8. All accidents on or around the site are reported and investigated by management. Any changes found necessary are quickly implemented. All accidents, reportable under RIDDOR 1995, are reported to the HSE on form F2508.
9. At least 1 qualified First Aider is available during working hours.
10. ?

Residual Risk i.e. after controls are in place.

Severity .... **Likelihood** .... **Residual Risk**...............

Information

Details of various relevant HSE and trade publications.

Comments from Line Manager	**Comments from the Risk Assessor**
Signed .... **Date**	**Signed** .... **Date** ....

Review Date

Likelihood	Severity		
	Slight 1	Serious 2	Major 3
Low 1	Low 1	Low 2	Medium 3
Medium 2	Low 2	Medium 4	High 6
High 3	Medium 3	High 6	High 9

Principles of control

6.1 Introduction

The control of risks is essential to secure and maintain a healthy and safe workplace which complies with the relevant legal requirements. Hazard identification and risk assessment are covered in Chapter 5 and these together with appropriate risk control measures form the core of HSG 65 'implementing and planning' section of the management model. Chapter 1 covers this in more detail.

This chapter concerns the principles that should be adopted when deciding on suitable measures to eliminate or control both acute and chronic risks to the health and safety of people at work. The principles of control can be applied to both health risks and safety risks, although health risks have some distinctive features that require a special approach.

Chapters 7 to 14 deal with specific workplace hazards and controls subject by subject. The principles of prevention now enshrined in the Management of Health and Safety at Work (MHSW) Regulations 1999 need to be used jointly with the hierarchy of control methods which give the preferred order of approach to risk control.

Figure 6.1 When controls break down. Reproduced with permission from the *Argus*, Brighton.

When risks have been analysed and assessed, decisions can be made about workplace precautions.

All final decisions about risk control methods must take into account the relevant legal requirements, which establish minimum levels of risk prevention or control. Some of the duties imposed by the HSW Act and the relevant statutory provisions are **absolute** and must be complied with. Many requirements are, however, qualified by the words, **so far as is reasonably practicable**, or **so far as is practicable**. These require an assessment of cost, along with information about relative costs, effectiveness and reliability of different control measures. Further guidance on the meaning of these three expressions is provided in Chapter 1.

6.2 Principles of prevention

The MHSW Regulations 1999 Schedule 1 specifies the general principles of prevention which are set out in article 6(2) of the European Council Directive 89/391/EEC. For the first time the principles have been enshrined directly in regulations which state, at Regulation 4, that *Where an employer implements any preventative measures he shall do so on the basis of the principles specified in Schedule 1.* These principles are:

1 Avoiding risks
   This means, for example, trying to stop doing the task or use different processes or doing the work in a different safer way.

2 Evaluating the risks which cannot be avoided
   This requires a risk assessment to be carried out.

3 Combating the risks at source
   This means that risks, such as a dusty work atmosphere, are controlled by removing the cause of the dust rather than providing special protection; or slippery floors are treated or replaced rather than putting up a sign.

4 Adapting the work to the individual, especially as regards the design of the workplace, the choice of work equipment and the choice of working and production methods, with a view, in particular, to alleviating monotonous work and work at a predetermined work-rate and to reducing their effect on health
   This will involve consulting those who will be affected when workplaces, methods of work and safety procedures are designed. The control individuals have over their work should be increased, and time spent working at predetermined speeds and in monotonous work should be reduced where it is reasonable to do so.

5 Adapting to technical progress
   It is important to take advantage of technological and technical progress, which often gives designers and employers the chance to improve both safety and working methods. With the Internet and other international information sources available a very wide knowledge, going beyond what is happening in the UK or Europe, will be expected by the enforcing authorities and the courts.

6 Replacing the dangerous by the non-dangerous or the less dangerous
This involves substituting, for example, equipment or substances with non-hazardous or less hazardous substances.

7 Developing a coherent overall prevention policy which covers technology, organization of work, working conditions, social relationships and the influence of factors relating to the working environment
Health and safety policies should be prepared and applied by reference to these principles.

8 Giving collective protective measures priority over individual protective measures
This means giving priority to control measures which make the workplace safe for everyone working there so giving the greatest benefit, for example, removing hazardous dust by exhaust ventilation rather than providing a filtering respirator to an individual worker. This is sometimes known as a 'Safe Place' approach to controlling risks.

9 Giving appropriate instruction to employees
This involves making sure that employees are fully aware of company policy, safety procedures, good practice, official guidance, any test results and legal requirements. This is sometimes known as a 'Safe Person' approach to controlling risks where the focus is on individuals. A properly set up health and safety management system should cover and balance both a Safe Place and Safe Person approach.

6.3 Hierarchy of risk control

When assessing the adequacy of existing controls or introducing new controls, a hierarchy of risk controls should be considered. The principles of prevention in the MHSW Regulations 1999 are not exactly a hierarchy, but must be considered alongside the usual hierarchy of risk controls, which is as follows:

➤ elimination
➤ substitution
➤ changing work methods/patterns
➤ reduced or limited time exposure
➤ engineering controls (e.g. isolation, insulation and ventilation)
➤ good housekeeping
➤ safe systems of work
➤ training and information
➤ personal protective equipment
➤ welfare
➤ monitoring and supervision
➤ review.

6.3.1 Elimination or substitution

This is the best and most effective way of avoiding a severe hazard and its associated risks. Elimination occurs when a process or activity is totally abandoned because the associated risk is too high. Substitution describes the use of a less hazardous form of the substance. There are many examples of substitution, such as the use of water-based rather than oil-based paints, the use of asbestos substitutes and the use of compressed air as a power source rather than electricity. Care must be taken not to introduce additional hazards and risks as a result of a substitution.

6.3.2 Changing work methods/patterns

In some cases it is possible to change the method of working so that exposures are reduced. For example use rods to clear drains instead of strong chemicals; use disposable hooks for holding articles being sprayed instead of exposing people during the cleaning of reusable hooks. Sometimes the pattern of work can be changed so that people can do things in a more natural way, for example when removing components and packing them consider whether people are right or left handed; encourage people in offices to take breaks from computer screens by getting up to photocopy, fetch files or printed documents.

6.3.3 Reduced time exposure

This involves reducing the time during the working day that the employee is exposed to the hazard, either by giving the employee other work or rest periods. It is only suitable for the control of health hazards associated with, for example, noise, display screens and hazardous substances. However, it is important to note that for many hazards there are short-term exposure limits as well as normal working occupational exposure limits over an 8-hour period (see Chapter 12). Short-term limits must not be exceeded during the reduced time exposure intervals.

6.3.4 Engineering controls

This describes the control of risks by means of engineering design rather than a reliance on preventative actions by the employee. There are several ways of achieving such controls:

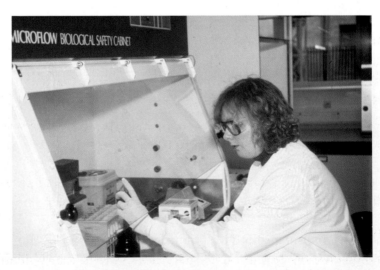

Figure 6.2 Proper control of gases and vapours in a laboratory.

1 control the risks at the source (e.g. the use of more efficient dust filters or the purchase of less noisy equipment)

2 control the risk of exposure by:

isolating the equipment by the use of an enclosure, a barrier or guard

insulating any electrical or temperature hazard

ventilate away any hazardous fumes or gases, either naturally or by the use of extractor fans and hoods.

6.3.5 Housekeeping

Housekeeping is a very cheap and effective means of controlling risks. It involves keeping the workplace clean and tidy at all times and maintaining good storage systems for hazardous substances and other potentially dangerous items. The risks most likely to be influenced by good housekeeping are fire and slips, trips and falls.

6.3.6 Systems of work

A safe system of work is a requirement of the HSW Act and is dealt with in detail later. The system of work describes the safe method of performing the job activity. If the risks involved are high or medium, the details of the system should be in writing and should be communicated to the employee formally in a training session. Systems for low risk activities may be conveyed verbally. There should be records that the employee (or contractor) has been trained or instructed in the safe system of work and that he or she understands it and will abide by it.

6.3.7 Training and information

Both these topics are important but should not be used in isolation. Information includes such items as signs, posters, systems of work and general health and safety arrangements. Details on the categories and design of safety signs are given in Chapter 17 (Health and Safety [Safety Signs and Signals] Regulations 1996).

6.3.8 Personal protective equipment

Personal protective equipment (PPE) should only be used as a last resort. There are many reasons for this. The most important limitations are that PPE:

➤ only protects the person wearing the equipment not others nearby;

➤ relies on people wearing the equipment at all times;

➤ must be used properly;

➤ must be replaced when it no longer offers the correct level of protection. This last point is particularly relevant when respiratory protection is used.

The benefits of PPE are:

➤ it gives immediate protection to allow a job to continue while engineering controls are put in place;

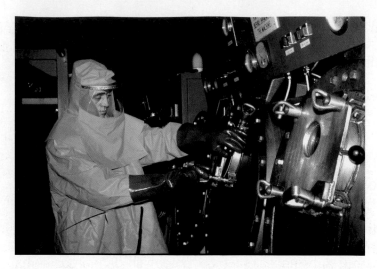

Figure 6.3 PPE used for loading a textile dye vessel.

➤ in an emergency it can be the only practicable way of effecting rescue or shutting down plant in hazardous atmospheres;

➤ it can be used to carry out work in confined spaces where alternatives are impracticable. But it should never be used to allow people to work in dangerous atmospheres, which are, for example, enriched with oxygen or explosive.

(See Chapter 12, Section 12.9.5 for more details on personal protective equipment.)

6.3.9 Welfare

Welfare facilities include general workplace ventilation, lighting and heating and the provision of drinking water, sanitation and washing facilities. There is also a requirement to provide eating and rest rooms. Risk control may be enhanced by the provision of eye

Figure 6.4 Welfare – washing facilities. Source HSE. Crown copyright material is reproduced with the permission of the Controller of HMSO and the Queen's Printer for Scotland.

washing and shower facilities for use after certain accidents. Within this area of welfare, first aid and health surveillance are important services that should be available.

6.3.10 Monitoring and supervision

All risk control measures, whether they rely on engineered or human behavioural controls, must be monitored for their effectiveness and supervised to ensure that they have been applied correctly. Competent people, who have a sound knowledge of the equipment or process, should undertake monitoring. Checklists are useful to ensure that no significant factor is forgotten. Any statutory inspection or insurance company reports should be checked to see whether any areas of concern were highlighted and if any recommendations were implemented. Details of any accidents, illnesses or other incidents will give an indication of the effectiveness of the risk control measures. Any emergency arrangements should be tested during the monitoring phase, including first aid provision.

It is crucial that the operator should be monitored to ascertain that all relevant procedures have been understood and followed. The operator may also be able to suggest improvements to the equipment or system of work. The supervisor is an important source of information during the monitoring process.

Where the organization is involved with shift work, it is essential that the risk controls are monitored on all shifts to ensure the uniformity of application.

The effectiveness and relevance of any training or instruction given should be monitored.

6.3.11 Review

Periodically the risk control measures should be reviewed. Monitoring and other reports are crucial for the review to be useful. Reviews often take place at safety committee and/or at management meetings. A serious accident or incident should lead to an immediate review of the risk control measures in place.

6.4 Controlling health risks

Figure 6.5   Health risks – checking on the contents.

6.4.1 Types of health risk

The principles of control for health risks are the same as those for safety. However, the nature of health risks can make the link between work activities and employee ill-health less obvious than in the case of injury from an accident. Guidance under the COSHH Regulations covers the principles in more detail.

Unlike safety risks, which can lead to immediate injury, the results of daily exposure to health risks may not manifest itself for months, years and, in some cases, decades. Irreversible health damage may occur before any symptoms are apparent. It is, therefore, essential to develop a preventive strategy to identify and control risks before anyone is exposed to them.

Risks to health from work activities include:

➤ skin contact with irritant substances, leading to dermatitis etc.
➤ inhalation of respiratory sensitizers, triggering immune responses such as asthma
➤ badly designed workstations requiring awkward body postures or repetitive movements, resulting in upper limb disorders, repetitive strain injury and other musculoskeletal conditions
➤ noise levels which are too high, causing deafness and conditions, such as tinnitus
➤ too much vibration, for example, from hand-held tools leading to hand–arm vibration syndrome and circulatory problems
➤ exposure to ionizing and non-ionizing radiation including ultraviolet in the sun's rays, causing burns, sickness and skin cancer
➤ infections ranging from minor sickness to life-threatening conditions, caused by inhaling or being contaminated by microbiological organisms
➤ stress causing mental and physical disorders.

Some illnesses or conditions, such as asthma and back pain, have both occupational and non-occupational causes and it may be difficult to establish a definite causal link with a person's work activity or their exposure to particular agents or substances. But, if there is evidence that shows the illness or condition is prevalent among the type of workers to which the person belongs or among workers exposed to similar agents or substances, it is likely that their work and exposure has contributed in some way.

6.4.2 Control hierarchy for exposure to substances hazardous to health

The following shows how the general principles and hierarchy can be applied to substances hazardous to health, which come under the COSHH Regulations:

➤ change the process or task so that there is no need for the hazardous substance;
➤ replace the substance with a safer material;
➤ use the substance in a safer form, for example, in liquid or pellets to prevent dust from powders. The HSE has produced a useful booklet on *7 steps to successful substitution of hazardous substances* HS (G) 110, HSE Books, which should be consulted;
➤ totally enclose the process;
➤ partially enclose the process and use local exhaust ventilation to extract the harmful substance;
➤ provide high quality general ventilation, if individual exposures do not breach the exposure limits;

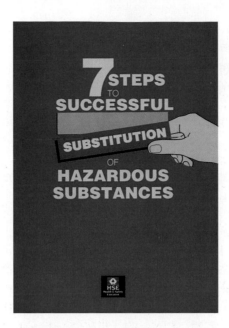

Figure 6.6 Guidance on substitution. Source HSE. Crown copyright material is reproduced with the permission of the Controller of HMSO and the Queen's Printer for Scotland.

➤ use safe systems of work and procedures to minimize exposures, spillage or leaks;
➤ reduce the number of people exposed or the duration of their exposure.

If none of the above control measures prove to be adequate on their own, PPE should be provided. This is a last resort and is only permitted by the COSHH Regulations in conjunction with other means of control if they prove inadequate. Special control requirements are needed for carcinogens, which are set out in the Carcinogen Approved Code of Practice.

6.4.3 Assessing exposure and health surveillance

Some aspects of health exposure will need input from specialist or professional advisers, such as occupational health hygienists, nurses and doctors. However, considerable progress can be made by taking straightforward measures such as:

➤ consulting the workforce on the design of workplaces;
➤ talking to manufacturers and suppliers of substances and work equipment about minimizing exposure;
➤ enclosing machinery to cut down dust, fume and noise;
➤ researching the use of less hazardous substances;
➤ ensuring that employees are given appropriate information and are trained in the safe handling of all the substances and materials to which they may be exposed.

To assess health risks and to make sure that control measures are working properly, it may be necessary, for example, to measure the concentration of substances in air to make sure that exposures remain within the assigned maximum exposure limits or occupational exposure standards. Sometimes health surveillance of workers who may be exposed will be needed. This will enable data to be collected to check control

113

measures and for early detection of any adverse changes to health. Health surveillance procedures available include biological monitoring for bodily uptake of substances, examination for symptoms and medical surveillance – which may entail clinical examinations and physiological or psychological measurements by occupationally qualified registered medical practitioners. The procedure chosen should be suitable for the case concerned. Sometimes a method of surveillance is specified for a particular substance, for example, in the COSHH ACOP. Whenever surveillance is undertaken, a health record has to be kept for the person concerned.

Health surveillance should be supervised by a registered medical practitioner or, where appropriate, it should be done by a suitably qualified person (e.g. an occupational nurse). In the case of inspections for easily detectable symptoms like chrome ulceration or early signs of dermatitis, health surveillance should be done by a suitably trained responsible person. If workers could be exposed to substances listed in Schedule 6 of the COSHH Regulations, medical surveillance, under the supervision of an HSE employment medical adviser or a doctor appointed by HSE, is required.

6.5 Safe systems of work

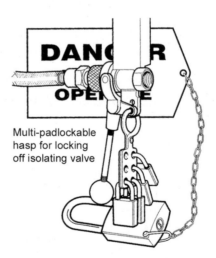

Figure 6.7 Multi-padlocked hasp for locking off an isolation valve. Source HSE. Crown copyright material is reproduced with the permission of the Controller of HMSO and the Queen's Printer for Scotland.

6.5.1 What is a safe system of work?

A safe system of work has been defined as:

> *The integration of personnel, articles and substances in a laid out and considered method of working which takes proper account of the risks to employees and others who may be affected, such as visitors and contractors, and provides a formal framework to ensure that all of the steps necessary for safe working have been anticipated and implemented.*

In simple terms, a safe system of work is a defined method for doing a job in a safe way. It takes account of all foreseeable hazards to health and safety and seeks to eliminate or minimize these. Safe systems of work are normally formal and documented, for example, in written operating procedures but, in some cases, they may be verbal.

The particular importance of safe systems of work stems from the recognition that most accidents are caused by a combination of factors (plant, substances, lack of training, supervision, etc.). Hence prevention must be based on an integral approach and not one which only deals with each factor in isolation. The adoption of a safe system of work provides this integral approach because an effective safe system:

➤ is based on looking at the job as a whole;
➤ starts from an analysis of all foreseeable hazards (physical, chemical, health, etc.);
➤ brings together all the necessary precautions, including design, physical precautions, training, monitoring, procedures and personal protective equipment.

It follows from this that the use of safe systems of work is in no way a replacement for other precautions, such as good equipment design, safe construction and the use of physical safeguards. However, there are many situations where these will not give adequate protection in themselves, and then a carefully thought-out and properly implemented safe system of work is especially important. The best example is maintenance and repair work, which will often involve as a first stage dismantling the guard or breaking through the containment, which exists for the protection of the ordinary process operator. In some of these operations, a permit to work procedure will be the most appropriate type of safe system of work.

The operations covered may be simple or complex, routine or unusual.

Whether the system is verbal or written, and whether the operation it covers is simple or complex, routine or unusual, the essential features are forethought and planning – to ensure that all foreseeable hazards are identified and controlled. In particular, this will involve scrutiny of:

➤ the sequence of operations to be carried out
➤ the equipment, plant, machinery and tools involved
➤ chemicals and other substances to which people might be exposed in the course of the work
➤ the people doing the work – their skill and experience
➤ foreseeable hazards (health, safety, environment), whether to the people doing the work or to others who might be affected by it
➤ practical precautions which, when adopted, will eliminate or minimize these hazards
➤ the training needs of those who will manage and operate under the procedure
➤ monitoring systems to ensure that the defined precautions are implemented effectively.

6.5.2 Legal requirements

The HSW Act Section 2 requires employers to provide safe plant and systems of work. In addition, many regulations made under the Act, such as the Provision and Use of Equipment Regulations 1998, require information and instruction to be provided to employees and others. In effect, this is also a more specific requirement to provide safe systems of work. Many of these safe systems, information and instructions will need to be in writing.

6.5.3 Assessment of what safe systems of work are required

Requirement

It is the responsibility of the management in each organization to ensure that its operations are assessed to determine where safe systems of work need to **be deve**loped.

This assessment must, at the same time, decide the **most appropriate** form for the safe system, that is:

➤ is a written procedure required?
➤ should the operation only be carried out under permit to **work?**
➤ is an informal system sufficient?

Factors to be considered

It is recognized that each organization must have the freedom to devise systems that match the risk potential of their operations and which are practicable in their situation. However, they should take account of the following factors in making their decision:

➤ types of risk involved in the operation
➤ magnitude of the risk, including consideration of the worst foreseeable loss
➤ complexity of the operation
➤ past accident and loss experience
➤ requirements and recommendations of the relevant health and safety authorities
➤ the type of documentation needed
➤ resources required to implement the safe system of work (including training and monitoring).

6.5.4 Development of safe systems

Role of competent person

Primarily management is responsible to provide safe systems of work. Managers and employees know the detailed way in which the task should be carried out. The competent person appointed under the MHSW Regulations should assist managers to draw up guidelines for safe systems of work with suitable forms and should advise management on the adequacy of the safe systems produced.

Analysis

The safe system of work should be based on a thorough analysis of the job or operation to be covered by the system. The way this analysis is done will depend on the nature of the job/operation.

If the operation being considered is a new one involving high loss potential, the use of formal hazard analysis techniques such as HAZOP (Hazard and Operability study), FTA (Fault Tree Analysis) or Failure Modes and Effects Analysis should be considered.

However, where the potential for loss is lower, a more simple approach, such as JSA (Job Safety Analysis), will be sufficient. This will involve three key stages:

➤ identification of the key steps in the job/operation – what activities will the work involve?
➤ analysis and assessment of the risks associated with each stage – what could go wrong?

➤ definition of the precautions or controls to be taken – what steps need to be taken to ensure the operation proceeds without danger, either to the people doing the work, or to anyone else?

The results of this analysis are then used to draw up the safe operating procedure. (See Appendix 6.2 for a suitable form.)

Consultation

Many people operating a piece of equipment or process are in the best position to help with the preparation of safe systems of work. Consultation with those employees who will be exposed to the risks, either directly or through their representatives, is also a legal requirement. The importance of discussing the proposed system with those who will have to work under it, and those who will have to supervise its operation, cannot be emphasized enough.

6.5.5 Preparation of safe systems

A checklist for use in the preparation of safe systems of work is set out as follows:

➤ what is the work to be done?
➤ what are the potential hazards?
➤ is the work covered by any existing instructions or procedures? If so, to what extent (if any) do these need to be modified?
➤ who is to do the work?
➤ what are their skills and abilities – is any special training needed?
➤ under whose control and supervision will the work be done?
➤ will any special tools, protective clothing or equipment (e.g. breathing apparatus) be needed? Are they ready and available for use?
➤ are the people who are to do the work adequately trained to use the above?
➤ what isolations and locking-off will be needed for the work to be done safely?
➤ is a permit to work required for any aspect of the work?
➤ will the work interfere with other activities? Will other activities create a hazard to the people doing the work?
➤ have other departments been informed about the work to be done, where appropriate?
➤ how will the people doing the work communicate with each other?
➤ have possible emergencies and the action to be taken been considered?
➤ should the emergency services be notified?
➤ what are the arrangements for handover of the plant/equipment at the end of the work? (For maintenance/project work etc.)
➤ do the planned precautions take account of all foreseeable hazards?
➤ who needs to be informed about or receive copies of the safe system of work?
➤ what arrangements will there be to see that the agreed system is followed and that it works in practice?
➤ what mechanism is there to ensure that the safe system of work stays relevant and up-to-date?

6.5.6 Documentation

Safe systems of work should be properly documented.

Wherever possible, they should be incorporated into normal process operating procedures. This is so that:

➤ health and safety are seen as an integral part of, and not add-on to, normal production procedures;
➤ the need for operators and supervisors to refer to separate manuals is minimized.

Whatever method is used, all written systems of work should be signed by the relevant managers to indicate approval or authorization. Version numbers should be included so that it can quickly be verified that the most up-to-date version is in use. Records should be kept of copies of the documentation, so that all sets are amended when updates and other revisions are issued.

As far as possible, systems should be written in a non-technical style and should specifically be designed to be as intelligible and user-friendly as possible. It may be necessary to produce simple summary sheets, which contain all the key points in an easy-to-read format.

6.5.7 Communication and training

People doing work or supervising work must be made fully aware of the laid-down safe systems that apply. The preparation of safe systems will often identify a training need that must be met before the system can be implemented effectively.

In addition, people should receive training in how the system is to operate. This applies not only to those directly involved in doing the work but also to supervisors/ managers who are to oversee it.

In particular, the training might include:

➤ why a safe system is needed
➤ what is involved in the work
➤ the hazards which have been identified
➤ the precautions which have been decided and, in particular
  ➤ the isolations and locking-off required, and how this is to be done
  ➤ details of the permit to work system, if applicable
  ➤ any monitoring (e.g. air testing) which is to be done during the work, or before it starts
  ➤ how to use any necessary personal protective equipment
  ➤ emergency procedures.

6.5.8 Monitoring safe systems

Safe systems of work should be monitored to ensure that they are effective in practice. This will involve:

➤ reviewing and revising the systems themselves, to ensure they stay up-to-date;
➤ inspection to identify how fully they are being implemented.

In practice, these two things go together, since it is likely that a system that is out of date will not be fully implemented by the people who are intended to operate it.

All organizations are responsible for ensuring that their safe systems of work are reviewed and revised as appropriate. Monitoring of implementation is part of all line managers' normal operating responsibilities, and should also take place during health and safety audits.

6.6 Lone workers

People who work by themselves without close or direct supervision are found in many work situations. In some cases they are the sole occupant of small workshops or warehouses; they may work in remote sections of a large site; they may work out of normal hours, like cleaners or security personnel; they may be working away from their main base as installers, or maintenance people; they could be people giving a service, like social workers, home helps, drivers and estate agents.

There is no general legal reason why people should not work alone but there may be special risks which require two or more people to be present, for example, during entry into a confined space in order to effect a rescue. It is important to ensure that a lone worker is not put at any higher risk than other workers. This is achieved by carrying out a specific risk assessment and introducing special protection arrangements for their safety. People particularly at risk, like young people or women, should also be considered. People's overall health and suitability to work alone should be taken into account.

Procedures may include:

➤ periodic visits from the supervisor to observe what is happening
➤ regular voice contact between the lone worker and the supervisor
➤ automatic warning devices to alert others if a specific signal is not received from the lone worker
➤ other devices to raise the alarm, which are activated by the absence of some specific action
➤ checks that the lone worker has returned safely home or to their base
➤ special arrangements for first aid to deal with minor injuries, this may include mobile first aid kits
➤ arrangements for emergencies should be established and employees trained.

6.7 Permits to work

6.7.1 Introduction

Safe systems of work are crucial in work such as the maintenance of chemical plant where the potential risks are high and the careful coordination of activities and precautions is essential to safe working. In this situation and others of similar risk potential, the safe system of work is likely to take the form of a permit to work procedure.

The permit to work procedure is a specialized type of safe system of work for ensuring that potentially very dangerous work (e.g. entry into process plant and other confined spaces) is done safely.

Although it has been developed and refined by the chemical industry, the principles of permit to work procedures are equally applicable to the management of complex risks in other industries.

Its fundamental principle is that certain defined operations are prohibited without the specific permission of a responsible manager, this permission being only granted once stringent checks have been made to ensure that all necessary precautions have been taken and that it is safe for work to go ahead.

The people doing the work take on responsibility for following and maintaining the safeguards set out in the permit, which will define the work to be done (no other work being permitted) and the timescale in which it must be carried out.

To be effective, the permit system requires the training needs of those involved to be identified and met, and monitoring procedures to ensure that the system is operating as intended.

Figure 6.8 Permit to work. Source HSE. Crown copyright material is reproduced with the permission of the Controller of HMSO and the Queen's Printer for Scotland.

6.7.2 Permit to work procedures

The permit to work procedure is a specialized type of safe system of work under which certain categories of high risk-potential work may only be done with the specific permission of an authorized manager. This permission (in the form of the permit to work) will only be given if the laid-down precautions are in force and have been checked.

The permit document will typically specify:

➤ what work is to be done
➤ the plant/equipment involved, and how they are identified
➤ who is authorized to do the work
➤ the steps which have already been taken to make the plant safe
➤ potential hazards which remain, or which may arise as the work proceeds
➤ the precautions to be taken against these hazards
➤ for how long the permit is valid
➤ that the equipment is released to those who are to carry out the work.

In accepting the permit, the person in charge of doing the authorized work normally undertakes to take/maintain whatever precautions are outlined in the permit. The permit will also include spaces for:

➤ signature certifying that the work is complete
➤ signature confirming re-acceptance of the plant/equipment.

6.7.3 Principles

Permit systems must adhere to the following eight principles:

1 wherever possible, and especially with routine jobs, hazards should be eliminated so that the work can be done safely without requiring a permit to work
2 although the Site Manager may delegate the responsibility for the operation of the permit system, the overall responsibility for ensuring safe operation rests with him/her
3 the permit must be recognized as the master instruction which, until it is cancelled, overrides all other instructions
4 the permit applies to everyone on site, including contractors
5 information given in a permit must be detailed and accurate. It must state:
   (a) which plant/equipment has been made safe and the steps by which this has been achieved
   (b) what work may be done
   (c) the time at which the permit comes into effect
6 the permit remains in force until the work has been completed and the permit is cancelled by the person who issued it, or by the person nominated by management to take over the responsibility (e.g. at the end of a shift or during absence)
7 no work other than that specified is authorized. If it is found that the planned work has to be changed, the existing permit should be cancelled and a new one issued
8 responsibility for the plant must be clearly defined at all stages.

6.7.4 Work requiring a permit

The main types of permit and the work to be covered by each are identified below. Appendix 6.2 illustrates the essential elements of a permit form with supporting notes on its operation.

General permit
The general permit should be used for work such as:

➤ alterations to or overhaul of plant or machinery where mechanical, toxic or electrical hazards may arise
➤ work on or near overhead crane tracks
➤ work on pipelines with hazardous contents
➤ repairs to railway tracks, tippers, conveyors
➤ work with asbestos-based materials
➤ work involving ionizing radiation
➤ roof work
➤ excavations to avoid underground services.

Figure 6.9 Entering a confined space.

Confined space permit

Confined spaces include chambers, tanks (sealed and open-top), vessels, furnaces, ducts, sewers, manholes, pits, flues, excavations, boilers, reactors and ovens.

Many fatal accidents have occurred where inadequate precautions were taken before and during work involving entry into confined spaces. The two main hazards are the potential presence of toxic or other dangerous substances and the absence of adequate oxygen. In addition, there may be mechanical hazards (entanglement on agitators) and raised temperatures. The work to be carried out may itself be especially hazardous when done in a confined space, for example, cleaning using solvents, cutting/welding work. Should the person working in a confined space get into difficulties for whatever reason, getting help in and getting the individual out may prove difficult and dangerous.

Stringent preparation, isolation, air testing and other precautions are therefore essential and experience shows that the use of a confined space entry permit is essential to confirm that all the appropriate precautions have been taken.

The Confined Spaces Regulations 1997 are summarized in Chapter17. They detail the specific controls that are necessary when people enter confined spaces.

Work on high voltage apparatus (including testing)

Work on high voltage apparatus (over about 600 volts) is potentially high risk. Hazards include:

➤ possibly fatal electric shock/burns to the people doing the work
➤ electrical fires/explosions
➤ consequential danger from disruption of power supply to safety-critical plant and equipment.

In view of the risk, this work must only be done by suitably trained and competent people acting under the terms of a high voltage permit.

Hot work

Hot work is potentially hazardous as a:

➤ source of ignition in any plant in which flammable materials are handled
➤ cause of fires in all processes, regardless of whether flammable materials are present.

Hot work includes cutting, welding, brazing, soldering and any process involving the application of a naked flame. Drilling and grinding should also be included where a flammable atmosphere is potentially present.

Hot work should therefore be done under the terms of a hot work permit, the only exception being where hot work is done in a designated maintenance area suitable for the purpose.

Figure 6.10 Hot work permit is usually essential except in designated areas.

6.7.5 Responsibilities

The effective operation of the permit system requires the involvement of many people. The following specific responsibilities can be identified:

(Note: all appointments, definitions of work requiring a permit etc. must be in writing. All the categories of people identified below should receive training in the operation of the permit system as it affects them.)

Site manager

➤ has overall responsibility for the operation and management of the permit system
➤ appoints a senior manager (normally the chief engineer) to act as senior authorized person.

Senior authorized person

➤ is responsible to the site manager for the operation of the permit system
➤ defines the work on the site which requires a permit
➤ ensures that people responsible for this work are aware that it must only be done under the terms of a valid permit
➤ appoints all necessary authorized persons
➤ appoints a deputy to act in his/her absence.

Authorized persons

➤ issue permits to competent persons and retain copies
➤ personally inspect the site to ensure that the conditions and proposed precautions are adequate and that it is safe for the work to proceed
➤ accompany the competent person to the site to ensure that the plant/equipment is correctly identified and that the competent person understands the permit
➤ cancel the permit on satisfactory completion of the work.

Competent persons

- receive permits from authorized persons
- read the permit and make sure they fully understand the work to be done and the precautions to be taken
- signify their acceptance of the permit by signing both copies
- comply with the permit and make sure those under their supervision similarly understand and implement the required precautions
- on completion of the work, return the permit to the authorized person who issued it.

Operatives

- read the permit and comply with its requirements, under the supervision of the competent person.

Specialists

A number of permits require the advice/skills of specialists in order to operate effectively. Such specialists may include chemists, electrical engineers, health and safety advisers and fire officers. Their role may involve:

- isolations within his/her discipline – e.g. electrical work
- using suitable techniques and equipment to monitor the working environment for toxic or flammable materials, or for lack of oxygen
- giving advice to managers on safe methods of working.

Specialists must not assume responsibility for the permit system. This lies with the site manager and the senior authorized person.

Engineers (and others responsible for work covered by permits)

- ensure that permits are raised as required.

Contractors

The permit system should be applied to contractors in the same way as to direct employees.

The contractor must be given adequate information and training on the permit system, the restrictions it imposes and the precautions it requires.

6.8 Emergency procedures

6.8.1 Introduction

Most of this chapter is about the principles of control to prevent accidents and ill-health. Emergency procedures, however, are about control procedures and equipment to limit the damage to people and property caused by an incident. Local fire authorities will often be involved and are normally prepared to give advice to employers.

Under Regulation 8 of the Management of Health and Safety at Work Regulations 1999, procedures must be established and set in motion when necessary to deal with serious and imminent danger to persons at work. Necessary links must be maintained

Figure 6.11 Emergency services at work.

with local authorities, particularly with regard to first aid, emergency medical care and rescue work.

Although fire is the most common emergency likely to be faced, there are many other possibilities, which should be considered including:

➤ gas explosion
➤ electrical burn or electrocution
➤ escape of toxic gases or fumes
➤ discovery of dangerous dusts like asbestos in the atmosphere
➤ bomb warning
➤ large vehicle crashing into the premises
➤ aircraft crash if near a flight path
➤ spread of highly infectious disease
➤ severe weather with high winds and flooding.

6.8.2 Fire routines and fire notices

Site managers must make sure that all employees are familiar with the means of escape in case of fire and their use, and with the routine to be followed in the event of fire.

To achieve this, routine procedures must be set up and made known to all employees, generally outlining the action to be taken in case of fire and specifically laying down the duties of certain nominated persons. Notices should be posted throughout the premises.

While the need in individual premises may vary, there are a number of basic components which should be considered when designing any fire routine procedures:

➤ the action to be taken on discovering a fire
➤ the method of operating the fire alarm
➤ the arrangements for calling the fire brigade
➤ the stopping of machinery and plant
➤ first-stage fire fighting by employees
➤ evacuation of the premises

125

➤ assembly of staff, customers and visitors and, carrying out a roll call to account for everyone on the premises.

The procedures must take account of those people who may have difficulty in escaping quickly from a building because of their location or a disability. Insurance companies and other responsible people may need to be consulted where special procedures are necessary to protect buildings and plant during or after people have evacuated. For example, there may be some special procedures necessary to ensure that a sprinkler system is operating in the event of fire. The duties of various managers should include any necessary actions.

6.8.3 Supervisory duties

A member of the staff should be nominated to supervise all fire and emergency arrangements. This person should be in a senior position or at least have direct access to a senior manager. Senior members of the staff should be appointed as departmental fire wardens, with deputies for every occasion of absence, however brief. In the event of fire or other emergency, their duties would be, while it remains safe to do so, to ensure that:

➤ the alarm has been raised
➤ the whole department, including toilets and small rooms, has been evacuated
➤ the fire brigade has been called
➤ fire doors are closed to prevent fire spread to adjoining compartments and to protect escape routes
➤ plant and machinery are shut down wherever possible and any other actions required to safeguard the premises are taken where they do not expose people to undue risks
➤ a roll call is carried out at the assembly point and the result reported to whoever is in control of the evacuation.

Under normal conditions fire wardens should check that good standards of housekeeping and preventive maintenance exist in their department, that exits and escape routes are kept free from obstruction, that all fire-fighting appliances are available for use and fire points are not obstructed, that smoking is rigidly controlled, and that all members of staff under their control are familiar with the emergency procedure and know how to use the fire alarm and fire fighting equipment.

6.8.4 Assembly and roll call

Assembly points should be established for use in the event of evacuation. It should be at a position, preferably undercover, which is unlikely to be affected at the time of fire. In some cases it may be necessary to make mutual arrangements with the occupiers of nearby premises.

In the case of small premises, a complete list of the names of all staff should be maintained so that a roll call can be made if evacuation becomes necessary.

In those premises where the number of staff would make a single roll call difficult, each departmental fire warden should maintain a list of the names of staff in their area. Roll call lists must be updated regularly.

6.8.5 Fire notices

Printed instructions for the action to be taken in the event of fire should be displayed throughout the premises. The information contained in the instructions should be stated briefly and clearly. The staff and their deputies to whom specific duties are allocated should be identified.

Instruction for the immediate calling of the fire brigade in case of fire should be displayed at telephone switchboards, exchange telephone instruments and security lodges.

A typical fire notice is given in Appendix 6.1.

6.8.6 Testing

The alarm system should be tested every week, while the premises are in normal use. The test should be carried out by activating a different call point each week, at a fixed time. (See Chapter 11 for more information on fire hazards and control.)

6.8.7 Fire drills

Once a fire routine has been established it must be tested at regular intervals in order to ensure that all staff are familiar with the action to be taken in an emergency.

The most effective way of achieving this is by carrying out fire drills at prescribed intervals. Drills should be held at least twice a year other than in areas dealing with hazardous processes where they should be more frequent. A programme of fire drills should be planned to ensure that all employees, including shift workers and part-time employees, are covered.

6.9 First aid at work

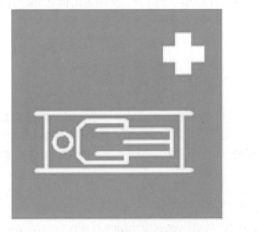 

Figure 6.12 (a) First aid and stretcher sign; (b) first aid sign

6.9.1 Introduction

People at work can suffer injuries or fall ill. It doesn't matter whether the injury or the illness is caused by the work they do or not. What is important is that they receive immediate attention and that an ambulance is called in serious cases. First aid at work covers the arrangements employers must make to ensure this happens. It can save lives and prevent minor injuries becoming major ones.

The Health and Safety (First-Aid) Regulations 1981 requires employers to provide adequate and appropriate equipment, facilities and personnel to enable first aid to be given to employees if they are injured or become ill at work.

What is adequate and appropriate will depend on the circumstances in a particular workplace.

The minimum first-aid provision on any worksite is:

➤ a suitably stocked first-aid box
➤ an appointed person to take charge of first-aid arrangements.

It is also important to remember that accidents can happen at any time. First-aid provision needs to be available at all times people are at work.

Many small firms will only need to make the minimum first-aid provision. However, there are factors which might make greater provision necessary. The following checklist covers the points that should be considered.

6.9.2 Aspects to consider

The risk assessments carried out under the MHSW and COSHH regulations should show whether there any specific risks in the workplace. The following should be considered:

➤ are there hazardous substances, dangerous tools and equipment; dangerous manual handling tasks, electrical shock risks, dangers from neighbours or animals?
➤ are there different levels of risk in parts of the premises or site?
➤ what is the accident and ill-health record, type and location of incidents?
➤ what is the total number of persons likely to be on site?
➤ are there young people, pregnant or nursing mothers on site, employees with disabilities or special health problems?
➤ are the facilities widely dispersed with several buildings or compact in a multi-storey building?
➤ what is the pattern of working hours, does it involve night work?
➤ is the site remote from emergency medical services?
➤ do employees travel a lot or work alone?
➤ do any employees work at sites occupied by other employers?
➤ are members of the public regularly on site?

6.9.3 Impact on first-aid provision if risks are significant

First aiders may need to be appointed if risks are significant.

This will involve a number of factors which must be considered, including:

> training for first aiders
> additional first-aid equipment and the contents of the first-aid box
> siting of first-aid equipment to meet the various demands in the premises. For example, provision of equipment in each building or on several floors. There needs to be first-aid provision at all times during working hours
> informing local medical services of the site and its risks
> any special arrangements that may be needed with the local emergency services.

If employees travel away from the site the employer needs to consider:

> issuing personal first-aid kits and providing training
> issuing mobile phones to employees
> making arrangements with employers on other sites.

Although there are no legal responsibilities for non-employees, the HSE strongly recommends that they are included in any first-aid provision.

6.9.4 Contents of the first-aid box

There is no standard list of items to put in a first-aid box. It depends on what the employer assesses the needs to be. Where there is no special risk in the workplace, a minimum stock of first-aid items is listed in Table 6.1.

Table 6.1   Contents of first aid box – low risk

Stock for up to 50 persons:	
A leaflet giving general guidance on first aid, e.g. HSE leaflet *Basic advice on first aid at work*.	
• Medical adhesive plasters	40
• Sterile eye pads	4
• Individually wrapped triangular bandages	6
• Safety pins	6
• Individually wrapped: medium sterile unmedicated wound dressings	8
• Individually wrapped: large sterile unmedicated wound dressings	4
• Individually wrapped wipes	10
• Paramedic shears	1
• Pairs of latex gloves	2
• Sterile eyewash if no clean running water	2

Tablets or medicines should not be kept in the first-aid box. Table 6.1 is a suggested contents list only; equivalent but different items will be considered acceptable.

6.9.5 Appointed persons

An appointed person is someone that is appointed by management to:

➤ take charge when someone is injured or falls ill. This includes calling an ambulance if required;
➤ look after the first-aid equipment, for example, keeping the first-aid box replenished;
➤ keeping records of treatment given.

Appointed persons should never attempt to give first aid for which they are not competent. Short emergency first-aid training courses are available. Remember that an appointed person should be available at all times when people are at work on site – this may mean appointing more than one. The training should be repeated every three years to keep up to date.

6.9.6 A first aider

A first aider is someone who has undergone an HSE approved training course in administering first aid at work and holds a current first aid at work certificate. Lists of local training organizations are available from the local environmental officer or HSE Offices. The training should be repeated every three years to maintain a valid certificate and keep the first aider up to date.

It is not possible to give hard and fast rules on when or how many first aiders or appointed persons might be needed. This will depend on the circumstances of each particular organization or worksite. Table 6.2 offers suggestions on how many first aiders or appointed persons might be needed in relation to categories of risk and number of employees. The details in the table are suggestions only, they are not definitive, nor are they a legal requirement.

Employees must be informed of the first-aid arrangements. Putting up notices telling staff who and where the first aiders or appointed persons are and where the first-aid box is will usually be enough. Special arrangements will be needed for employees with reading or language difficulties.

6.9.7 Suggested numbers of first-aid personnel

To ensure cover at all times when people are at work and where there are special circumstances, such as remoteness from emergency medical services, shift-work, or sites with several separate buildings, there may need to be more first-aid personnel than set out in Table 6.2. Provision must be sufficient to cover for absences.

Table 6.2 Number of first-aid personnel

Category of risk	Numbers employed at any location	Suggested number of first-aid personnel
Lower risk		
e.g. shops and offices, libraries	Fewer than 50 50–100 More than 100	At least one appointed person At least one first aider One additional first aider for every 100 employed
Medium risk		
e.g. light engineering and assembly work, food processing, warehousing	Fewer than 20 20–100 More than 100	At least one appointed person At least one first aider for every 50 employed (or part thereof) One additional first aider for every 100 employed
Higher risk		
e.g. most construction, slaughterhouses, chemical manufacture, extensive work with dangerous machinery or sharp instruments	Fewer than 5 5–50 More than 50	At least one appointed person At least one first aider One additional first aider for every 50 employed

Source HSE

6.10 Practice NEBOSH questions for Chapter 6

1. Outline the precautions that should be taken in order to ensure the safety of employees undertaking maintenance work in an underground storage vessel. (March 2001)
2. State the shape and colour, and give a relevant example, of each of the following types of safety sign:
   (i) prohibition
   (ii) warning
   (iii) mandatory
   (iv) emergency escape or first-aid. (March 2001)
3. (a) Identify the two main functions of first-aid treatment.
   (b) Outline the factors to consider when making an assessment of first-aid provision in a workplace. (December 2001)

4. Identify eight sources of information that might usefully be consulted when developing a safe system of work. (June 2001)

5. (a) Identify two situations where a permit-to-work system might be considered appropriate.
   (b) Outline the key elements of a permit-to-work system. (December 2000)

6. Outline the factors that should be considered when developing a safe system of work. (June 1999)

7. Outline the factors that should be considered when preparing a procedure to deal with a workplace emergency. (June 1998)

8. Outline the reasons why employees may fail to comply with safety procedures at work. (June 1998)

9. Outline the factors to consider when making an assessment of first-aid provision in a workplace. (June 1998)

10. (a) Give two examples of a confined space.
    (b) Outline specific hazards associated with working in confined spaces. (June 1997)

Appendix 6.1 – Fire notice

FIRE INSTRUCTIONS

IF YOU DISCOVER A FIRE
● Operate the nearest fire alarm call point immediately.
● Attack the fire with a suitable extinguisher if it is safe to do so.

IF THE FIRE ALARM SOUNDS
● The brigade will be called by
● Leave the building immediately.
● Do not stop to collect personal belongings.
● Do not use a lift.
● Close all doors behind you.
● Report to the assembly point at

● Do not re-enter the building until instructed to do so.

TO CALL THE FIRE BRIGADE
● Lift the receiver and dial   999.
● Ask the operator for the fire brigade.
● When the brigade answer, state that the fire is at

● Do not replace the receiver until the details have been repeated.

Appendix 6.2 – Job safety analysis form

JOB SAFETY ANALYSIS						
Job			Date			
Department			Carried out by			
Description of job						
Legal requirements and guidance						
Task steps	Hazards	Consequence	Severity	Risk C × S	Controls	
Safe system of work						
Job Instruction						
Training requirements						
Review date						

Appendix 6.3 – Essential elements of a permit-to-work form

2. Should have any reference to other relevant permits or isolation certificates.	1 Permit Title	2 Permit Number
	3 Job Title	
	4 Plant identification	
5. Description of work to be done and its limitations.	5 Description of Work	
6. Hazard identification including residual hazards and hazards introduced by the work.	6 Hazard Identification	
7. Precautions necessary – person(s) who carry out precautions, e.g. isolations, should sign that precautions have been taken.	7 Precautions necessary	7 Signatures ....
8. Protective equipment needed for the task.	8 Protective Equipment	
9. Authorization signature confirming that isolations have been made and precautions taken, except where these can only be taken during the work. Date and time duration of permit.	9 Authorization	
10. Acceptance signature confirming understanding of work to be done, hazards involved and precautions required. Also confirming permit information has been explained to all workers involved.	10 Acceptance	
11. Extension/Shift handover signature confirming checks have been made that plant remains safe to be worked on, and new acceptance/workers made fully aware of hazards/precautions. New time expiry given.	11 Extension/Shift handover	
12. Hand back signed by acceptor certifying work completed. Signed by issuer certifying work completed and plant ready for testing and re-commissioning.	12 Hand back	
13. Cancellation certifying work tested and plant satisfactorily re-commissioned.	12 Cancellation	

Source HSE

Movement of people and vehicles – hazards and control

7.1 Introduction

People are most often involved in accidents as they walk around the workplace or when they come into contact with vehicles in or around the workplace. It is therefore important to understand the various common accident causes and the control strategies that can be employed to reduce them. Slips, trips and falls account for the majority of accidents to pedestrians and the more serious accidents between pedestrians and vehicles can often be traced back to excessive speed or other unsafe vehicle practices, such as lack

Figure 7.1 Tripping hazards. Source HSE. Crown copyright material is reproduced with the permission of the Controller of HMSO and the Queen's Printer for Scotland.

of driver training. Many of the risks associated with these hazards can be significantly reduced by an effective management system.

7.2 Hazards to pedestrians

The most common hazards to pedestrians at work are slips, trips and falls on the same level, falls from height, collisions with moving vehicles, being struck by moving, falling or flying objects and striking against fixed or stationary objects. Each of these will be considered in turn, including the conditions and environment in which the particular hazard may arise.

7.2.1 Slips, trips and falls on the same level

These are the most common of the hazards facing pedestrians and accounted for 31% of all the major accidents and 20% of over three-day injuries reported to the HSE in 1999/2000. It has been estimated that the annual cost of these accidents to the nation is £750m and a direct cost to employers of £300m. The highest reported injuries are reported in the food and related industries. Older workers, especially women, are the most severely injured group from falls resulting in fractures of the hips and/or femur. Civil compensation claims are becoming more common and costly to employers and such claims are now being made by members of the public who have tripped on uneven paving slabs on pavements or in shopping centres.

The Health and Safety Commission has been so concerned at the large number of such accidents that it has identified slips, trips and falls on the same level as a key risk area. The costs of slips, trips and falls on the same level are high to the injured employee (lost income and pain), the employer (direct and indirect costs including lost production) and to society as a whole in terms of health and social security costs.

Figure 7.2   Cleaning must be done carefully to prevent slipping.

Slip hazards are caused by:

> wet or dusty floors
> the spillage of wet or dry substances – oil, water, flour dust and plastic pellets used in plastic manufacture
> loose mats on slippery floors
> wet and/or icy weather conditions
> unsuitable footwear or floor coverings or sloping floors.

Trip hazards are caused by:

> loose floorboards or carpets
> obstructions, low walls, low fixtures on the floor
> cables or trailing leads across walkways or uneven surfaces. Leads to portable electrical hand tools and other electrical appliances (vacuum cleaners and overhead projectors). Raised telephone and electrical sockets are also a serious trip hazard (this can be a significant problem when the display screen workstations are re-orientated in an office)
> rugs and mats – particularly when worn or placed on a polished surface
> poor housekeeping – obstacles left on walkways, rubbish not removed regularly
> poor lighting levels – particularly near steps or other changes in level
> sloping or uneven floors – particularly where there is poor lighting or no handrails
> unsuitable footwear – shoes with a slippery sole or lack of ankle support.

The vast majority of major accidents involving slips, trips and falls on the same level result in dislocated or fractured bones.

7.2.2 Falls from a height

These are the most common cause of serious injury or death in the construction industry and the topic is covered in Chapter 14. These accidents are usually concerned with falls of greater than 2 m and often result in fractured bones, serious head injuries, loss of consciousness and death. Twenty-five per cent of all deaths at work and 19% of all major accidents are due to falls from a height. Falls down staircases and stairways, through fragile roofs, off landings and stepladders and from vehicles, all come into this category. Injury, sometimes serious, can also result from falls below 2 m, for example, using swivel chairs for access to high shelves.

7.2.3 Collisions with moving vehicles

These can occur within the workplace premises or on the access roads around the building. It is a particular problem where there is no separation between pedestrians and vehicles or where vehicles are speeding. Poor lighting, blind corners, the lack of warning signs and barriers at road crossing points also increase the risk of this type of accident. Eighteen per cent of fatalities at work are caused by collisions between pedestrians and moving vehicles with the greatest number occurring in the service sector (primarily in retail and warehouse activities).

7.2.4 Being struck by moving, falling or flying objects

This causes 18% of fatalities at work and is the second highest cause of fatality in the construction industry. It also causes 15% of all major and 14% of over three-day accidents. Moving objects include, articles being moved, moving parts of machinery or conveyor belt systems, and flying objects are often generated by the disintegration of a moving part or a failure of a system under pressure. Falling objects are a major problem in construction (due to careless working at height) and in warehouse work (due to careless stacking of pallets on racking). The head is particularly vulnerable to these hazards. Items falling off high shelves and moving loads are also significant hazards in many sectors of industry.

7.2.5 Striking against fixed or stationary objects

This accounts for between 1200 and 1400 major accidents each year. Injuries are caused to a person either by colliding with a fixed part of the building structure, work in progress, a machine member or a stationary vehicle or by falling against such objects. The head appears to be the most vulnerable part of the body to this particular hazard and this is invariably caused by the misjudgement of the height of an obstacle. Concussion in a mild form is the most common outcome and a medical check-up is normally recommended. It is a very common injury during maintenance operations when there is, perhaps, less familiarity with particular space restrictions around a machine. Effective solutions to all these hazards need not be expensive, time consuming or complicated. Employee awareness and common sense combined with a good house-keeping regime will solve many of the problems.

7.3 Control strategies for pedestrian hazards

7.3.1 Slips, trips and falls on the same level

These may be prevented or, at least, reduced by several control strategies. These and all the other pedestrian hazards discussed should be included in the workplace risk assessments required under the Management of Health and Safety at Work Regulations 1999 by identifying slip or trip hazards, such as poor or uneven floor/pavement surfaces, badly lit stairways and puddles from leaking roofs. There is also a legal requirement in the Workplace (Health, Safety and Welfare) Regulations 1992 for all floors to be suitable, in good condition and free from obstructions. Traffic routes must be so organized that people can move around the workplace safely.

The key elements of a health and safety management system are as relevant to these as to any other hazards:

> **planning** – remove or minimize the risks by using appropriate control measures and defined working practices (e.g. covering all trailing leads)
> **organization** – involve employees and supervisors in the planning process by defining responsibility for keeping given areas tidy and free from trip hazards

➤ **control** – record all cleaning and maintenance work. Ensure that anti-slip covers and cappings are placed on stairs, ladders, catwalks, kitchen floors and smooth walkways. Use warning signs when floor surfaces have been recently washed

➤ **monitoring and review** – carry out regular safety audits of cleaning and housekeeping procedures and include trip hazards in safety surveys. Check on accident records to see either if there has been an improvement or if an accident black spot may be identified.

7.3.2 Falls from a height

These may be controlled by the use of suitable guardrails and barriers and also by the application of the hierarchy of controls discussed in Chapters 6 and 14 which is:

➤ remove the possibility of falling more than 2 m (e.g. by undertaking the work at ground level);
➤ protect against the hazard of falling more than 2 m (e.g. by using handrails);
➤ stop the person from falling more than 2 m (e.g. by the provision of safety harnesses);
➤ mitigate the consequences of falling more than 2 m (e.g. by the use of air bags).

Figure 7.3 Falling from a height – tower scaffold.

The principal means of preventing falls of people or materials includes the use of fencing, guardrails, toe boards, working platforms, access boards and ladder hoops. Safety nets and safety harnesses should only be used when all other possibilities are not practical. The use of banisters on open sides of stairways and handrails fitted on adjacent walls will also help to prevent people from falling. Holes in floors and pits should always be fenced or adequately covered. Precautions should be taken when working on fragile roofs (see Chapter 14 for details).

Staircases are a source of accidents included within this category of falling from a height and the following design and safety features will help to reduce the risk of such accidents:

➤ adequate width of the stairway, depth of the tread and provision of landings and banisters or handrails and intermediate rails. The treads and risers should always be of uniform size throughout the staircase and designed to meet Building Regulations requirements for angle of incline (i.e. steepness of staircase)
➤ provision of non-slip surfaces and reflective edging
➤ adequate lighting
➤ adequate maintenance
➤ special or alternative provision for disabled people (for example, personnel elevator at the side of the staircase).

Great care should be used when people are loading or unloading vehicles, as far as possible people should avoid climbing onto vehicles or their loads. For example sheeting of lorries should be carried out in designated places using properly designed access equipment.

7.3.3 Collisions with moving vehicles

These are best prevented by completely separating pedestrians and vehicles, providing well marked, protected and laid out pedestrian walkways. People should cross roads by designated and clearly marked pedestrian crossings. Suitable guardrails and barriers should be erected at entrances and exits from buildings and at 'blind' corners at the end

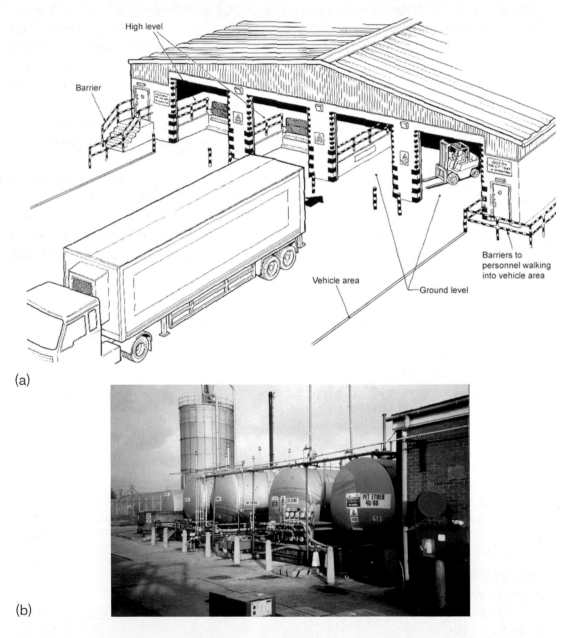

(a)

(b)

Figure 7.4   (a) Typical warehouse vehicle loading/unloading area with separate pedestrian access. Source HSE. Crown copyright material is reproduced with the permission of the Controller of HMSO and the Queen's Printer for Scotland; (b) barriers to prevent collision with tank surrounds/bunds.

of racking in warehouses. Particular care must be taken in areas where lorries are being loaded or unloaded. It is important that separate doorways are provided for pedestrians and vehicles and all such doorways should be provided with a vision panel and an indication of the safe clearance height, if used by vehicles. Finally, the enforcement of a sensible speed limit, coupled where practicable, with speed governing devices, is another effective control measure.

7.3.4 Being struck by moving, falling or flying objects

These may be prevented by guarding or fencing the moving part (as discussed in Chapter 9) or by adopting the measures outlined for construction work (Chapter 14). Both construction workers and members of the public need to be protected from the hazards associated with falling objects. Both groups should be protected by the use of covered walkways or suitable netting to catch falling debris where this is a significant hazard. Waste material should be brought to ground level by the use of chutes or hoists. Waste should not be thrown from a height and only minimal quantities of building materials should be stored on working platforms. Appropriate personal protective equipment, such as hard hats or safety glasses, should be worn at all times when construction operations are taking place.

It is often possible to remove high-level storage in offices and provide driver protection on lift truck cabs in warehouses. (See the section on fork lift trucks later in this chapter.) Storage racking is particularly vulnerable and should be strong and stable enough for the loads it has to carry. Damage from vehicles in warehouse can easily weaken the structure and cause collapse. Uprights need protection, particularly at corners.

The following action can be taken to keep racking serviceable:

> inspect them regularly and encourage workers to report any problems;
> post notices with maximum permissible loads and never exceed the loading;
> use good pallets and safe stacking methods;
> band, box or wrap articles to prevent items falling;
> set limits on the height of stacks and regularly inspect to make sure that limits are being followed;
> provide instruction and training for staff and special procedures for difficult objects.

7.3.5 Striking against fixed or stationary objects

This can only be effectively controlled by:

> having good standards of lighting and housekeeping;
> defining walkways and making sure they are used;
> the use of awareness measures, such as training and information in the form of signs or distinctive colouring;
> the use of appropriate personal protective equipment, in some cases, as discussed previously.

Figure 7.5 (a) Internal roadway with appropriate markings; (b) unsafe stacks of heavy boxes.

7.3.6 General preventative measures for pedestrian hazards

Minimizing pedestrian hazards and promoting good work practices requires a mixture of sensible planning, good housekeeping and common sense. Few of the required measures are costly or difficult to introduce and, although they are mainly applicable to slips, trips and falls on the same level and collisions with moving vehicles, they can be adapted to all types of pedestrian hazard. Typical measures include:

> develop a safe workplace as early as possible and ensure that suitable floor surfaces and lighting are selected and vehicle and pedestrian routes are carefully planned. Lighting should not dazzle approaching vehicles nor should pedestrians be obscured by stored products. Lighting is very important where there are changes of level or stairways. Any physical hazards, such as low beams, vehicular movements or pedestrian crossings, should be clearly marked. Staircases need particular attention to ensure that they are slip resistant and the edges of the stairs marked to indicate a trip hazard;

> consider pedestrian safety when re-orientating the workplace layout (e.g. the need to reposition lighting and emergency lighting);

> adopt and mark designated walkways;

> apply good housekeeping principles by keeping all areas, particularly walkways, as tidy as possible and ensure that any spillages are quickly removed;

> ensure that all workers are suitably trained in the correct use of any safety devices (such as machine guarding or personal protective equipment) or cleaning equipment provided by the employer;

> only use cleaning materials and substances that are effective and compatible with the surfaces being cleaned, so that additional slip hazards are not created;

> ensure that a suitable system of maintenance, cleaning, fault reporting and repair are in place and working effectively. Areas that are being cleaned must be fenced and warning

signs erected. Care must also be taken with trailing electrical leads used with the cleaning equipment. Records of cleaning, repairs and maintenance should be kept;

➤ ensure that all workers are wearing appropriate footwear with the correct type of anti-slip soles suitable for the type of flooring;

➤ consider whether there are significant pedestrian hazards present in the area when any workplace risk assessments are being undertaken.

7.4 Hazards in vehicle operations

Many different kinds of vehicle are used in the workplace, including dumper trucks, heavy goods vehicles, all terrain vehicles and, perhaps the most common, the fork lift truck. Approximately 70 persons are killed annually following vehicle accidents in the workplace. There are also over 1000 major accidents (involving serious fractures, head injuries and amputations) caused by:

➤ collisions between pedestrians and vehicles
➤ people falling from vehicles
➤ people being struck by objects falling from vehicles
➤ people being struck by an overturning vehicle

Figure 7.6   Industrial counter-balanced lift truck. Source HSE. Crown copyright material is reproduced with the permission of the Controller of HMSO and the Queen's Printer for Scotland.

➤ communication problems between vehicle drivers and employees or members of the public.

A key cause of these accidents is the lack of competent and documented driver training. HSE investigations, for example, have shown that in over 30% of dumper truck accidents on construction sites, the drivers had little experience and no training. Common forms of these accidents include driving into excavations, overturning while driving up steep inclines and runaway vehicles which have been left unattended with the engine running.

Risks of injuries to employees and members of the public involving vehicles could arise due to the following occurrences:

➤ collision with pedestrians
➤ collision with other vehicles
➤ overloading of vehicles
➤ overturning of vehicles
➤ general vehicle movements and parking
➤ dangerous occurrences or other emergency incidents (including fire)
➤ access and egress from the buildings and the site.

There are several other more general hazardous situations involving pedestrians and vehicles. These include the following:

➤ reversing of vehicles, especially inside buildings
➤ poor road surfaces and/or poorly drained road surfaces
➤ roadways too narrow with insufficient safe parking areas
➤ roadways poorly marked out and inappropriate or unfamiliar signs used
➤ too few pedestrian crossing points
➤ the non-separation of pedestrians and vehicles
➤ lack of barriers along roadways
➤ lack of directional and other signs
➤ poor environmental factors, such as lighting, dust and noise
➤ ill-defined speed limits and/or speed limits which are not enforced
➤ poor or no regular maintenance checks
➤ vehicles used by untrained and/or unauthorized personnel
➤ poor training or lack of refresher training.

Vehicle operations need to be carefully planned so that the possibility of accidents is minimized.

7.5 Control strategies for safe vehicle operations

Any control strategy involving vehicle operations will involve a risk assessment to ascertain where, on traffic routes, accidents are most likely to happen. It is important that the risk assessment examines both internal and external traffic routes, particularly when goods are loaded and unloaded from lorries. It should also assess whether designated traffic routes are suitable for the purpose and sufficient for the volume of traffic.

The following needs to be addressed:

Figure 7.7 Separate doors for vehicles and pedestrians. Source HSE. Crown copyright material is reproduced with the permission of the Controller of HMSO and the Queen's Printer for Scotland.

> traffic routes, loading and storage areas need to be well designed with enforced speed limits, good visibility and the separation of vehicles and pedestrians whenever reasonably practicable;

> environmental considerations, such as visibility, road surface conditions, road gradients and changes in road level, must also be taken into account;

> the use of one-way systems and separate site access gates for vehicles and pedestrians may be required;

> the safety of members of the public must be considered, particularly where vehicles cross public footpaths;

> all external roadways must be appropriately marked, particularly where there could be doubt on right of way, and suitable direction and speed limit signs erected along the roadways. While there may well be a difference between internal and external speed limits, it is important that all speed limits are observed;

> induction training for all new employees must include the location and designation of pedestrian walkways and crossings and the location of areas in the factory where pedestrians and fork lift trucks use the same roadways;

> the identification of recognized and prohibited parking areas around the site should also be given during these training sessions.

7.6 The management of vehicle movements

The movement of vehicles must be properly managed, as must vehicle maintenance and driver training. The development of an agreed code of practice for drivers, to which all drivers should sign up, and the enforcement of site rules covering all vehicular movements are essential for effective vehicle management.

All vehicles should be subject to appropriate regular preventative maintenance programmes with appropriate records kept and all vehicle maintenance procedures properly documented. Consideration must be given to driver protection by fitting driver restraint (seat belts), falling object protective structures (FOPS) and roll-over or tip-over protective structures known as ROPS or TOPS. (See the summary of the Provision and Use of Work Equipment Regulations in Chapter 17.) Many vehicles, such as mobile cranes, require regular inspection by a competent person and test certificates.

Certain vehicle movements, such as reversing, are more hazardous than others and particular safe systems should be set up. The reversing of lorries, for example, must be kept to a minimum (and then restricted to particular areas). Vehicles should be fitted with reversing warning systems as well as being able to give warning of approach. Refuges, where pedestrians can stand to avoid reversing vehicles are a useful safety measure. Banksmen, who direct reversing vehicles, should also be alert to the possibility of pedestrians crossing in the path of the vehicle. Where there are many vehicle movements, consideration should be given to the provision of high visibility clothing.

Fire is often a hazard which is associated with many vehicular activities, such as battery charging and the storage of warehouse pallets. All batteries should be recharged in a separate well-ventilated area.

As mentioned earlier, driver training, given by competent people, is essential. Only trained drivers should be allowed to drive vehicles and the training should be relevant to the particular vehicle (e.g. fork lift truck, dumper truck, lorry, etc.). All drivers must receive specific training and instruction before they are permitted to drive vehicles. They must also be given refresher training at regular intervals. This involves a management system for assuring driver competence, which must include detailed records of all drivers with appropriate training dates and certification in the form of a driving licence or authorization.

The HSE publications *Workplace Transport Safety. Guidance for Employers* HSG136, and *Managing Vehicle Safety at the Workplace* INDG199, provide useful checklists of relevant safety requirements that should be in place when vehicles are used in a workplace.

7.7 Practice NEBOSH questions for Chapter 7

1. (a) Identify the types of hazard that may cause slips or trips at work.
   (b) Outline how slip and trip hazards in the workplace might be controlled. (December 2001)
2. Outline measures to be taken to prevent accidents when pedestrians are required to work in vehicle manoeuvring areas. (June 2001)
3. A cleaner is required to polish floors using a rotary floor polisher.
   (i) Identify the hazards that might be associated with this operation.
   (ii) Outline suitable control measures that might be used to minimize the risk. (June 2001)
4. Outline the precautionary measures to be taken to avoid accidents involving reversing vehicles within a workplace. (June 2000)
5. List eight design features and/or safe practices intended to reduce the risk of accidents on staircases used as internal pedestrian routes within work premises. (March 2000)

Manual and mechanical handling hazards and control

8.1 Introduction

Until a few years ago, accidents caused by the manual handling of loads were the largest single cause of over three-day accidents reported to the HSE. The Manual Handling Operations Regulations 1992 recognized this fact and helped to reduce the number of these accidents. However, accidents due to poor manual handling technique still accounts for over 25% of all reported accidents and in some occupational sectors, such as the health service, the figure rises above 50%. An understanding of the factors causing some of these accidents is essential if they are to be further reduced. Mechanical handling methods should always be used whenever possible, but they are not without their hazards, many of which have been outlined in Chapter 7. Much mechanical handling involves the use of lifting equipment, such as cranes and lifts which present specific hazards to both the users and bystanders. The risks from these hazards are reduced by thorough examinations and inspections as required by the Lifting Operations and Lifting Equipment Regulations 1998 (LOLER).

Figure 8.1 Handling goods onto a truck in a typical docking bay.

8.2 Manual handling hazards and injuries

The term 'manual handling' is defined as the movement of a load by human effort alone. This effort may be applied directly or indirectly using a rope or a lever. Manual handling may involve the transportation of the load or the direct support of the load including pushing, pulling, carrying, moving using bodily force and, of course, straightforward lifting. Back injuries due to the lifting of heavy loads is very common and several million working days are lost each year as a result of such injuries.

Typical hazards of manual handling include:

> lifting a load which is too heavy or too cumbersome resulting in back injury
> poor posture during lifting or poor lifting technique resulting in back injury
> dropping a load, resulting in foot injury
> lifting sharp-edged or hot loads resulting in hand injuries.

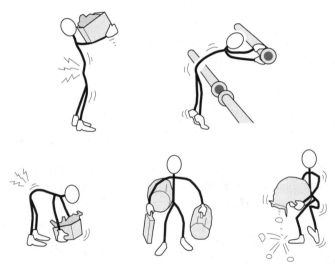

Figure 8.2 Manual handling – there are many potential hazards.

8.2.1 Injuries caused by manual handling

Manual handling operations can cause a wide range of acute and chronic injuries to workers. Acute injuries normally lead to sickness leave from work and a period of rest during which time the damage heals. Chronic injuries build up over a long period of time and are usually irreversible producing illnesses such as arthritic and spinal disorders. There is considerable evidence to suggest that modern lifestyles, such as a lack of exercise and regular physical effort, have contributed to the long-term serious effects of these injuries.

The most common injuries associated with poor manual handling techniques are all musculoskeletal in nature and are:

> muscular sprains and strains – caused when a muscular tissue (or ligament or tendon) is stretched beyond its normal capability leading to a weakening, bruising and painful inflammation of the area affected. Such injuries normally occur in the back or in the arms and wrists

148

➤ back injuries – include injuries to the discs situated between the spinal vertebrae (i.e. bones) and can lead to a very painful prolapsed disc lesion (commonly known as a slipped disc). This type of injury can lead to other conditions known as lumbago and sciatica (where pain travels down the leg)

➤ trapped nerve – usually occurring in the back as a result of another injury but aggravated by manual handling

➤ hernia – this is a rupture of the body cavity wall in the lower abdomen causing a protrusion of part of the intestine. This condition eventually requires surgery to repair the damage

➤ cuts, bruising and abrasions – caused by handling loads with unprotected sharp corners or edges

➤ fractures – normally of the feet due to the dropping of a load. Fractures of the hand also occur but are less common

➤ work-related upper limb disorders (WRULDs) – cover a wide range of musculoskeletal disorders which are discussed in detail in Chapter 13

➤ rheumatism – this is a chronic disorder involving severe pain in the joints. It has many causes, one of which is believed to be the muscular strains induced by poor manual handling lifting technique.

The sites on the body of injuries caused by manual handling accidents are shown in Figure 8.3.

In general, pulling a load is much easier for the body than pushing one. If a load can only be pushed, then pushing backwards using the back is less stressful on body muscles. Lifting a load from a surface at waist level is easier than lifting from floor level and most injuries during lifting are caused by lifting and twisting at the same time. If a load has to be carried, it is easier to carry it at waist level and close to the body trunk. A firm grip is essential when moving any type of load.

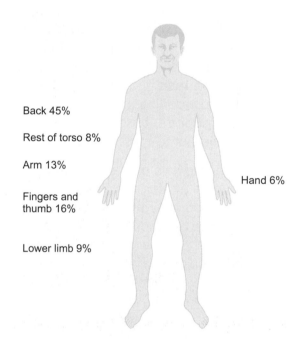

Back 45%

Rest of torso 8%

Arm 13%

Hand 6%

Fingers and thumb 16%

Lower limb 9%

Figure 8.3 Main injury sites caused by manual handling accidents.

8.3 Manual handling risk assessments

8.3.1 Hierarchy of measures for manual handling operations

With the introduction of the Manual Handling Operations Regulations 1992, the emphasis during the assessment of lifting operations changed from a simple reliance on safe lifting techniques to an analysis, using risk assessment, of the need for manual handling. The Regulations established a clear hierarchy of measures to be taken when an employer is confronted with a manual handling operation:

> avoid manual handling operations so far as is reasonably practicable by either redesigning the task to avoid moving the load or by automating or mechanizing the operations;
> if manual handling cannot be avoided, then a suitable and sufficient risk assessment should be made;
> reduce the risk of injury from those operations so far as is reasonably practicable, either by the use of mechanical handling or making improvements to the task, the load and the working environment.

The guidance given to the Manual Handling Operations Regulations 1992 (available in the HSE Legal series – L23 (revised in 1998)) is a very useful document. It gives very helpful advice on manual handling assessments and manual handling training. The advice is applicable to all occupational sectors. Appendix 8.1 gives an example of manual handling assessment forms.

8.3.2 Manual handling assessments

The Regulations specify four main factors which must be taken into account during the assessment. These are the task, the load, the working environment and the capability of the individual who is expected to do the lifting.

The **task** should be analysed in detail so that all aspects of manual handling are covered including the use of mechanical assistance. The number of people involved and the cost of the task should also be considered. Some or all of the following questions are relevant to most manual handling tasks:

> is the load held or manipulated at a distance from the trunk? The further from the trunk, the more difficult it is to control the load and the stress imposed on the back is greater;
> is a satisfactory body posture being adopted? Feet should be firmly on the ground and slightly apart and there should be no stooping or twisting of the trunk. It should not be necessary to reach upwards since this will place additional stresses on the arms, back and shoulders. The effect of these risk factors is significantly increased if several are present while the task is being performed;
> are there excessive distances to carry or lift the load? Over distances greater than 10 m, the physical demands of carrying the load will dominate the operation. The frequency of lifting, and the vertical and horizontal distances the load needs to be carried (particularly if it has to be lifted from the ground and/or placed on a high shelf) are very important considerations;
> is there excessive pulling and pushing of the load? The state of floor surfaces and the footwear of the individual should be noted so that slips and trips may be avoided;
> is there a risk of a sudden movement of the load? The load may be restricted or jammed in some way;
> is frequent or prolonged physical effort required? Frequent and prolonged tasks can lead to fatigue and a greater risk of injury;
> are there sufficient rest or recovery periods? Breaks and/or the changing of tasks enables the body to recover more easily from strenuous activity;
> is there an imposed rate of work on the task? This is a particular problem with some automated production lines and can be addressed by spells on other operations away from the line;
> are the loads being handled while the individual is seated? In these cases, the legs are not used during the lifting processes and stress is placed on the arms and back;

> does the handling involve two or more people? The handling capability of an individual reduces when he becomes a member of a team (e.g. for a 3 person team, the capability is half the sum of the individual capabilities). Visibility, obstructions and the roughness of the ground must all be considered when team handling takes place.

The **load** must be carefully considered during the assessment and the following questions asked:

> is the load too heavy? The maximum load that an individual can lift will depend on the capability of the individual and the position of the load relative to the body. There is therefore no safe load, but Figure 8.4 is reproduced from the HSE guidance, which does give some advice on loading levels. It does not recommend that loads in excess of 25 kg should be lifted or carried by a man (and this is only permissible when the load is at the level of and adjacent to the thighs). For women, the guideline figures should be reduced by about one third;
> is the load too bulky or unwieldy? In general, if any dimension of the load exceeds 0.75 m (2 ft), its handling is likely to pose a risk of injury. Visibility around the load is important. It may hit obstructions or become unstable in windy conditions. The position of the centre of gravity is very important for stable lifting – it should be as close to the body as possible;
> is the load difficult to grasp? Grip difficulties will be caused by slippery surfaces, rounded corners or a lack of foot room;
> are the contents of the load likely to shift? This is a particular problem when the load is a container full of smaller items, such as a sack full of nuts and bolts. The movement of people (in a nursing home) or animals (in a veterinary surgery) are loads which fall into this category;
> is the load sharp, hot or cold? Personal protective equipment may be required.

The **working environment** in which the manual handling operation is to take place, must be considered during the assessment. The following areas will need to be assessed:

> any space constraints which might inhibit good posture. Such constraints include lack of headroom, narrow walkways and items of furniture

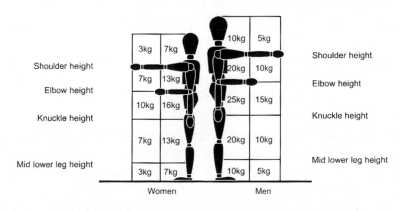

Lifting and lowering

Figure 8.4 HSE guidance for manual lifting – recommended weights. Source HSE. Crown copyright material is reproduced with the permission of the Controller of HMSO and the Queen's Printer for Scotland.

> slippery, uneven or unstable floors
> variations in levels of floors or work surfaces, possibly requiring the use of ladders
> extremes of temperature and humidity. These effects are discussed in detail in Chapter 13
> ventilation problems or gusts of wind
> poor lighting conditions.

Finally, the capability of the **individual** to lift or carry the load must be assessed. The following questions will need to be asked:

> does the task require unusual characteristics of the individual (e.g. strength or height)? It is important to remember that the strength and general manual handling ability depends on age, gender, state of health and fitness;
> are employees who might reasonably be considered to be pregnant or to have a health problem, put at risk by the task? Particular care should be taken to protect pregnant women or those who have recently given birth from handling loads. Allowance should also be given to any employee who has a health problem, which could be exacerbated by manual handling.

The assessment must be reviewed if there is reason to suspect that it is no longer valid or there has been a significant change to the manual handling operations to which it relates.

8.3.3 Reducing the risk of injury

This involves the introduction of control measures resulting from the manual handling risk assessment. The guidance to the Regulations (L 23) and the HSE publication *Manual Handling – solutions you can handle* (HSG115) contain many ideas to reduce the risk of injury from manual handling operations. An ergonomic approach, discussed in detail in Chapter 13, is generally required to design and develop the manual handling operation as a whole. The control measures can be grouped under five headings. However, the first consideration, when it is reasonably practicable, is mechanical assistance.

Mechanical assistance involves the use of mechanical aids to assist the manual handling operation such as hand-powered hydraulic hoists, specially adapted trolleys, hoist for lifting patients and roller conveyors.

The **task** can be improved by changing the layout of the workstation by, for example, storing frequently used loads at waist level. The removal of obstacles and the use of a better lifting technique that relies on the leg rather than back muscles should be encouraged. When pushing, the hands should be positioned correctly. The work routine should also be examined to see whether job rotation is being used as effectively as it could be. Special attention should be paid to seated manual handlers to ensure that loads are not lifted from the floor while they are seated. Employees should be encouraged to seek help if a difficult load is to be moved so that a team of people can move the load. Adequate and suitable personal protective equipment should be provided where there is a risk of loss of grip or injury. Care must be taken to ensure that the clothing does not become a hazard in itself (e.g. the snagging of fasteners and pockets).

The **load** should be examined to see whether it could be made lighter, smaller or easier to grasp or manage. This could be achieved by splitting the load, the positioning

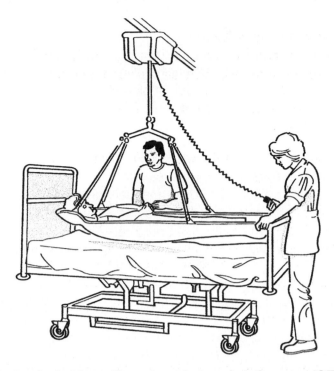

Figure 8.5 Mechanical aids to lift patients in hospital. Source HSE. Crown copyright material is reproduced with the permission of the Controller of HMSO and the Queen's Printer for Scotland.

of handholds or a sling, or ensuring that the centre of gravity is brought closer to the handler's body. Attempts should be made to make the load more stable and any surface hazards, such as slippery deposits or sharp edges, should be removed. It is very important to ensure that any improvements do not, inadvertently, lead to the creation of additional hazards.

The **working environment** can be improved in many ways. Space constraints should be removed or reduced. Floors should be regularly cleaned and repaired when damaged. Adequate lighting is essential and working at more than one level should be minimized so that hazardous ladder work is avoided. Attention should be given to the need for suitable temperatures and ventilation in the working area.

The **capability of the individual** is the fifth area where control measures can be applied to reduce the risk of injury. The state of health of the

Figure 8.6 A pallet truck. Source HSE. Crown copyright material is reproduced with the permission of the Controller of HMSO and the Queen's Printer for Scotland.

employee and his/her medical record will provide the first indication as to whether the individual is capable of undertaking the task. A period of sick leave or a change of job can make an individual vulnerable to manual handling injury. The Regulations require that the employee be given information and training. The information includes the provision, where it is reasonably practicable to do so, of precise information on the weight of each load and the heaviest side of any load whose centre of gravity is not centrally positioned. The training requirements are given in the next section.

8.3.4 Manual handling training

Training alone will not reduce manual handling injuries – there still need to be safe systems of work in place and the full implementation of the control measures highlighted in the manual handling assessment. The following topics should be addressed in a manual handling training session:

➤ types of injuries associated with manual handling activities
➤ the findings of the manual handling assessment
➤ the recognition of potentially hazardous manual handling operations
➤ the correct use of mechanical handling aids
➤ the correct use of personal protective equipment
➤ features of the working environment which aid safety in manual handling operations
➤ good housekeeping issues
➤ factors which affect the capability of the individual
➤ good lifting or manual handling technique as shown in Figure 8.7.

Finally, it needs to be stressed that if injuries involving manual handling operations are to be avoided, planning, control and effective supervision are essential.

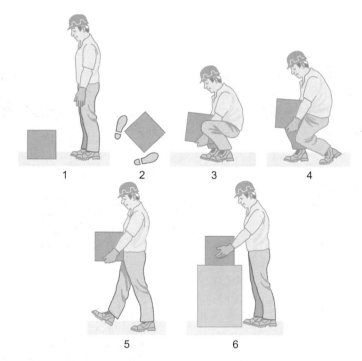

1. Check suitable clothing and assess load. Heaviest side to body.

2. Place feet apart – bend knees – straight back.

3. Firm grip – close to body.

4. Back straight – lift smoothly to knee level and then waist level.

5. With clear visibility move forward without twisting.

6. Set load down at waist level or to knee level and then floor.

Figure 8.7   The main elements of a good lifting technique.

8.4 Types of mechanical handling and lifting equipment

There are four elements to mechanical handling, each of which can present hazards. These are handling equipment, the load, the workplace and the employees involved.

The **mechanical handling equipment** must be capable of lifting and/or moving the load. It must be fault-free, well maintained and inspected on a regular basis. The hazards related to such equipment include collisions between people and the equipment and personal injury from being trapped in moving parts of the equipment (such as belt and screw conveyors).

The **load** should be prepared for transportation in such a way as to minimize the possibility of accidents. The hazards will be related to the nature of the load (e.g. substances which are flammable or hazardous to health) or the security and stability of the load (e.g. collapse of bales or incorrectly stacked pallets).

The **workplace** should be designed so that, whenever possible, workers and the load are kept apart. If, for example, an overhead crane is to be used, then people should be segregated away or barred from the path of the load.

The **employees** and others and any other people who are to use the equipment must be properly trained and competent in its safe use.

8.4.1 Conveyors and elevators

Conveyors transport loads along a given level which may not be completely horizontal, whereas elevators move loads from one level or floor to another.

There are three common forms of **conveyor** – belt, roller and screw conveyors. The most common hazards and preventative measures are:

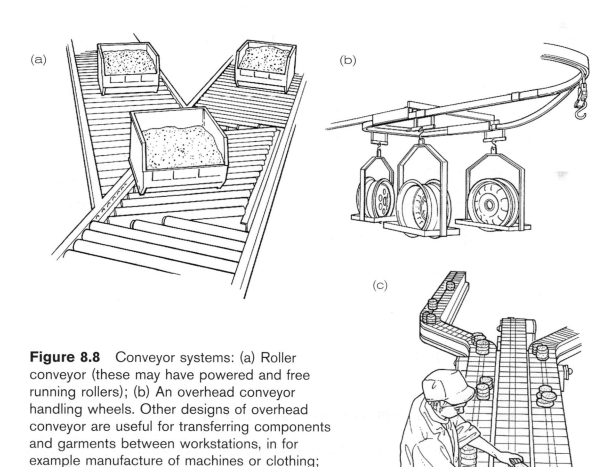

(a)

(b)

(c)

Figure 8.8 Conveyor systems: (a) Roller conveyor (these may have powered and free running rollers); (b) An overhead conveyor handling wheels. Other designs of overhead conveyor are useful for transferring components and garments between workstations, in for example manufacture of machines or clothing; (c) A slat conveyor in use in the food industry. Source HSE. Crown copyright material is reproduced with the permission of the Controller of HMSO and the Queen's Printer for Scotland.

➤ the in-running nip, where a hand is trapped between the rotating rollers and the belt. Protection from this hazard can be provided by nip guards and trip devices

➤ entanglement with the power drive requiring the fitting of fixed guards and the restriction of loose clothing which could become caught in the drive

➤ loads falling from the conveyor. This can be avoided by edge guards and barriers

➤ impact against overhead systems. Protection against this hazard may be given by the use of bump caps, warning signs and restricted access

➤ contact hazards prevented by the removal of sharp edges, conveyor edge protection and restricted access

➤ manual handling hazards

➤ noise and vibration hazards.

Screw conveyors, often used to move very viscous substances, must be provided with either fixed guards or covers to prevent accidental access. Man riding on belt conveyors should be prohibited and emergency trip wires or stop buttons must be fitted and operational at all times.

Elevators are used to transport goods between floors, such as the transportation of building bricks to upper storeys during the construction of a building or the transportation of grain sacks into the loft of a barn. Guards should be fitted at either end of the elevator and around the power drive. The most common hazard is injury due to loads falling from elevators. There are also potential manual handling problems at both the feed and discharge ends of the elevator.

Figure 8.9 A brick elevator. Source HSE. Crown copyright material is reproduced with the permission of the Controller of HMSO and the Queen's Printer for Scotland.

8.4.2 Fork lift trucks

The most common form of mobile handling equipment is the fork lift truck (Figures 7.6 and 8.10). It comes from the group of vehicles, known as lift trucks, and can be used in factories, on construction sites and on farms. The term fork lift truck is normally applied to the counterbalanced lift truck, where the load on the forks is counterbalanced by the weight of the vehicle over the rear wheels. The reach truck is designed to operate in narrower aisles in warehouses and enables the load to be retracted within the wheelbase. The very narrow aisle (VNA) truck does not turn within the aisle to deposit or retrieve a load. It is often guided by guides or rails on the floor. Other forms of lift truck include the pallet truck and the pallet stacker truck, both of which may be pedestrian or rider controlled.

There are many hazards associated with the use of fork lift trucks. These include:

> overturning – manoeuvring at too high a speed (particularly cornering), wheels hitting an obstruction such as a kerb, sudden braking, poor tyre condition leading to skidding, driving forwards down a ramp, movement of the load, insecure, excessive or uneven loading, incorrect tilt or driving along a ramp

> overloading – exceeding the rated capacity of the machine

> collisions – particularly with warehouse racking which can lead to a collapse of the whole racking system

> silent operation of the electrically powered fork lift truck – can make pedestrians unaware of its presence

> uneven road surface – can cause the vehicle to overturn and/or cause musculoskeletal problems for the driver

> overhead obstructions – a particular problem for inexperienced drivers

> loss of load – shrink wrapping or sheeting will reduce this hazard

> inadequate maintenance leading to mechanical failure

> use as a work platform

> speeding – strict enforcement of speed limits is essential

> poor vision around the load

> pedestrians – particularly when pedestrians and vehicles use the same roadways. Warning signs, indicating the presence of fork lift trucks, should be posted at regular intervals

> dangerous stacking or de-stacking technique – this can destabilize a complete racking column

> carrying passengers – this should be a disciplinary offence

> battery charging – presents an explosion and fire risk

> fire – often caused by poor maintenance resulting in fuel leakages or engine/motor burn-out, or through using a fork lift truck in areas where flammable liquids or gases are used and stored

> lack of driver training.

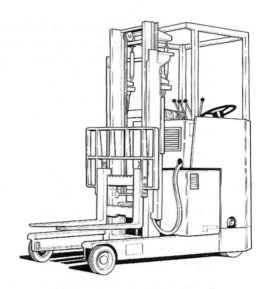

Figure 8.10 Reach truck designed so that the load retracts within the wheelbase to save space. Source HSE. Crown copyright material is reproduced with the permission of the Controller of HMSO and the Queen's Printer for Scotland.

There are also the following physical hazards:

> noise – caused by poor silencing of the power unit

> exhaust fumes – should only be a problem when the maintenance regime is poor

> vibrations – often caused by a rough road surface or wide expansion joints. Badly inflated tyres will exacerbate this problem

> manual handling – resulting from manoeuvring the load by hand or lifting batteries or gas cylinders

> ergonomic – musculoskeletal injuries caused by soft tyres and/or undulating road surface or holes or cracks in the road surface (e.g. expansion joints).

Regular and documented maintenance by competent mechanics is essential. However, the driver should undertake the following checks at the beginning of each shift:

- condition of tyres and correct tyre pressures
- effectiveness of all brakes
- audible reversing horn and light working properly
- lights, if fitted, working correctly
- mirrors, if fitted, in good working order and properly set
- secure and properly adjusted seat
- correct fluid levels, when appropriate
- fully charged batteries, when appropriate
- correct working of all lifting and tilting systems.

A more detailed inspection should be undertaken by a competent person within the organization on a weekly basis to include the mast and the steering gear. Driver training is essential and should be given by a competent trainer. The training session must include the site rules covering items such as the fork lift truck driver code of practice for the organization, speed limits, stacking procedures and reversing rules. Refresher training should be provided at regular intervals and a detailed record kept of all training received. Table 8.1 illustrates some key requirements of fork lift truck drivers and the points listed should be included in most codes of practice.

Table 8.1   Safe driving of lift trucks

Drivers must:

- drive at a suitable speed to suit road conditions and visibility
- use the horn when necessary (at blind corners and doorways)
- always be aware of pedestrians and other vehicles
- take special care when reversing (do not rely on mirrors)
- take special care when handling loads which restrict visibility
- travel with the forks (or other equipment fitted to the mast) lowered
- use the prescribed lanes
- obey the speed limits
- take special care on wet and uneven surfaces
- use the handbrake, tilt and other controls correctly
- take special care on ramps
- always leave the truck in a state which is safe and discourages unauthorized use (brake on, motor off, forks down, key out)

Drivers must not:

- operate in conditions in which it is not possible to drive and handle loads safely (e.g. partially blocked aisles)
- travel with the forks raised
- use the forks to raise or lower persons unless a purpose-built working cage is used
- carry passengers
- park in an unsafe place (e.g. obstructing emergency exits)
- turn round on ramps
- drive into areas where the truck would cause a hazard (flammable substance store)
- allow unauthorized use.

Finally, care must be taken with the selection of drivers, including relevant health checks and previous experience. Drivers should be at least 18 years of age and their fitness to drive should be reassessed at three-yearly intervals.

8.4.3 Other forms of lifting equipment

The other types of lifting equipment to be considered are cranes (mobile overhead and jib), lifts and hoists and lifting tackle.

Figure 8.11   Manoeuvring a yacht using a large overhead travelling gantry and slings in a marina.

Cranes may be either a jib crane or an overhead gantry travelling crane. The safety requirements are similar for each type. All cranes need to be properly designed, constructed, installed and maintained. They must also be operated in accordance with a safe system of work. They should only be driven by authorized persons who are fit and trained. Each crane is issued with a certificate by its manufacturer giving details of the safe working load (SWL). The safe working load must **never** be exceeded and should be marked on the crane structure. If the safe working load is variable, as with a jib crane (the safe working load decreases as the operating radius increases), a safe working load indicator should be fitted. Care should be taken to avoid sudden shock loading since this will impose very high stresses to the crane structure. It is also very important that the load is properly shackled and all eyebolts tightened. Safe slinging should be included in any training programme. All controls should be clearly marked and be of the 'hold to run' type.

Large cranes, which incorporate a driving cab, often work in conjunction with a banksman who will direct the lifting operation from the ground. It is important that banksmen are trained so that they understand recognized crane signals.

The basic principles for the safe operation of cranes are as follows. **For all cranes**, the driver must:

➤ undertake a brief inspection of the crane and associated lifting tackle each time before it is used;
➤ ensure that loads are not left suspended when the crane is not in use;
➤ before a lift is made, ensure that nobody can be struck by the crane or the load;
➤ ensure that loads are never carried over people;
➤ ensure good visibility and communications;
➤ lift loads vertically – cranes must not be used to drag a load;
➤ travel with the load as close to the ground as possible;
➤ switch off power to the crane when it is left unattended.

For mobile jib cranes, the following points should be considered:

➤ each lift must be properly planned, with the maximum load and radius of operation known
➤ overhead obstructions or hazards must be identified, it may be necessary to protect the crane from overhead power lines by using goal posts and bunting to mark the safe headroom
➤ the ground on which the crane is to stand should be assessed for its load-bearing capacity
➤ if fitted, outriggers should be used.

The principal reasons for crane failure, including loss of load, are:

➤ overloading
➤ poor slinging of load
➤ insecure or unbalanced load
➤ loss of load
➤ overturning
➤ collision with another structure or overhead power lines
➤ foundation failure
➤ structural failure of the crane
➤ operator error
➤ lack of maintenance and/or regular inspections.

During lifting operations using cranes, it must be ensured that:

➤ the driver has good visibility
➤ there are no pedestrians below the load by using barriers, if necessary
➤ an audible warning is given prior to the lifting operation.

If lifting takes place in windy conditions, tag lines may need to be attached to the load to control its movement.

A **lift or hoist** incorporates a platform or cage and is restricted in its movement by guides. Hoists are generally used in industrial settings (e.g. construction sites and garages), whereas lifts are normally used inside buildings. Lifts and hoists may be designed to carry passengers and/or goods alone. They should be of sound mechanical construction and have interlocking doors or gates, which must be completely closed before the lift or hoist moves. Passenger carrying lifts must be fitted with an automatic braking system to prevent overrunning, at least two suspension ropes, each capable alone of supporting the maximum working load and a safety device which could support the lift in the event of suspension rope failure. Maintenance procedures must be rigorous, recorded and only undertaken by competent persons. It is very important that a safe system of work is employed during maintenance operations to protect others, such as members of the public, from falling down the lift shaft and other hazards.

Other **items of lifting tackle**, usually used with cranes, include chain slings and hooks, wire and fibre rope slings, eyebolts and shackles. Special care should be taken when slings are used to ensure that the load is properly secured and balanced. Lifting hooks should be checked for signs of wear and any distortion of the hook. Shackles and eyebolts must be correctly tightened. Slings should always be checked for any damage before they are used and only competent people should use them. Training and instruction in the use of lifting tackle is essential and should include regular inspections of the tackle, in addition to the mandatory thorough examinations. Finally, care should be taken when these items are being stored between use.

Figure 8.12   Hoist for lifting a car.

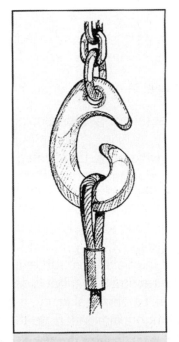

A hook designed to prevent displacement of the load.

A hook with a safety catch.

Figure 8.13   Specially designed safety hooks. Source HSE. Crown copyright material is reproduced with the permission of the Controller of HMSO and the Queen's Printer for Scotland.

8.5 Requirements for the statutory examination of lifting equipment

The Lifting Operations and Lifting Equipment Regulations 1998 specify examinations required for lifting equipment by using two terms – an inspection and a thorough examination. Both terms are defined by the HSE in guidance accompanying various regulations.

An **inspection** is used to identify whether the equipment can be operated, adjusted and maintained safely so that any defect, damage or wear can be detected before it results in unacceptable risks. It is normally performed by a competent person appointed by the employer (often an employee).

A **thorough examination** is a detailed examination, which may involve a visual check, a disassembly and testing of components and/or an equipment test under operating conditions. Such an examination must normally be carried out by a competent person who is independent of the employer. The examination is usually carried out according to a written scheme and a written report is submitted to the employer.

A detailed summary of the thorough examination and inspection requirements of the Lifting Operations and Lifting Equipment Regulations 1998 is given in Chapter 17.

A thorough examination of lifting equipment should be undertaken at the following times:

➤ before the equipment is used for the first time
➤ after it has been assembled at a new location
➤ at least every 6 months for equipment for lifting persons or a lifting accessory
➤ at least every 12 months for all other lifting equipment
➤ in accordance with a particular examination scheme drawn up by an independent competent person
➤ each time that exceptional circumstances, which are likely to jeopardize the safety of the lifting equipment, have occurred (such as severe weather).

The person making the thorough examination of lifting equipment must:

➤ notify the employer forthwith of any defect which, in their opinion, is or could become dangerous;
➤ as soon as practicable (within 28 days) write an authenticated report to the employer and any person who leased or hired the equipment.

The Regulations specify the information that should be included in the report.

The initial report should be kept for as long as the lifting equipment is used (except for a lifting accessory which need only be kept for two years). For all other examinations, a copy of the report should be kept until the next thorough examination is made or for two years (whichever is the longer). If the report shows that a defect exists, which could lead to an existing or imminent risk of serious personal injury, a copy of the report must be sent, by the person making the thorough examination, to the appropriate enforcing authority.

The equipment should be inspected at suitable intervals between thorough examinations. The frequency and the extent of the inspection is determined by the level of risk presented by the lifting equipment. A report or record should be made of the inspection which should be kept until the next inspection. Unless stated otherwise, lifts and hoists should be inspected every week.

8.6 Practice NEBOSH questions for Chapter 8

1. (a) Outline the main requirements of the Manual Handling Operations Regulations 1992.
   (b) List the possible indications of a manual handling problem in a workplace. (December 2001)

2. Outline a procedure for the safe lifting of a load by a crane, having ensured that the crane has been correctly selected and positioned for the job. (June 2001)

3. Outline the health and safety considerations when a fork lift truck is to be used to unload palletized goods from a vehicle parked in a factory car park. (December 2000)

4. List eight items to be included on a checklist for the routine inspection of a fork lift truck at the beginning of a shift. (December 2001)

5. Outline the factors to be considered when undertaking a manual handling assessment of a task that involves lifting buckets of water out of a sink. (June 2000)

6. An engineering workshop uses an overhead gantry crane to transport materials.
   (i) Identify two reasons why loads may fall from this crane.
   (ii) Outline precautions to prevent accidents to employees working at ground level when overhead cranes are in use. (March 2000)

7. (a) List two types of injury that may be caused by the incorrect manual handling of loads.
   (b) Outline a good manual handling technique that could be adopted by a person required to lift a load from the ground. (June 1999)

8. List the ways in which a fork lift truck may become unstable while in operation. (June 1998)

Appendix 8.1 – Manual handling of loads assessment checklist

Section A – Preliminary assessment	
Task description: **Factors beyond the limits of the guidelines?**	**Is an assessment needed?** (i.e. is there a potential risk of injury, and are the factors beyond the limits of the guidelines?) Yes / No
If 'YES' continue. If 'NO' no further assessment required.	
Tasks covered by this assessment: **Location:** **People involved:** **Date of assessment:**	**Diagram and other information**
Section B – See detailed analysis form	
Section C – Overall assessment of the risk of injury? Low / Med / High	
Section D – Remedial action needed:	
Remedial action, in priority order: a b c d f g h	
Date action should be completed:	**Date for review:**
Assessor's name:	**Signature:**

Manual Handling Risk Assessment

Task description...   Employee...............................

....

....   I.D. Number...............................

Risk Factors

A. Task characteristics	Yes/No	Risk Level			Current Controls
		H	M	L	See guidance
1. Loads held away from trunk?					
2. Twisting?					
3. Stooping?					
4. Reaching upwards?					
5. Extensive vertical movements?					
6. Long carrying distances?					
7. Strenuous pushing or pulling?					
8. Unpredictable movements of loads?					
9. Repetitive handling operations?					
10. Insufficient periods of rest/recovery?					
11. High work rate imposed?					
B. Load characteristics					
1. Heavy?					
2. Bulky?					
3. Difficult to grasp?					
4. Unstable/unpredictable?					
5. Harmful (sharp/hot)?					
C. Work environment characteristics					
1. Postural constraints?					
2. Floor suitability?					
3. Even surface?					
4. Thermal/humidity suitability?					
5. Lighting suitability?					

D. Individual characteristics

1. Unusual capability required?

2. Hazard to those with health problems?

3. Hazard to pregnant workers?

4. Special information/training required?

Any further action needed?   **Yes/No**

Details:

9

Work equipment hazards and control

9.1 Introduction

This chapter covers the scope and main requirements for work equipment as covered by Parts II and III of the Provision and Use of Work Equipment Regulations (PUWER)1998. The requirements for the supply of new machinery are also included. Summaries of PUWER and The Supply of Machinery (Safety) Regulations 1992 are given in Chapter 17. The safe use of hand tools, hand-held power tools and the proper safeguarding of a small range of machinery used in industry and commerce is included.

Figure 9.1 Typical machinery safety notice.

Any equipment used by an employee at work is generally covered by the term 'Work Equipment'. The scope is extremely wide and includes, hand tools, power tools, ladders, photocopiers, laboratory apparatus, lifting equipment, fork lift trucks, and motor vehicles (which are not privately owned). Virtually anything used to do a job of work, including employees' own equipment, is covered. The uses covered include starting or stopping the equipment, repairing, modifying, maintaining, servicing, cleaning and transporting.

Employers and the self-employed must ensure that work equipment is suitable; maintained; inspected if necessary; provided with adequate information and instruction; and only used by people who have received sufficient training.

9.2 Suitability of work equipment and CE marking

When work equipment is provided it has to conform to standards which cover its supply as a new or second-hand piece of equipment and its use in the workplace. This involves:

> its initial integrity
> the place where it will be used
> the purpose for which it will be used.

There are two groups of law that deal with the provision of work equipment:

➤ one deals with what manufacturers and suppliers have to do. This can be called the 'supply' law. One of the most common of these is the Supply of Machinery (Safety) Regulations 1992, which requires manufacturers and suppliers to ensure that machinery is safe when supplied and has CE marking. Its **primary** purpose is to prevent barriers to trade across the EU, and not to protect people at work;

➤ the other deals with what the users of machinery and other work equipment have to do. This can be called the 'user' law, PUWER 98 and applies to most pieces of work equipment. Its **primary** purpose is to protect people at work.

Under 'user' law employers have to provide safe equipment of the correct type; ensure that it is correctly used; and maintain it in a safe condition. When buying new equipment the 'user' has to check that the equipment complies with all the 'supply' law that is relevant. **The user must check that the machine is safe before it is used.**

Most new equipment, including machinery in particular, should have 'CE' marking when purchased. 'CE' marking is only a claim by the manufacturer that the equipment is safe and that they have met relevant supply law. If this is done properly manufacturers will have to do the following:

CE

Figure 9.2 CE mark.

➤ find out about the health and safety hazards (trapping, noise, crushing, electrical shock, dust, vibration, etc.) that are likely to be present when the machine is used;

➤ assess the likely risks;

➤ design out the hazards that result in risks or, if that is not possible;

➤ provide safeguards (e.g. guarding dangerous parts of the machine, providing noise enclosures for noisy parts) or, if that is not possible;

➤ use warning signs on the machine to warn of hazards that cannot be designed out or safeguarded (e.g. 'noisy machine' signs).

Manufacturers also have to:

➤ keep information, explaining what they have done and why, in a technical file;

➤ fix CE marking to the machine where necessary, to show that they have complied with all the relevant supply laws;

➤ issue a 'Declaration of Conformity' for the machine (see Figure 9.3). This is a statement that the machine complies with the relevant essential health and safety requirements or with the example that underwent type-examination. A declaration of conformity must

   ➤ state the name and address of the manufacturer or importer into the EU

   ➤ contain a description of the machine, its make, type and serial number

   ➤ indicate all relevant European directives with which the machinery complies

   ➤ state details of any notified body that has been involved

   ➤ specify which standards have been used in the manufacture (if any)

EC Declaration of Conformity

We declare that units:
BD710,BD711,BD713,BD725,BD735E,
BD750,SR600

conform to:
89/392/EEC,EN50144

A weighted sound pressure   103dB (A)
A weighted sound power   116dB (A)
Hand/arm weighted vibration   <2.5m/s²

Brian Cooke

Director of Engineering
Spennymoor, County Durham DL16 6JG
United Kingdom

Figure 9.3 Typical certificate of conformity.

> be signed by a person with authority to do so;
> provide the buyer with instructions to explain how to install, use and maintain the machinery safely.

Before buying new equipment the buyer will need to think about:

> where and how it will be used
> what it will be used for
> who will use it (skilled employees, trainees)
> what risks to health and safety might result
> how well health and safety risks are controlled by different manufacturers.

This can help to decide which equipment may be suitable, particularly if buying a standard piece of equipment 'off the shelf'.

If buying a more complex or custom-built machine the buyer should discuss their requirements with potential suppliers. For a custom-built piece of equipment, there is the opportunity to work with the supplier to design out the causes of injury and ill-health. Time spent now on agreeing the necessary safeguards, to control health and safety risks, could save time and money later.

Note: Sometimes equipment is supplied via another organization, for example, an importer, rather than direct from the manufacturer, so this other organization is referred to as the supplier. It is important to realize that the supplier may not be the manufacturer.

When the equipment has been supplied the buyer should look for CE marking, check for a copy of the Declaration of Conformity and that there is a set of instructions in English on how the machine should be used and most important of all, **check to see if they think that it is safe**.

9.3 Use and maintenance of equipment with specific risks

Some pieces of work equipment involve specific risks to health and safety where it is not possible to control adequately the hazards by physical measures alone, for example, the use of a bench-mounted circular saw or an abrasive wheel. In all cases the hierarchy of controls should be adopted to reduce the risks by:

> eliminating the risks or, if this is not possible;
> taking physical measures to control the risks such as guards, but if the risks cannot be adequately controlled;
> taking appropriate software measures, such as a safe system of work.

Figure 9.4 Using a bench-mounted abrasive wheel. (Courtesy of Draper.)

PUWER 98 Regulation 7 restricts the use of such equipment to the persons designated to use it. These people need to have received sufficient information, instruction and training so that they can carry out the work using the equipment safely.

Repairs, modifications, maintenance or servicing is also restricted to designated persons. A designated person may be the operator if they have the necessary skills and have received specific instruction and training. Another person specifically trained to carry out a particular maintenance task, for example, dressing an abrasive wheel, may not be the operator but designated to do this type of servicing task on a range of machines.

9.4 Information, instruction and training

People using and maintaining work equipment, where there are residual risks that cannot be sufficiently reduced by physical means, require enough information, instruction and training to operate safely.

The information and instructions are likely to come from the manufacturer in the form of operating and maintenance manuals. It is up to the employer to ensure that what is provided is easily understood, set out logically with illustrations and standard symbols where appropriate. The information should normally be in good plain English but other languages may be necessary in some cases.

The extent of the information and instructions will depend on the complexity of the equipment and the specific risks associated with its use. They should cover:

> all safety and health aspects
> any limitations on the use of the equipment
> any foreseeable problems that could occur
> safe methods to deal with the problems
> any relevant experience with the equipment that would reduce the risks or help others to work more safely, should be recorded and circulated to everyone concerned.

Everyone who uses and maintains work equipment needs to be adequately trained. The amount of training required will depend on:

> the complexity and level of risk involved in using or maintaining the equipment
> the experience and skills of the person doing the work, whether it is normal use or maintenance.

Training needs will be greatest when a person is first recruited but will also need to be considered:

> when working tasks are changed, particularly if the level of risk changes
> if new technology or new equipment is introduced
> where a system of work changes
> when legal requirements change
> periodically to update and refresh peoples' knowledge and skills.

Supervisors and managers also require adequate training to carry out their function, particularly if they only supervise a particular task occasionally. The training and supervision of young persons is particularly important because of their relative immaturity, unfamiliarity with a working environment and awareness of existing or potential risks. Some Approved Codes of Practice, for example on the *Safe Use of Woodworking*

Machinery restrict the use of high-risk machinery, so that only young persons with sufficient maturity and competence, who have finished their training, may use the equipment unsupervised.

9.5 Maintenance and inspection

9.5.1 Maintenance

Work equipment needs to be properly maintained so that it continues to operate safely and in the way it was designed to perform. The amount of maintenance will be stipulated in the manufacturers' instructions and will depend on the amount of use, the working environment and the type of equipment. High speed, high risk machines, which are heavily used in an adverse environment like salt water, may require very frequent maintenance, whereas a simple hand tool, like a shovel, may require very little.

Figure 9.5 Typical maintenance notice.

Maintenance management schemes can be based around a number of techniques designed to focus on those parts which deteriorate and need to be maintained to prevent health and safety risks. These techniques include:

➤ **preventative planned maintenance** – which involves replacing parts and consumables or making necessary adjustments at preset intervals, normally set by the manufacturer, so that there are no hazards created by component deterioration or failure. Vehicles are normally maintained on this basis

➤ **condition based maintenance** – this involves monitoring the condition of critical parts and carrying out maintenance whenever necessary to avoid hazards which could otherwise occur

➤ **breakdown based maintenance** – here maintenance is only carried out when faults or failures have occurred. This is only acceptable if the failure does not present an immediate hazard and can be corrected before the risk is increased. If, for example, a bearing overheating can be detected by a monitoring device, it is acceptable to wait for the overheating to occur as long as the equipment can be stopped and repairs carried out before the fault becomes dangerous to persons employed.

In the context of health and safety, maintenance is not concerned with operational efficiency but only with avoiding risks to people. It is essential to ensure that maintenance work can be carried out safely. This will involve:

➤ competent well-trained maintenance people;
➤ the equipment being made safe for the maintenance work to be carried out. In many cases the normal safeguards for operating the equipment may not be sufficient as maintenance sometimes involves going inside guards to observe and subsequently adjust, lubricate or repair the equipment. Careful design allowing adjustments, lubrication and observation from outside the guards, for example, can often eliminate the hazard. Making equipment safe will usually involve disconnecting the power supply and then preventing anything moving, falling or starting during the work. It may also involve waiting for equipment to cool or warm up to room temperature;

➤ a safe system of work being used to carry out the necessary procedures to make and keep the equipment safe and perform the maintenance tasks. This can often involve a formal 'permit to work' scheme to ensure that the correct sequence of safety critical tasks has been performed and all necessary precautions taken;

➤ correct tools and safety equipment being available to perform the maintenance work without risks to people. For example special lighting or ventilation may be required.

9.5.2 Inspection under PUWER

Complex equipment and/or high-risk equipment will probably need a maintenance log and may require a more rigid inspection regime to ensure continued safe operation. This is covered by PUWER 98, Regulation 6.

PUWER requires that where safety is dependent on the installation conditions and/or the work equipment is exposed to conditions causing deterioration, which may result in a significant risk and a dangerous situation developing, that the equipment is inspected by a competent person. In this case the competent person would normally be an employee, but there may be circumstances like fairground equipment, where an outside competent person would be used.

The inspection must be done:

➤ after installation for the first time
➤ after assembly at a new site or in a new location and thereafter
➤ at suitable intervals and
➤ each time exceptional circumstances occur which could affect safety.

The inspection under PUWER will vary from a simple visual inspection to a detailed comprehensive inspection, which may include some dismantling and/or testing. The level of inspection required would normally be less rigorous and intrusive than thorough examinations under the Lifting Operations and Lifting Equipment Regulations 1998(LOLER) for certain lifting equipment. In the case of boilers and air receivers the inspection is covered by the Pressure Systems Safety Regulations 2000 which involves a thorough examination (see the summary in Chapter 17). The inspection under PUWER would only be needed in these cases if the examinations did not cover all the significant health and safety risks that are likely to arise.

9.5.3 Examination of boilers and air receivers

Under the Pressure Systems Safety Regulations 2000, a wide range of pressure vessels and systems require thorough examination by a competent person to an agreed specifically written scheme. This includes steam boilers, pressurized hot water plants and air receivers which are contained in the Certificate syllabus.

The Regulations place duties on designers and manufacturers but this section is only concerned with the duty on users to have the vessels examined. An employer who operates a steam boiler and/or a pressurized hot water plant and/or an air receiver must ensure:

➤ that it is supplied with the correct written information and markings
➤ that the equipment is properly installed

➤ that it is used within its operating limits
➤ there is a written scheme for periodic examination of the equipment certified by a competent person (in the case of standard steam boilers and air receivers the scheme is likely to be provided by the manufacturer)
➤ that the equipment is examined in accordance with the written scheme by a competent person within the specified period
➤ that a report of the periodic examination is held on file giving the required particulars
➤ the actions required by the report are carried out
➤ any other safety critical maintenance work is carried out, whether or not covered by the report.

Figure 9.6 Typical compressor and air receiver. (Courtesy of Draper.)

In these cases the competent person is usually a specialist inspector from an external inspection organization. Many of these organizations are linked to the insurance companies who cover the financial risks of the pressure vessel.

9.6 Operation and working environment

To operate work equipment safely it must be fitted with easily reached and operated controls; kept stable; properly lit; kept clear; and provided with adequate markings and warning signs. These are covered by PUWER 98, which applies to all types of work equipment.

Equipment controls should:
➤ be easily reached from the operating positions
➤ not permit accidental starting of equipment
➤ move in the same direction as the motion being controlled
➤ vary in mode, shape and direction of movement to prevent inadvertent operation of the wrong control
➤ incorporate adequate red stop buttons
➤ incorporate adequate red emergency stop buttons, of the mushroom headed type with lock-off
➤ have shrouded or sunk green start buttons to prevent accidental starting of the equipment
➤ be clearly marked to show what they do.

Equipment should be operated in clear unobstructed work areas, particularly if they are high-risk machines. Debris and rubbish given off from the equipment should be removed quickly. Sometimes there will be need for special systems to remove waste to prevent accumulations of dust and/or substances hazardous to health, for example, wood dust from wood working machinery.

Equipment should be provided with efficient means of isolating it from all sources of energy. In many cases this source is the mains electrical energy but it must cover all sources such as steam, compressed air, hydraulic, batteries, heat, etc. In some cases special consideration is necessary where, for example, hydraulic power is switched off, not to let heavy pieces of equipment like the ram, fall under gravity.

Stability is important and is normally achieved by bolting equipment in place or if this is not possible by using clamps. Some equipment can be tied down, counterbalanced or weighted, so that it remains stable under all operating conditions. If portable equipment is weighted or counterbalanced it should be reappraised when the equipment is moved to another position. If outriggers are needed for stability in certain conditions, for example, to stabilize mobile access towers, they should be employed whenever conditions warrant the additional support. In severe weather conditions it may be necessary to stop using the equipment or reappraise the situation to ensure stability is maintained.

The quality of general and local lighting will need to be considered to ensure the safe operation of the equipment. The level of lighting and its position relative to the working area are often critical to the safe use of work equipment. Poor levels of lighting, glare and shadows can be dangerous when operating equipment. Some types of lighting, for example, sodium lights, can change the colour of equipment, which may increase the level of risk. This is particularly important if the colour coding of pipe work or cables is essential for safety.

Markings on equipment must be clearly visible and durable. They should follow international conventions for some hazards like radiation and lasers and, as far as possible, conform to the Health and Safety (Safety Signs and Signals) Regulations 1996 (see Chapter 17 for a summary). The contents, or the hazards of the contents, as well as controls, will need to be marked on some equipment. Warnings or warning devices are required in some cases to alert operators or people nearby to any dangers, for example, 'wear hardhats'; a flashing light on an airport vehicle; or a reversing horn on a truck.

9.7 User responsibilities

The responsibilities on users of work equipment are covered in section 7 of HSW Act and regulation 14 of the Management Regulations 1999. Section 7 requires employees to take reasonable care for themselves and others who may be affected and to cooperate with the employer. Regulation 12 requires employees to use equipment properly in accordance with instructions and training. They must also inform employers of dangerous situations and shortcomings in protection arrangements. The self-employed who are users have similar responsibilities to employers and employees combined.

9.8 Hand-held tools

Work equipment includes hand tools and hand-held power tools. These tools need to be correct for the task, well maintained and properly used.

Hazards from the misuse or poor maintenance of hand tools include:

➤ broken handles on files/chisels/screwdrivers/hammers which can cause cut hands or hammer heads to fly off
➤ incorrect use of knives, saws and chisels with hands getting injured in the path of the cutting edge
➤ tools that slip causing stab wounds

- poor quality uncomfortable handles that damage hands
- splayed spanners that slip and damage hands or faces
- chipped or loose hammer heads that fly off or slip
- incorrectly sharpened chisels that slip and cut hands
- flying particles that damage eyes from breaking up stone or concrete
- electrocution or burns by using incorrect or damaged tools for electrical work
- use of poorly insulated tools for hot work in catering or food industry
- use of pipes or similar equipment as extension handles for a spanner which is likely to slip causing hand or face injury
- mushroomed headed chisels or drifts which can damage hands or cause hammers (not suitable for chisels) and mallets to slip
- use of spark producing or percussion tools in flammable atmospheres.

Figure 9.7 Typical range of hand tools.

9.8.1 Hand tools

Hand tools should be properly controlled including those tools owned by employees. The following controls are important.

Suitability – all tools should be suitable for the purpose and location in which they are to be used. This will include:

- specially protected and insulated tools for electricians
- non-sparking tools for flammable atmospheres and the use of non-percussion tools and cold cutting methods
- tools made of suitable quality materials which will not chip or splay in normal use
- the correct tools for the job, for example, using the right-sized spanner and the use of mallets on chisel heads
- safety knives with enclosed blades for regular cutting operations
- scissors with blunted ends.

Inspection – all tools should be maintained in a safe and proper condition. This can be achieved through:

- the regular inspection of hand tools
- discarding or prompt repair of defective tools
- taking time to keep tools in the proper condition and ready for use
- proper storage to prevent damage and corrosion
- locking tools away when not in use to prevent them being used by unauthorized people.

Training – all users of hand tools should be properly trained in their use. This may well have been done through apprenticeships and similar training. This will be particularly important with specialist working conditions or work involving young people.

9.8.2 Hand-held power tools

Hazards

The electrical hazards of portable hand-held tools are covered in Chapter 10. This section deals with other physical hazards relating to this type of equipment, particularly a drill and sander.

Other hazards involve:

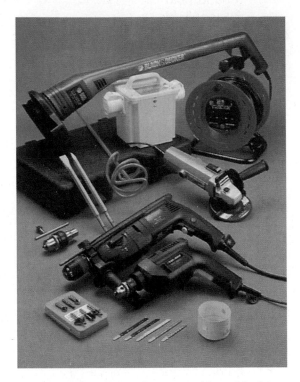

➤ mechanical entanglement in rotating spindles or sanding discs
➤ waste material flying out of the cutting area
➤ coming into contact with the cutting blades or drill bits
➤ risk of hitting electrical, gas or water services when drilling into building surfaces
➤ manual handling problem with a risk of injury if the tool is heavy or very powerful
➤ hand–arm vibration especially with petrol strimmers and chainsaws
➤ tripping hazard from trailing cables or power supplies

Figure 9.8 Typical range of hand-held power tools.

➤ eye hazard from flying particles
➤ explosion risk with petrol driven tools or when used near flammable liquids or gases
➤ high noise levels with routers, planes and saws in particular (see Chapter 13)
➤ dust levels (but see Chapter 12).

Typical safety instructions and controls

When using power tools, the following basic safety measures should be observed to protect against electrical shock, personal injury, ill-health and risk of fire. Operators should read these instructions before using the equipment and ensure that they are followed:

➤ maintain a clean and tidy working area that is well lit and clear of obstructions;
➤ never expose power tools to rain. Do not use power tools in damp or wet surroundings;
➤ do not use power tools in the vicinity of combustible fluids or gases;
➤ protect against electrical shock (if tools are electrically powered) by avoiding body contact with grounded objects such as pipes, radiators, stoves and refrigerators;
➤ keep children away;
➤ do not let other persons handle the tool or the cable. Keep them away from the working area;

➤ store tools in a safe place when not in use where they are in a dry, locked area, which is inaccessible to children;

➤ tools should not be overloaded as they operate better and safer in the performance range for which they were intended;

➤ use the right tool. Do not use small tools or attachments for heavy work. Do not use tools for purposes and tasks for which they were not intended; for example, do not use a hand-held circle saw to cut down trees or cut off branches;

➤ wear suitable work clothes. Do not wear loose fitting clothing or jewelry. They can get entangled in moving parts. For outdoor work, rubber gloves and non-skid footwear are recommended. Long hair should be protected with a hair net;

➤ use safety glasses;

➤ also use filtering respirator mask for work that generates dust;

➤ do not abuse the power cable;

➤ do not carry the tool by the power cable and do not use the cable to pull the plug out of the power socket. Protect the cable from heat, oil and sharp edges;

➤ secure the work piece. Use clamps or a vice to hold work piece. It is safer than using hands and it frees both hands for operating the tool;

➤ do not overreach the work area. Avoid abnormal body postures. Maintain a safe stance and maintain a proper balance at all times;

➤ maintain tools with care. Keep your tools clean and sharp for efficient and safe work. Follow the maintenance regulations and instructions for the changing of tools. Check the plug and cable regularly and in case of damage, have repaired by a qualified service engineer. Also inspect extension cables regularly and replace if damaged;

➤ keep the handle dry and free of oil or grease;

➤ disconnect the power plug when not in use, before servicing and when changing the tool, i.e. blade, bits, cutter, sanding disc, etc.;

➤ do not forget to remove key. Check before switching on that the key and any tools for adjustment are removed;

➤ avoid unintentional switch-on. Do not carry tools that are connected to power with your finger on the power switch. Check that the switch is turned off before connecting the power cable;

➤ outdoors use extension cables. When working outdoors, use only extension cables which are intended for such use and marked accordingly;

➤ stay alert, keep eyes on the work. Use common sense. Do not operate tools when there are significant distractions;

➤ check the equipment for damage. Before further use of a tool, check carefully the protection devices or lightly damaged parts for proper operation and performance of their intended functions. Check movable parts for proper function, whether there is binding or for damaged parts. All parts must be correctly mounted and meet all conditions necessary to ensure proper operation of the equipment;

➤ damaged protection devices and parts should be repaired or replaced by a competent service centre unless otherwise stated in the operating instructions. Damaged switches must be replaced by a competent service centre. Do not use any tool which cannot be turned on and off with the switch;

➤ only use accessories and attachments that are described in the operating instructions or are provided or recommended by the tool manufacturer. The use of tools other than those described in the operating instructions or in the catalogue of recommended tool inserts or accessories can result in a risk of personal injury.

9.9 | Mechanical machinery hazards

Most machinery has the potential to cause injury to people, and machinery accidents figure prominently in official accident statistics. These injuries may range in severity from a minor cut or bruise, through various degrees of wounding and disabling mutilation, to crushing, decapitation or other fatal injury. Nor is it solely powered machinery that is hazardous, for many manually operated machines (e.g. hand-operated guillotines and fly presses) can still cause injury if not properly safeguarded.

Machinery movement basically consists of rotary, sliding or reciprocating action, or a combination of these. These movements may cause injury by entanglement, friction or abrasion, cutting, shearing, stabbing or puncture, impact, crushing, or by drawing a person into a position where one or more of these types of injury can occur.

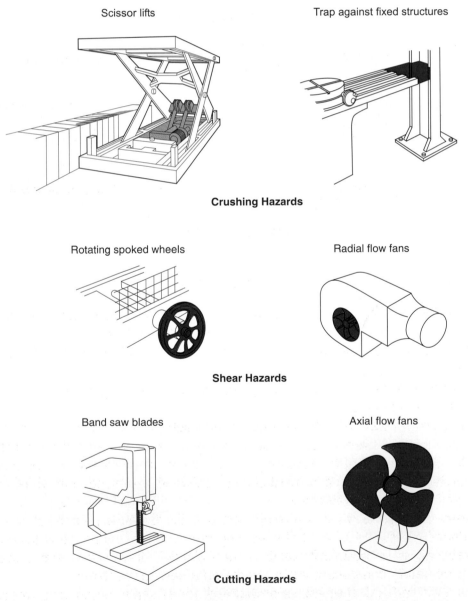

Scissor lifts   Trap against fixed structures

Crushing Hazards

Rotating spoked wheels   Radial flow fans

Shear Hazards

Band saw blades   Axial flow fans

Cutting Hazards

Figure 9.9 Range of mechanical hazards. Reproduced with permission from *Safety with Machinery* (2002) by Ridley and Pearce, Butterworth-Heinemann.

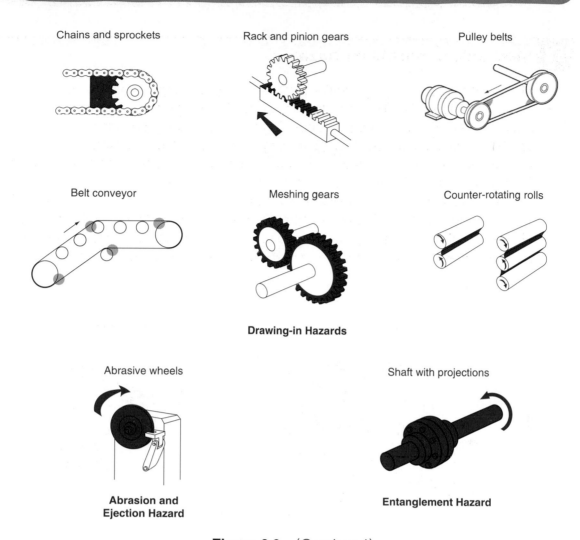

Figure 9.9 (*Continued.*)

A person may be injured at machinery as a result of:

> a **crushing hazard** through being trapped between a moving part of a machine and a fixed structure, such as a wall or any material in a machine;

> a **shearing hazard** which traps part of the body, typically a hand or fingers, between moving and fixed parts of the machine;

> a **cutting or severing hazard** through contact with a cutting edge, such as a band saw or rotating cutting disc;

> **entanglement hazard** with the machinery which grips loose clothing, hair or working material, such as emery paper, around revolving exposed parts of the machinery. The smaller the diameter of the revolving part the easier it is to get a wrap or entanglement;

> a **drawing-in or trapping hazard** such as between in-running gear wheels or rollers or between belts and pulley drives;

> an **impact hazard** when a moving part directly strikes a person, such as with the accidental movement of a robot's working arm when maintenance is taking place;

> a **stabbing or puncture hazard** through ejection of particles from a machine or a sharp operating component like a needle on a sewing machine;

> contact with a **friction or abrasion hazard,** for example, on grinding wheels or sanding machines;

➤ **high pressure fluid ejection hazard**, for example, from a hydraulic system leak.

In practice, injury may involve several of these at once, for example, contact, followed by entanglement of clothing, followed by trapping.

9.10 Non-mechanical machinery hazards

Non-mechanical hazards include:

➤ access: slips, trips and falls; falling and moving objects; obstructions and projections
➤ lifting and handling
➤ electricity (including static electricity): shock, burns
➤ fire and explosion
➤ noise and vibration
➤ pressure and vacuum
➤ high/low temperature
➤ inhalation of dust/fume/mist
➤ suffocation
➤ radiation: ionizing and non-ionizing
➤ biological: viral or bacterial.

In many cases it will be practicable to install safeguards which protect the operator from both mechanical and non-mechanical hazards.

For example, a guard may prevent access to hot or electrically live parts as well as to moving ones. The use of guards which reduce noise levels at the same time is also common.

As a matter of policy machinery hazards should be dealt with in this integrated way instead of dealing with each hazard in isolation.

9.11 Examples of machinery hazards

The following examples are given to demonstrate a small range of machines found in industry and commerce, which are included in the Certificate syllabus.

Office – photocopier
The hazards are:

➤ contact with moving parts when clearing a jam
➤ electrical – when clearing a jam, maintaining the machine or through poorly maintained plug and wiring
➤ heat through contact with hot parts when clearing a jam
➤ health hazard from ozone or lack of ventilation in the area.

Office – document shredder
The hazards are:

➤ drawing in between the rotating cutters when feeding paper into the shredder
➤ contact with the rotating cutters when emptying the waste container or clearing a jam
➤ electrical through faulty plug and wiring or during maintenance

➤ noise from the cutting action of the machine
➤ possible dust from the cutting action.

Manufacturing and maintenance – bench top grinding machine
The hazards are:

➤ contact with the rotating wheel causing abrasion
➤ drawing in between the rotating wheel and a badly adjusted tool rest
➤ bursting of the wheel, ejecting fragments which puncture the operator
➤ electrical through faulty wiring and/or earthing or during maintenance
➤ fragments given off during the grinding process causing eye injury
➤ hot fragments given off which could cause a fire
➤ noise produced during the grinding process
➤ possible health hazard from dust/particles/fumes given off during grinding.

Manufacturing and maintenance – pedestal drill
The hazards are:

➤ entanglement around the rotating spindle and chuck
➤ contact with the cutting drill or work-piece
➤ being struck by the work-piece if it rotates
➤ being cut or punctured by fragments ejected from the rotating spindle and cutting device
➤ drawing in to the rotating drive belt and pulley
➤ contact or entanglement with the rotating motor
➤ electrical from faulty wiring and/or earthing or during maintenance
➤ possible health hazard from cutting fluids or dust given off during the process.

Agriculture/horticultural – cylinder mower
The hazards are:

➤ trapping, typically hands or fingers, in the shear caused by the rotating cutters
➤ contact and entanglement with moving parts of the drive motor
➤ drawing in between chain and sprocket drives
➤ impact and cutting injuries from the machine starting accidentally
➤ burns from hot parts of the engine
➤ fire from the use of highly flammable petrol as a fuel
➤ possible noise hazard from the drive motor
➤ electrical if electrically powered, but this is unlikely
➤ possible sensitization health hazard from cutting grass, for example, hay fever
➤ possible health hazard from exhaust fumes.

Agriculture/horticultural – brush cutter/strimmer
The hazards are:

➤ entanglement with rotating parts of motor and shaft
➤ cutting from contact with cutting head/line
➤ electric shock, if electrically powered but this is unlikely
➤ burns from hot parts of the engine
➤ fire from the use of highly flammable petrol as a fuel
➤ possible noise hazard from the drive motor and cutting action
➤ eye and face puncture wounds from ejected particles

➤ health hazard from vibration causing white finger and other problems
➤ back strain from carrying the machine while operating
➤ health hazards from animal faeces.

Retail – compactor
The hazards are:

➤ crushing hazard between the ram and the machine sides
➤ trapping between the ram and machine frame (shear action)
➤ crushing when the waste unit is being changed if removed by truck
➤ entanglement with moving parts of pump motor
➤ electrical from faulty wiring and/or earthing or during maintenance
➤ failure of hydraulic hoses with liquid released under pressure causing puncture to eyes or other parts of body
➤ falling of vertical ram under gravity if the hydraulic system fails
➤ handling hazards during loading and unloading.

Retail – checkout conveyor system
The hazards are:

➤ entanglement with belt fasteners if fitted
➤ drawing in between belt and rollers if under tension
➤ drawing in between drive belt and pulley
➤ contact or entanglement with motor drive
➤ electrical from faulty wiring and/or earthing or during maintenance.

Construction – cement/concrete mixer
The hazards are:

➤ contact and entanglement with moving parts of the drive motor
➤ crushing between loading hopper (if fitted) and drum
➤ drawing-in between chain and sprocket drives
➤ electrical, if electrically powered
➤ burns from hot parts of engine
➤ fire if highly flammable liquids used as fuel
➤ possible noise hazards from the motor and dry mixing of aggregates
➤ eye injury from splashing cement slurry
➤ possible health hazard from cement dust and cement slurry while handling.

Construction – bench-mounted circular saw
The hazards are:

➤ contact with the cutting blade above and below the bench
➤ ejection of the work-piece or timber as it closes after passing the cutting blade
➤ drawing-in between chain and sprocket or V belt drives
➤ contact and entanglement with moving parts of the drive motor
➤ likely noise hazards from the cutting action and motor
➤ health hazards from wood dust given off during cutting
➤ electric shock from faulty wiring and/or earthing or during maintenance.

9.12 Practical safeguards

PUWER requires that access to dangerous parts of machinery should be prevented in a preferred order or hierarchy of control methods. The standard required is a 'practicable' one, so that the only acceptable reason for non-compliance is that there is no technical solution. Cost is not a factor. (See Chapter 1 for more details on standards of compliance.)

The levels of protection required are, in order of implementation:

➤ fixed enclosing guarding
➤ other guards or protection devices, such as interlocked guards and pressure sensitive mats
➤ protection appliances, such as jigs, holders and push-sticks and
➤ the provision of information, instruction, training and supervision.

Since the mechanical hazard of machinery arises principally from someone coming into contact or entanglement with dangerous components, risk reduction is based on preventing this contact occurring.

This may be by means of:

➤ a physical barrier between the individual and the component (e.g. a fixed enclosing guard)
➤ a device which only allows access when the component is in a safe state (e.g. an interlocked guard which prevents the machine starting unless a guard is closed and acts to stop the machine if the guard is opened) or
➤ a device which detects that the individual is entering a risk area and then stops the machine (e.g. certain photoelectric guards and pressure-sensitive mats).

The best method should, ideally, be chosen by the designer as early in the life of the machine as possible. It is often found that safeguards which are 'bolted on' instead of 'built in' are not only less effective in reducing risk, but are also more likely to inhibit the normal operation of the machine. In addition, they may in themselves create hazards and are likely to be difficult and hence expensive to maintain.

9.12.1 Fixed guards

Fixed guards have the advantage of being simple, always in position, difficult to remove and almost maintenance free. Their disadvantage is that they do not always properly prevent access, they are often left off by maintenance staff and can create difficulties for the operation of the machine.

A fixed guard has no moving parts and should, by its design, prevent access to the dangerous parts of the machinery. It must be of robust construction and sufficient to withstand the stresses of the process and environmental conditions. If visibility or free air flow (e.g. for cooling) are necessary, this must be allowed for in the design and construction of the guard. If the guard can be opened or removed, this must only be possible with the aid of a tool.

An alternative fixed guard is the distance fixed guard, which does not completely enclose a hazard, but which reduces access by virtue of its dimensions and its distance from the hazard. Where perimeter-fence guards are used, the guard must follow the

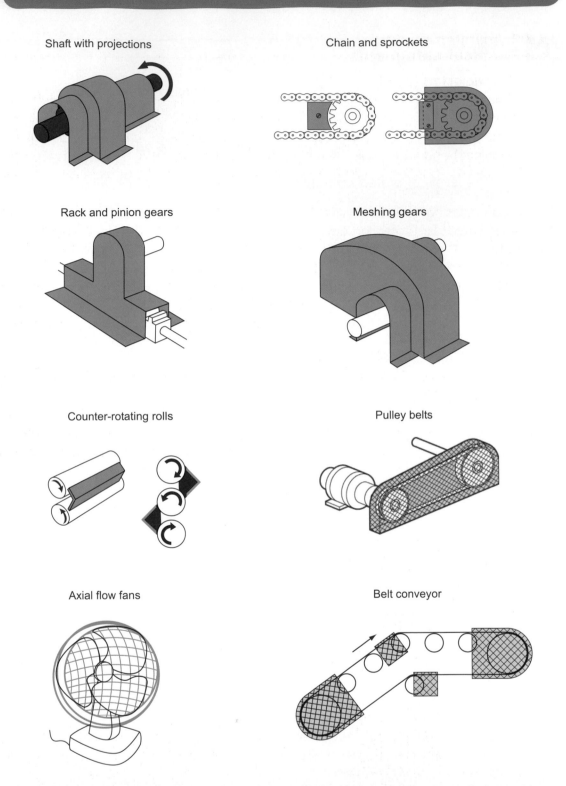

Shaft with projections

Chain and sprockets

Rack and pinion gears

Meshing gears

Counter-rotating rolls

Pulley belts

Axial flow fans

Belt conveyor

Figure 9.10   Range of fixed guards. Reproduced with permission from *Safety with Machinery* (2002) by Ridley and Pearce, Butterworth-Heinemann.

contours of the machinery as far as possible, thus minimizing space between the guard and the machinery. With this type of guard it is important that the safety devices and operating systems prevent the machinery being operated with the guards closed and someone inside the guard, i.e. in the danger area. Figure 9.10 shows a range of fixed guards for some of the examples shown in Figure 9.9.

9.12.2 Adjustable guards

User adjusted guard

These are fixed or moveable guards, which are adjustable for a particular operation during which they remain fixed. They are particularly used with machine tools where some access to the dangerous part is required (e.g. drills, circular saws, milling machines) and where the clearance required will vary (e.g. with the size of the cutter in use on a horizontal milling machine or with the size of the timber being sawn on a circular saw bench).

Adjustable guards may be the only option with cutting tools, which are otherwise very difficult to guard, but they have the disadvantage of requiring frequent re-adjustment. By the nature of the machines on which they are most frequently used, there will still be some access to the dangerous parts, so these machines must only be used by suitably trained operators. Jigs, push sticks and false tables must be used wherever possible to minimize hazards during the feeding of the work-piece. The working area should be well lit and kept free of anything which might cause the operator to slip or trip.

Self-adjusting guard

A self-adjusting guard is one which adjusts itself to accommodate, for example, the passage of material.

A good example is the spring-loaded guard fitted to many portable circular saws.

As with adjustable guards (see above) they only provide a partial solution in that they may well still allow access to the dangerous part of the machinery. They require careful maintenance to ensure they work to the best advantage.

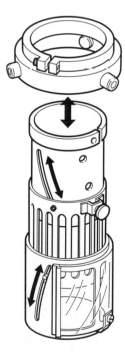

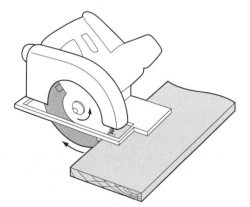

Figure 9.11   Adjustable guard for a rotating shaft, such as a pedestal drill. (Courtesy EJA Ltd, Nelsa Machine Guarding Systems.)

Figure 9.12   Self-adjusting guard on a wood saw. Reproduced with permission from *Safety with Machinery* (2002) by Ridley and Pearce, Butterworth-Heinemann.

9.12.3 Interlocking guard

The advantages of interlocked guards are that they allow safe access to operate and maintain the machine without dismantling the safety devices. Their disadvantage stems from the constant need to ensure that they are operating correctly and designed to fail safe. Maintenance and inspection procedures must be very strict.

This is a guard which is movable (or which has a movable part) whose movement is connected with the power or control system of the machine.

An interlocking guard must be so connected to the machine controls such that:

➤ until the guard is closed the interlock prevents the machinery from operating by interrupting the power medium
➤ either the guard remains locked closed until the risk of injury from the hazard has passed or opening the guard causes the hazard to be eliminated before access is possible.

A passenger lift or hoist is a good illustration of these principles: the lift will not move unless the doors are closed, and the doors remain closed and locked until the lift is stationary and in such a position that it is safe for the doors to open.

Special care is needed with systems which have stored energy. This might be the momentum of a heavy moving part, stored pressure in a hydraulic or pneumatic system, or even the simple fact of a part being able to move under gravity even though the power is disconnected. In these situations, dangerous movement may continue or be possible with the guard open, and these factors need to be considered in the overall design. Braking devices (to arrest movement when the guard is opened) or delay devices (to prevent the guard opening until the machinery is safe) may be needed. All interlocking systems must be designed to minimize the risk of failure-to-danger and should not be easy to defeat.

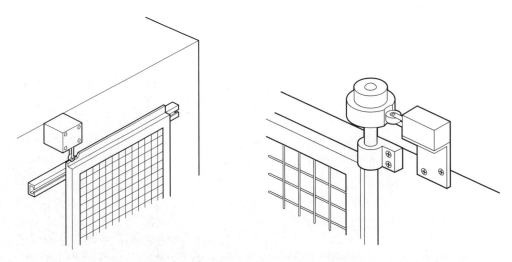

Figure 9.13 Typical sliding and hinged interlocking guards. Reproduced with permission from *Safety with Machinery* (2002) by Ridley and Pearce, Butterworth-Heinemann.

9.13 Other safety devices

9.13.1 Trip devices

A trip device does not physically keep people away but detects when a person approaches close to a danger point. It should be designed to stop the machine before injury occurs. A trip device depends on the ability of the machine to stop quickly and in some cases a brake may need to be fitted. Trip devices can be:

➤ mechanical in the form of a bar or barrier
➤ electrical in the form of a trip switch on an actuator rod, wire or other mechanism
➤ photoelectric or other type of presence-sensing device
➤ pressure-sensitive mat.

They should be designed to be self-resetting so that the machine must be restarted using the normal procedure.

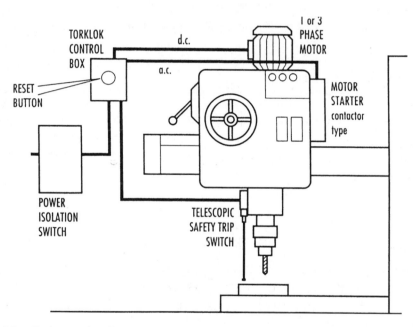

Figure 9.14   Schematic diagram of a telescopic trip device fitted to a redial drill. (Courtesy EJA Ltd, Nelsa Machine Guarding Systems.)

9.13.2 Two-handed control devices

These are devices which require the operator to have both hands in a safe place (the location of the controls) before the machine can be operated. They are an option on machinery that is otherwise very difficult to guard but they have the drawback that they only protect the operator's hands. It is therefore essential that the design does not allow any other part of the operator's body to enter the danger zone during operation. More significantly, they give no protection to anyone other than the operator.

Where two-handed controls are used, the following principles must be followed:

➤ the controls should be so placed, separated and protected as to prevent spanning with one hand only,
being operated with one hand and another part of the
body, or being readily bridged;

➤ it should not be possible to set the dangerous parts in
motion unless the controls are operated within approximately 0.5 seconds of each other. Having set the dangerous parts in motion, it should not be possible to do so
again until both controls have been returned to their off
position;

➤ movement of the dangerous parts should be arrested
immediately or, where appropriate, arrested and reversed
if one or both controls are released while there is still
danger from movement of the parts;

➤ the hand controls should be situated at such a distance
from the danger point that, on releasing the controls, it is
not possible for the operator to reach the danger point
before the motion of the dangerous parts has been
arrested or, where appropriate, arrested and reversed.

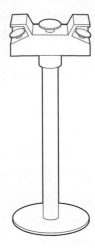

Figure 9.15 Pedestal
mounted free-standing
two-hand control device.
Reproduced with
permission from *Safety
with Machinery* (2002) by
Ridley and Pearce,
Butterworth-Heinemann.

9.13.3 Hold-to-run control

This is a control which allows movement of the machinery only as long as the control is
held in a set position. The control must return automatically to the stop position when
released. Where the machinery runs at crawl speed, this speed should be kept as low
as practicable.

Hold-to-run controls give even less protection to the operator than two-handed
controls and have the same main drawback in that they give no protection to anyone
other than the operator.

However, along with limited movement devices (systems which permit only a
limited amount of machine movement on each occasion that the control is operated
and are often called 'inching devices'), they are extremely relevant to operations such as
setting, where access may well be necessary and safeguarding by any other means is
difficult to achieve.

9.14 Application of safeguards to the range of machines

The application of the safeguards to the Certificate range of machines is as follows.

Office – photocopier
Application of safeguards

➤ The machines are provided with an all-enclosing case which prevents access to the
internal moving, hot or electrical parts.

➤ The access doors are interlocked so that the machine is automatically switched off
when gaining access to clear jams or maintain the machine. It is good practice to switch
off when opening the machine.

Figure 9.16 Typical photocopier. (Courtesy of Canon.)

➤ Internal electrics are insulated and protected to prevent contact.
➤ Regular inspection and maintenance should be carried out.
➤ The machine should be on the PAT (potable electrical testing) schedule.
➤ Good ventilation in the machine room should be maintained.

Office – document shredder
Application of safeguards

➤ Enclosed fixed guards surround the cutters with restricted access for paper only, which prevents fingers reaching the dangerous parts.
➤ Interlocks are fitted to the cutter head so that the machine is switched off when the waste bin is emptied.

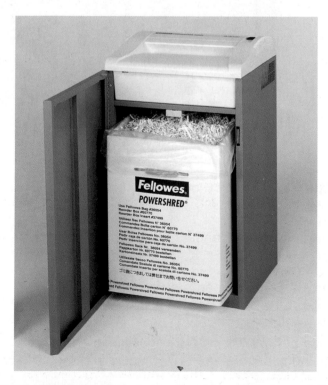

Figure 9.17 Typical office paper shredder. (Courtesy of Fellowes.)

➤ A trip device is used to start the machine automatically when paper is fed in.
➤ Machine should be on PAT schedule and regularly checked.
➤ General ventilation will cover most dust problems except for very large machines where dust extraction may be necessary.
➤ Noise levels should be checked and the equipment perhaps placed on a rubber mat if standing on a hard reflective floor.

Manufacturing and maintenance – bench top grinder
Application of safeguards

➤ Wheel should be enclosed as much as possible in a strong casing capable of containing a burst wheel.
➤ An adjustable tool rest should be adjusted as close as possible to the wheel.
➤ Adjustable screen should be fitted over the wheel to protect the eyes of the operator. Goggles should also be worn.
➤ Only properly trained people should mount an abrasive wheel.
➤ The maximum speed should be marked on the machine so that the abrasive wheel can be matched to the machine speed.
➤ Noise levels should be checked and attenuating screens used if necessary.
➤ The machine should be on the PAT schedule and regularly checked.
➤ If necessary extract ventilation should be fitted to the wheel encasing to remove dust at source.

Figure 9.18 Typical bench-mounted grinder. (Courtesy of Draper.)

Manufacturing and maintenance – pedestal drill
Application of safeguards

➤ Motor and drive should be fitted with fixed guard.
➤ The spindle should be guarded by an adjustable guard, which is fixed in position during the work.
➤ A clamp should be available on the pedestal base to secure work-pieces.
➤ The machine should be on the PAT schedule and regularly checked.

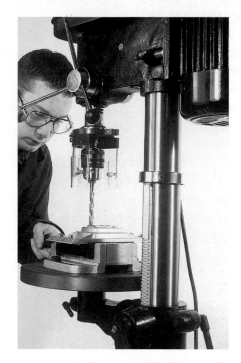

Figure 9.19 Typical pedestal drill. (Courtesy of Draper.)

➤ Cutting fluid, if used, should be contained and not allowed to get onto clothing or skin. A splash guard may be required but is unlikely.

➤ Goggles should be worn by the operator.

Agricultural/horticultural – cylinder mower
Application of safeguards

➤ Machine should be designed to operate with the grass collection box in position to restrict access to the bottom blade trap. A warning sign should be fitted on the machine.

Figure 9.20 Typical large cylinder mower. (Courtesy of Atco-Qualcast.)

191

➤ On pedestrian-controlled machines the control handle should automatically stop the blade rotation when the operators' hands are removed. It should take two separate actions to restart.

➤ Ride on machines should be fitted with a device to automatically stop the blades when the operator leaves the operating position. This is normally a switch under the seat and it should be tested to ensure that it is functioning correctly and has not been defeated.

➤ Drives and motor should be completely encased with a fixed guard.

➤ The machine should only be refuelled in the open air with a cool engine, using a proper highly flammable fuel container with pourer to restrict spillage. No smoking should be allowed.

➤ Hot surfaces like the exhaust should be covered.

➤ Engine must only be run in the open air to prevent a build up of fumes.

➤ Noise levels should be checked and an improved silencer fitted to the engine and, if necessary, hearing protection used.

➤ Hay-fever-like problems from grass cutting are difficult to control. A suitable dust mask may be required to protect the user.

Agricultural/horticultural – brush cutter/strimmer
Application of safeguards

➤ Moving engine parts should be enclosed.

➤ Rotating shafts should be encased in a fixed drive shaft cover.

➤ Rotating cutting head should have a fixed top guard, which extends out on the user side of the machine.

➤ Line changes must only be done either automatically or with the engine switched off.

➤ Engine should only be run in the open air.

➤ Refuelling should only be done in the open air using a proper highly flammable liquid container with pouring spout.

➤ Boots with steel toe cap and good grip, stout trousers and non-snagging upper garments should be worn plus hard hat fitted with full face screen and safety glasses.

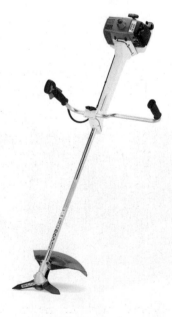

Figure 9.21 STIHL petrol-driven brush cutter for professional users. Picture supplied courtesy of STIHL GB.

➤ If the noise levels are sufficiently high (normally they are with petrol driven units) suitable hearing protection should be worn.

➤ Low vibration characteristics should be balanced with engine power and speed of work to achieve the minimum overall vibration exposure. Handles should be of an anti-vibration type. Engines should be mounted on flexible mountings. Work periods should be limited to allow recovery.

➤ Washing arrangements should be provided and warm impervious gloves to guard against health risks.

➤ Properly constructed harness should be worn which comfortably balances the weight of the machine.

Retail – compactor
Application of safeguards

➤ Access doors to the loading area, which gives access to the ram, should be positively interlocked with electrical and/or hydraulic mechanisms.

➤ The ram pressure should be dumped if it is hydraulic.

➤ Drives of motors should be properly guarded.

➤ If the waste unit is removed by truck, the ram mechanism should be interlocked with the unit so that it cannot operate when the unit is changed for an empty one.

➤ The machine should be regularly inspected and tested by a competent person.

➤ If the hydraulic ram can fall under gravity, mechanical restraints should automatically move into position when the doors are opened.

➤ Emergency stop buttons should be fitted on each side.

Figure 9.22 Typical retail compactor. (Courtesy of Pakawaste.)

Retail – checkout conveyor
Application of safeguards

➤ All traps between belt and rollers are either provided with fixed guards or interlocked guards.
➤ Motor and drive unit should be provided with a fixed guard and access to underside of conveyor is prevented by enclosure.
➤ Adequate emergency stop buttons must be provided by the checkout operator.
➤ Machine should be on the PAT electrical inspection schedule and regularly checked.
➤ Auto stop system should be fitted so that conveyor does not push products into the operator's working zone.

Construction – cement mixer
Application of safeguards

➤ Operating position for the hopper hoist should be designed so that anyone in the trapping area is visible to the operator. The use of the machine should be restricted to designated operators only. As far as possible the trapping point should be designed out. The hoist operating location should be fenced off just allowing access for barrows etc., to the unloading area.
➤ Drives and rotating parts of engine should be enclosed.
➤ The drum gearing should be enclosed and persons kept away from the rotating drum, which is normally fairly high on large machines.
➤ No one should be allowed to stand on the machine while it is in motion.
➤ Goggles should be worn to prevent cement splashes.
➤ If petrol driven, care is required with flammable liquids and refuelling.
➤ Engines must only be run in the open air.
➤ Electric machines should be regularly checked and be on the PAT schedule.

Figure 9.23   Typical cement/concrete mixer. (Courtesy of Winget.)

➤ Noise levels should be checked and noise attenuation, for example, silencers and damping, fitted if necessary.

Construction – bench-mounted circular saw
Application of safeguards

➤ A fixed guard should be fitted to the blade below the bench.
➤ Fixed guards should be fitted to the motor and drives.
➤ An adjustable top guard should be fitted to the blade above the bench which encloses as much of the blade as possible. An adjustable front section should also be fitted.
➤ A riving knife should be fitted behind the blade to keep the cut timber apart and prevent ejection.
➤ Noise attenuation should be applied to the machine, for example, damping, special quiet saw blades and if necessary fitting in an enclosure. Hearing protection may have to be used.
➤ Protection against wet weather should be provided.
➤ The electrical parts should be regularly checked plus all the mechanical guards.
➤ Extraction ventilation will be required for the wood dust and shavings.
➤ Suitable dust masks should be worn.

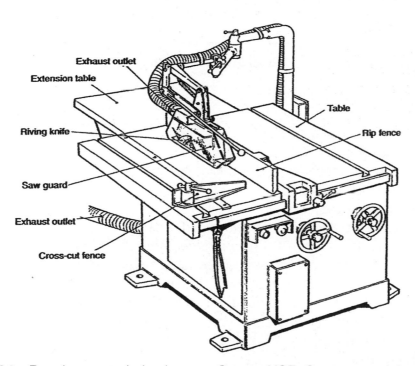

Figure 9.24 Bench-mounted circular saw. Source HSE. Crown copyright material is reproduced with the permission of the Controller of HMSO and the Queen's Printer for Scotland.

9.15 Guard construction

The design and construction of guards must be appropriate to the risks identified and the mode of operation of the machinery in question.

The following factors should be considered:

➤ strength – guards should be adequate for the purpose, able to resist the forces and vibration involved, and able to withstand impact (where applicable)
➤ weight and size – in relation to the need to remove and replace the guard during maintenance
➤ compatibility with materials being processed and lubricants etc.
➤ hygiene and the need to comply with food safety regulations
➤ visibility – it may be necessary to see through the guard for both operational and safety reasons
➤ noise attenuation – guards can often be used to reduce the noise levels produced by a machine. Conversely the resonance of large undamped panels may exacerbate the noise problem
➤ enabling a free flow of air – where necessary (e.g. for ventilation)
➤ avoidance of additional hazards – for example, free of sharp edges
➤ ease of maintenance and cleanliness
➤ openings – the size of openings and their distance from the dangerous parts should not allow anyone to be able to reach into a danger zone. These values can be determined by experiment or by reference to standard tables. If doing so by experiment, it is essential that the machine is first stopped and made safe (e.g. by isolation). The detailed information on openings is contained in EN 294:1992, EN 349:1993 and EN 811:1997.

9.16 Practice NEBOSH questions for Chapter 9

1. Outline the sources and possible effects of four non-mechanical hazards commonly encountered in a woodworking shop. (March 2001)
2. Identify four mechanical hazards presented by pedestal drills and outline in each case how injury may occur. (June 2000)
3. (a) In relation to machine safety, outline the principles of:
      (i) interlocked guards
      (ii) trip devices
   (b) Other than contact with dangerous parts, identify four types of danger against which fixed guards on machines may provide protection. (March 2000)
4. Outline eight factors that may be important in determining the maintenance requirements for an item of work equipment. (June 1999)
5. Provide sketches to show clearly the nature of the following mechanical hazards from moving parts of machinery:
      (i) entanglement
      (ii) crushing
      (iii) drawing-in
      (iv) shear. (June 1998)
6. List the main requirements of the Provision and Use of Work Equipment Regulations 1992. *(Note: now 1998)*. (June 1998)
7. With reference to an accident involving an operator who comes into contact with a dangerous part of a machine, describe:
      (i) the possible immediate causes
      (ii) the possible root (underlying) causes. (June 1997)

Electrical hazards and control

10.1 Introduction

Electricity is a widely used, efficient and convenient, but potentially hazardous method of transmitting and using energy. It is in use in every factory, workshop, laboratory and office in the country. Any use of electricity has the potential to be very hazardous with possible fatal results. Legislation has been in place for many years to control and regulate the use of electrical energy and the activities associated with its use. Such legislation provides a framework for the standards required in the design, installation, maintenance and use of electrical equipment and systems and the supervision of these activities to

Figure 10A Beware of electricity – typical sign.

minimize the risk of injury. Electrical work from the largest to the smallest installation must be carried out by people known to be competent to undertake such work. New installations always require expert advice at all appropriate levels to cover both design aspects of the system and its associated equipment. Electrical systems and equipment must be properly selected, installed, used and maintained.

Approximately 8% of all fatalities at work are caused by electric shock. Over the last few years, there have been between 12 and 16 employee deaths due to electricity, between 210 and 258 major accidents and about 500 over three-day accidents each year. The majority of the fatalities occur in the agriculture, extractive and utility supply and service industries, while the majority of the major accidents happen in the manufacturing, construction and service industries.

Only voltages up to and including **mains voltage** (220/240V) will be considered in detail in this chapter and the three principal electrical hazards – electric shock, electric burns and electrical fires and explosions.

10.2 Principles of electricity and some definitions

10.2.1 Basic principles and measurement of electricity

In simple terms, electricity is the flow or movement of electrons through a substance which allows the transfer of electrical energy from one position to another. The substance through which the electricity flows is called a **conductor**. This flow or movement of electrons is known as the **electric current**. There are two forms of electric current – direct and alternating. **Direct current (dc)** involves the flow of electrons along a conductor from one end to the other. This type of current is mainly restricted to batteries and similar devices. **Alternating current (ac)** is produced by a rotating alternator and causes an oscillation of the electrons rather than a flow of electrons so that energy is passed from one electron to the adjacent one and so on through the length of the conductor.

It is sometimes easier to understand the basic principles of electricity by comparing its movement with that of water in a pipe flowing downhill. The flow rate of water through the pipe (measured in litres/s) is similar to the current flowing through the conductor which is measured in amperes, normally abbreviated to **amps**. Sometimes very small currents are used and these are measured in milliamps (mA).

The higher the pressure drop is along the pipeline, the greater will be the flow rate of water and, in a similar way, the higher the electrical 'pressure difference' along the conductor, the higher the current will be. This electrical 'pressure difference' or potential difference is measured in **volts**.

The flow rate through the pipe will also vary for a fixed pressure drop as the roughness on the inside surface of the pipe varies – the rougher the surface, the slower the flow and the higher the resistance to flow becomes. Similarly, for electricity, the poorer the conductor, the higher the resistance is to electrical current and the lower the current becomes. Electrical resistance is measured in **ohms**.

The voltage (V), the current (I) and the resistance (R) are related by the following formula, known as Ohms law:

$$V = I \times R \text{ (Volts)}$$

and, electrical power (*P*) is given by:

$$P = V \times I \text{ (Watts)}$$

These basic formulae enable simple calculations to be made so that, for example, the correct size of fuse may be ascertained for a particular piece of electrical equipment.

Conductors and insulators

Conductors are nearly always metals, copper being a particularly good conductor, and are usually in wire form but they can be gases or liquids, water being a particularly good conductor of electricity. Superconductors is a term given to certain metals which have a very low resistance to electricity at low temperatures.

Very poor conductors are known as **insulators** and include materials such as rubber, timber and plastics. Insulating material is used to protect people from some of the hazards associated with electricity.

Short circuit

Electrical equipment components and an electrical power supply (normally the mains or a battery) are joined together by a conductor to form a **circuit**. If the circuit is broken in some way so that the current flows directly to earth rather than to a piece of equipment, a **short circuit** is made. Since the resistance is greatly reduced but the voltage remains the same, a rapid increase in current occurs which could cause significant problems if suitable protection were not available.

Earthing

The electricity supply company has one of its conductors solidly connected to the earth and every circuit supplied by the company must have one of its conductors connected to earth. This means that if there is a fault, such as a break in the circuit, the current, known as the earth fault current, will return directly to earth, which forms the circuit of least resistance, thus maintaining the supply circuit. This process is known as **earthing**. Other devices, such as fuses and residual current devices, which will be described later, will also be needed within the circuit to interrupt the current flow to earth so as to protect people from electric shock and equipment from overheating. Good and effective earthing is absolutely essential and must be connected and checked by a competent person. Where a direct contact with earth is not possible, for example, in a motor car, a common voltage reference point is used, such as the vehicle chassis.

Where other potential metallic conductors exist near to electrical conductors in a building, they must be connected to the main earth terminal to ensure **equipotential bonding** of all conductors to earth. This applies to gas, water and central heating pipes and other devices such as lightning protection systems. **Supplementary bonding** is required in bathrooms and kitchens where, for example, metal sinks and other metallic equipment surfaces are present. This involves the connection of a conductor from the sink to a water supply pipe which has been earthed by equipotential bonding. There have been several fatalities due to electric shocks from 'live' service pipes or kitchen sinks.

10.2.2 Some definitions

Certain terms are frequently used with reference to electricity and the more common ones are defined here.

Low voltage – a voltage normally not exceeding 600 volts ac between conductors and earth or 1000 volts ac between phases. Mains voltage falls into this category.

High voltage – a voltage normally exceeding 600 volts ac between conductors and earth or 1000 volts ac between phases.

Mains voltage – this is the common voltage available in domestic premises and many workplaces and is normally taken from three pin socket points. In the UK it is distributed by the national grid and is usually supplied between 220V and 240V, alternating current and at 50 cycles/s.

Maintenance – a combination of any actions carried out to retain an item of electrical equipment in, or restore it to, an acceptable and safe condition.

Testing – a measurement carried out to monitor the conditions of an item of electrical equipment without physically altering the construction of the item or the electrical system to which it is connected.

Inspection – a maintenance action involving the careful scrutiny of an item of electrical equipment, using, if necessary, all the senses to detect any failure to meet an acceptable and safe condition. An inspection does not include any dismantling of the item of equipment.

Examination – an inspection together with the possible partial dismantling of an item of electrical equipment, including measurement and non-destructive testing as required, in order to arrive at a reliable conclusion as to its condition and safety.

Isolation – involves cutting off the electrical supply from all or a discrete section of the installation by separating the installation or section from every source of electrical energy. This is the normal practice so as to ensure the safety of persons working on or in the vicinity of electrical components which are normally live and where there is a risk of direct contact with live electricity.

Competent electrical person – a person possessing sufficient electrical knowledge and experience to avoid the risks to health and safety associated with electrical equipment and electricity in general.

10.3 Electrical hazards and injuries

Electricity is a safe, clean and quiet method of transmitting energy. However, this apparently benign source of energy when accidentally brought into contact with conducting material, such as people, animals or metals, permits releases of energy which may result in serious damage or loss of life. Constant awareness is necessary to avoid and prevent danger from accidental releases of electrical energy.

The principal hazards associated with electricity are:

➤ electric shock
➤ electric burns
➤ electrical fires and explosions
➤ arcing
➤ portable electrical equipment
➤ secondary hazards.

10.3.1 Electric shock and burns

Electric shock is the convulsive reaction by the human body to the flow of electrical current through it. This sense of shock is accompanied by pain and, in more severe cases, by burning. The shock can be produced by low voltages, high voltages or lightning. Most incidents of electric shock occur when the person becomes the route to earth for a live conductor. The effect of electrical shock and the resultant severity of injury depend upon the size of the electrical current passing through the body which, in turn, depends on the voltage and the electrical resistance of the skin. If the skin is wet, a shock from mains voltage (220/240V) could well be fatal. The effect of shock is very dependent on conditions at the time but it is always dangerous and must be avoided. Electrical burns are usually more severe than those caused by heat, since they can penetrate deep into the tissues of the body.

The effect of electric current on the human body depends on its pathway through the body (e.g. hand to hand or hand to foot), the frequency of the current, the length of time of the shock and the size of the current. Current size is dependent on the duration of contact and the electrical resistance of body tissue. The electrical resistance of the body is greatest in the skin and is approximately 100 000 ohm, however, this may be reduced by a factor of 100 when the skin is wet. The body beneath the skin offers very little resistance to electricity due to its very high water content and, while the overall body resistance varies considerably between people and during the lifetime of each person, it averages at 1000 ohm. Skin that is wounded, bruised or damaged will considerably reduce human electrical resistance and work should not be undertaken on electrical equipment if damaged skin is unprotected.

An electrical current of 1 mA is detectable by touch and one of 10 mA will cause muscle contraction which may prevent the person from being able to release the conductor, and if the chest is in the current path, respiratory movement may be prevented causing asphyxia. Current passing through the chest may also cause fibrillation of the heart (vibration of the heart muscle) and disrupt the normal rhythm of the heart, though this is likely only within a particular range of currents. The shock can also cause the heart to stop completely (cardiac arrest) and this will lead to the cessation of breathing. Current passing through the respiratory centre of the brain may cause respiratory arrest that does not quickly respond to the breaking of the electrical contact. These effects on the heart and respiratory system can be caused by currents as low as 25 mA. It is not possible to be precise on the threshold current because it is dependent on the environmental conditions at the time, as well as the age, sex, body weight and health of the person.

Burns of the skin occur at the point of electrical contact due to the high resistance of skin. These burns may be deep, slow to heal and often leave permanent scars. Burns may also occur inside the body along the path of the electric current causing damage to muscle tissue and blood cells. Burns associated with radiation and microwaves are dealt with in Chapter 13.

10.3.2 Treatment of electric shock and burns

There are many excellent posters available which illustrate a first-aid procedure for treating electric shock and such posters should be positioned close to electrical junction

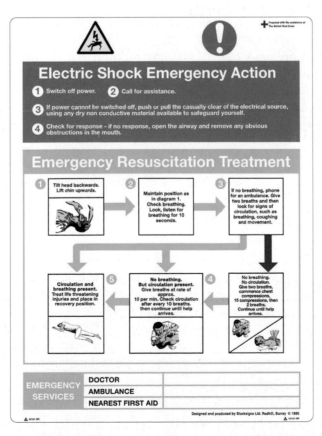

Figure 10.1   Typical electric shock treatment poster. (Courtesy of Stocksigns.)

boxes or isolation switches. The recommended procedure for treating an unconscious person who has received a **low voltage** electric shock is as follows:

1 On finding a person suffering from electric shock, raise the alarm by calling for help from colleagues (including a trained first aider).
2 Switch off the power if it is possible and/or the position of the emergency isolation switch is known.
3 Call for an ambulance.
4 If it is not possible to switch off the power, then push or pull the person away from the conductor using an object made from a good insulator, such as a wooden chair or broom. Remember to stand on dry insulating material, for example, a wooden pallet, rubber mat or wooden box. If these precautions are not taken, then the rescuer will also be electrocuted.
5 If the person is breathing, place them in the recovery position so that an open airway is maintained and the mouth can drain if necessary.
6 If the person is not breathing apply mouth-to-mouth resuscitation and, in the absence of a pulse, chest compressions. When the person is breathing normally place them in the recovery position.
7 Treat any burns by placing a sterile dressing over the burn and secure with a bandage. Any loose skin or blisters should not be touched nor any lotions or ointments applied to the burn wound.
8 If the person regains consciousness, treat for normal shock.
9 Remain with the person until they are taken to a hospital or local surgery.

It is important to note that electrocution by high voltage electricity is normally instantly fatal. On discovering a person who has been electrocuted by high voltage electricity, the police and electricity supply company should be informed. If the person remains in contact with or within 18 m of the supply, they should not be approached to within 18 m by others until the supply has been switched off and clearance has been given by the emergency services. High voltage electricity can 'arc' over distances less than 18 m, thus electrocuting the would-be rescuer.

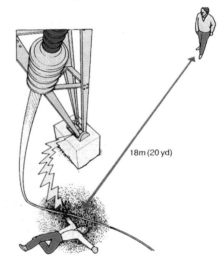

18m (20 yd)

Figure 10.2   Keep 18 m clear on high voltage lines.

10.3.3 Electrical fires and explosions

Over 25% of all fires have a cause linked to a malfunction of either a piece of electrical equipment or wiring or both. Electrical fires are often caused by a lack of reasonable care in the maintenance and use of electrical installations and equipment. The electricity that provides heat and light and drives electric motors is capable of igniting insulating or other combustible material if the equipment is misused, is not adequate to carry the electrical load, or is not properly installed and maintained. The most common causes of fire in electrical installations are short circuits, overheating of cables and equipment, the ignition of flammable gases and vapours, and the ignition of combustible substances by static electrical discharges.

Short circuits happen, as mentioned earlier, if insulation becomes faulty, and an unintended flow of current between two conductors or between one conductor and earth occurs. The amount of the current depends, among other things, upon the voltage, the condition of the insulating material and the distance between the conductors. At first the current flow will be low but as the fault develops the current will increase and the area surrounding the fault will heat up. In time, if the fault persists, a total breakdown of insulation will result and excessive current will flow through the fault. If the fuse fails to operate or is in excess of the recommended fuse rating, overheating will occur and a fire will result. A fire can also be caused if combustible material is in close proximity to the heated wire or hot sparks are ejected. Short circuits are most likely to occur where electrical equipment or cables are susceptible to damage by water leaks or mechanical damage. Twisted or bent cables can also cause breakdowns in insulation materials.

Inspection covers and cable boxes are particular problem areas. Effective steps should be taken to prevent the entry of moisture as this will reduce or eliminate the risk. Covers can themselves be a problem especially in dusty areas where the dust can accumulate on flat insulating surfaces resulting in tracking between conductors at different voltages and a subsequent insulation failure. The interior of inspection panels should be kept clean and dust free by using a suitable vacuum cleaner.

Overheating of cables and equipment will occur if they become overloaded. Electrical equipment and circuits are normally rated to carry a given safe current which will keep the temperature rise of the conductors in the circuit or appliance within permissible limits and avoid the possibility of fire. These safe currents define the maximum size of the fuse (the fuse rating) required for the appliance. A common cause of circuit overloading is the use of equipment and cables which are too small for the imposed electrical load. This is often caused by the addition of more and more equipment to the circuit thus taking it beyond its original design specification. In offices, the overuse of multisocket unfused outlet adaptors can create overload problems (sometimes known as the Christmas tree effect). The more modern multiplugs are much safer as they lead to one fused plug and cannot be easily overloaded (see Figure 10.3). Another cause of overloading is mechanical breakdown or wear of an electric motor and the driven machinery. Motors must be maintained in good condition with particular

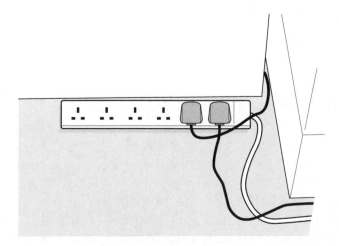

Figure 10.3 Modern multiplug.

attention paid to bearing surfaces. Fuses do not always provide total protection against the overloading of motors and, in some cases, severe heating may occur without the fuses being activated.

Loose cable connections are one of the most common causes of overheating and may be readily detected (as well as overloaded cables) by a thermal imaging survey (a technique which indicates the presence of hot spots). The bunching of cables together can also cause excessive heat to be developed within the inner cable leading to a fire risk. This can happen with cable extension reels, which have only been partially unwound, used for high-energy appliances like an electric heater.

Ventilation is necessary to maintain safe temperatures in most electrical equipment and overheating is liable to occur if ventilation is in any way obstructed or reduced. All electric equipment must be kept free of any obstructions that restrict the free supply of air to the equipment and, in particular, to the ventilation apertures.

Most electrical equipment either sparks in normal operation or is liable to spark under fault conditions. Some electrical appliances such as electric heaters, are specifically designed to produce high temperatures. These circumstances create fire and explosion hazards, which demand very careful assessment in locations where processes capable of producing flammable concentrations of gas or vapour are used, or where flammable liquids are stored.

It is likely that many fires are caused by static electrical discharges. Static electricity can, in general, be eliminated by the careful design and selection of materials used in equipment and plant, and the materials used in products being manufactured. When it is impractical to avoid the generation of static electricity, a means of control must be devised. Where flammable materials are present, especially if they are gases or dusts, then there is a great danger of fire and explosion, even if there is only a small discharge of static electricity. The control and prevention of static electricity is considered in more detail later.

The use of electrical equipment in potentially flammable atmospheres should be avoided as far as possible. However, there will be many cases where electrical equipment must be used and, in these cases, the standards for the construction of the equipment should comply with the Equipment and Protective Systems Intended for Use in Potentially Explosive Atmospheres Regulations 1996 (known as ATEX which come fully into force in June 2003) and details on the classification or zoning of areas are published by the British Standards Institution and the Health and Safety Executive.

Before electrical equipment is installed in any location where flammable vapours or gases may be present, the area must be zoned in accordance with the relevant standard and records of the zoned areas must be marked on building drawings and revised when any zoned area is changed. The installation and maintenance of electrical equipment in potentially flammable atmospheres is a specialized task. It must only be undertaken by electricians or instrument mechanics who have an understanding of the techniques involved.

In the case of a fire involving electrical equipment, the first action must be the isolation of the power supply so that the circuit is no longer live. This is achieved by switching off the power supply at the mains isolation switch or at another appropriate point in the system. Where it is not possible to switch off the current, the fire must be attacked in a way which will not cause additional danger. The use of non-conducting extinguishing media, such as carbon dioxide or powder, is necessary. After extinguishing such a fire careful watch should be kept for renewed outbreaks until the fault has been rectified. Re-ignition is a particular problem when carbon dioxide extinguishers are used, although less equipment may be damaged than is the case when powder is used.

Finally, the chances of electrical fires occurring are considerably reduced if the original installation was undertaken by competent electricians working to recognized standards, such as the Institution of Electrical Engineers Regulations. It is also important to have a system of regular testing and inspection in place so that any remedial maintenance can take place.

10.3.4 Electric arcing

A person who is standing on earth too close to a high voltage conductor may suffer flash burns as a result of arc formation. Such burns may be extensive and lower the resistance of the skin so that electric shock may add to the ill effects. Electrical arc faults can cause temporary blindness by burning the retina of the eye and this may lead to additional secondary hazards. The quantity of electrical energy is as important as the size of the voltage since the voltage will determine the distance over which the arc will travel. The risk of arcing can be reduced by the insulation of live conductors.

Strong electromagnetic fields induce surface charges on people. If these charges accumulate, skin sensation is affected and spark discharges to earth may cause localized pain or bruising. Whether prolonged exposure to strong fields has any other significant effects on health has not been proven. However, the action of an implanted cardiac pacemaker may be disturbed by the close proximity of its wearer to a powerful electromagnetic field. The health effects of arcing and other non-ionizing radiation are covered in Chapter 13.

10.3.5 Static electricity

Static electricity is produced by the build-up of electrons on weak electrical conductors or insulating materials. These materials may be gaseous, liquid or solid and may include flammable liquids, powders, plastic films and granules. The generation of static may be caused by the rapid separation of highly insulated materials by friction or by transfer from one highly charged material to another in an electric field by induction.

Discharges of static electricity may be sufficient to cause serious electric shock and are always a potential source of ignition when flammable liquid, dusts or powders are present. Flour dust in a mill, for example, has been ignited by static electricity.

Static electricity may build up on both materials and people. When a charged person approaches flammable gases or vapours and a spark ignites the substance, the resulting explosion or fire often causes serious injury.

Figure 10.4 Prevention of static discharge – container connected to earthed drum.

Lightning strikes are a natural form of static electricity and result in large amounts of electrical energy being dissipated in a short time in a limited space with a varying degree of damage. The current produced in the vast majority of strikes exceeds 3000 amps over a short period of time. Before a strike, the electrical potential between the cloud and earth might be about 100 million volts and the energy released at its peak might be about 100 million watts per metre of strike.

The need to provide lightning protection depends on a number of factors, which include:

> the risk of a strike occurring
> the number of people likely to be affected
> the location of structure and the nearness of other tall structures in the vicinity
> the type of construction, including the materials used
> the contents of structure or building (including any flammable substances)
> the value of the building and its contents.

Expert advice will be required from a specialist company in lightning protection, especially when flammable substances are involved. Lightning strikes can also cause complete destruction and/or significant disruption of electronic equipment.

10.3.6 Portable electrical equipment

Portable and transportable electrical equipment is defined by the Health and Safety Executive as 'not part of a fixed installation but may be connected to a fixed installation by means of a flexible cable and either a socket and plug or a spur box or similar means. It may be hand held or hand operated while connected to the supply, or is intended or likely to be moved while connected to the supply'. The auxiliary equipment, such as extension leads, plugs and sockets, used with portable tools is also classified as portable equipment. The term 'portable' means both portable and transportable.

Almost 25% of all reportable electrical accidents involve portable electrical equipment (known as portable appliances). While most of these accidents were caused by electric shock, over 2000 fires in 1991 were started by faulty cables used by portable appliances, caused by a lack of effective maintenance. Portable electrical tools often present a high risk of injury, which is frequently caused by the conditions under which they are used. These conditions include the use of defective or unsuitable equipment and, indeed, the misuse of equipment. There must be a system to record the inspection, maintenance and repair of these tools.

Where plugs and sockets are used for portable tools, sufficient sockets must be provided for all the equipment and adaptors should not be used. Many accidents are caused by faulty flexible cables, extension leads, plugs and sockets, particularly when these items become damp or worn. Accidents often occur when contact is made with some part of the tool which has become live (probably at mains voltage), while the user is standing on, or in contact with, an earthed conducting surface. If the electrical supply is at more than 50 volts ac, then the electric shock that a person may receive from such defective equipment is potentially lethal. In adverse environmental conditions, such as humid or damp atmospheres, even lower voltages can be dangerous. Portable electrical equipment should not be used in flammable atmospheres if it can be avoided and it

Figure 10.5 Portable hand-held electric power tools. (Courtesy of DeWalt.)

must also comply with any standard relevant to the particular environment. Air operated equipment should also be used as an alternative whenever it is practical.

Some portable equipment requires substantial power to operate and may require voltages higher than those usually used for portable tools, so that the current is kept down to reasonable levels. In these cases, power leads with a separate earth conductor and earth screen must be used. Earth leakage relays and earth monitoring equipment must also be used, together with substantial plugs and sockets designed for this type of system.

Electrical equipment is safe when properly selected, used and maintained. It is important, however, that the environmental conditions are always carefully considered. The hazards associated with portable appliances increase with the frequency of use and the harshness of the environment (construction sites are often particularly hazardous in this respect). These factors must be considered when inspection, testing and maintenance procedures are being developed.

10.3.7 Secondary hazards

It is important to note that there are other hazards associated with portable electrical appliances, such as abrasion and impact, noise and vibration. Trailing leads used for portable equipment and raised socket points offer serious trip hazards and both should be used with great care near pedestrian walkways. Power drives from electric motors should always be guarded against entanglement hazards.

Secondary hazards are those additional hazards which present themselves as a result of an electrical hazard. It is very important that these hazards are considered during a risk assessment. An electric shock could lead to a fall from height if the shock occurred on a scaffold or it could lead to a collision with a vehicle if the victim collapsed on to a roadway.

Similarly an electrical fire could lead to all the associated fire hazards outlined in Chapter 11 (e.g. suffocation, burns and structural collapse) and electrical burns can easily lead to infections.

10.4 General control measures for electrical hazards

The principal control measures for electrical hazards are contained in the statutory precautionary requirements covered by the Electricity at Work Regulations 1989, the main provisions of which are outlined in Chapter 17. They are applicable to all electrical equipment and systems found at the workplace and impose duties on employers, employees and the self-employed.

The regulations cover the following topics:

> the design, construction and maintenance of electrical systems, work activities and protective equipment
> the strength and capability of electrical equipment
> the protection of equipment against adverse and hazardous environments
> the insulation, protection and placing of electrical conductors
> the earthing of conductors and other suitable precautions
> the integrity of referenced conductors
> the suitability of joints and connections used in electrical systems
> means for protection from excess current
> means for cutting off the supply and for isolation
> the precautions to be taken for work on equipment made dead
> working on or near live conductors
> adequate working space, access and lighting
> the competence requirements for persons working on electrical equipment to prevent danger and injury.

Detailed safety standards for designers and installers of electrical systems and equipment are given a code of practice published by the Institution of Electrical Engineers, known as the IEE Regulations. While these regulations are not legally binding, they are recognized as a code of good practice and widely used as an industry standard.

The risk of injury and damage inherent in the use of electricity can only be controlled effectively by the introduction of employee training, safe operating procedures (safe systems of work) and guidance to cover specific tasks.

Training is required at all levels of the organization ranging from simple on-the-job instruction to apprenticeship for electrical technicians and supervisory courses for experienced electrical engineers. First aid training related to the need for cardiovascular resuscitation and treatment of electrical burns should be available to all people working on electrical equipment and their supervisors.

A **management system** should be in place to ensure that the electrical systems are installed, operated and maintained in a safe manner. All managers should be responsible for the provision of adequate resources of people, material and advice to ensure that the safety of electrical systems under their control is satisfactory and that **safe systems of work** are in place for all electrical equipment. (Chapter 6 gives more information on both safe systems of work and permits-to-work.)

For small factories and office or shop premises where the system voltages are normally at mains voltage, it may be necessary for an external competent person to be available to offer the necessary advice. Managers must set up a high voltage permit-to-work system for all work at and above 600 volts. The system should be appropriate to the extent of the electrical system involved. Consideration should also be given to the introduction of a permit system for voltages under 600 volts when appropriate and for all work on live conductors.

The additional control measures that should be taken when working with electrical or using electrical equipment are summarized by the following topics:

➤ the selection of suitable equipment
➤ the use of protective systems
➤ inspection and maintenance strategies.

These three groups of measures will be discussed in detail.

10.5 The selection and suitability of equipment

Many factors which affect the selection of suitable electrical equipment, such as flammable, explosive and damp atmospheres and adverse weather conditions, have already been considered. Other issues include high or low temperatures, dirty or corrosive processes or problems associated with vegetation or animals (for example, tree roots touching and displacing underground power cables, farm animals urinating near power supply lines and rats gnawing through cables). Temperature extremes will affect, for example, the lubrication of motor bearings and corrosive atmospheres can lead to the breakdown of insulating materials. The equipment selected must be suitable for the task demanded else either it will become overloaded or running costs will be too high.

The equipment should be installed to a recognized standard and capable of being isolated in the event of an emergency. It is also important that the equipment is effectively and safely earthed. Electrical supply failures may affect process plant and equipment. These are certain to happen at some time and the design of the installation should be such that a safe shut-down can be achieved in the event of a total mains failure. This may require the use of a battery backed shut-down systems or emergency stand-by electrical generators (assuming that this is cost effective).

Finally, it is important to stress that electrical equipment must only be used within the rating performance given by the manufacturer and any accompanying instructions from the manufacturer or supplier must be carefully followed.

10.5.1 The advantages and limitations of protective systems

There are several different types of protective systems and techniques that may be used to protect people, plant and premises from electrical hazards, some of which, for example, earthing, have already been considered earlier in this chapter. However, only the more common types of protection will be considered here.

Fuse

A fuse will provide protection against faults and continuous marginal current overloads. It is basically a thin strip of conducting wire which will melt when the rated current passes through it, thus breaking the circuit. A fuse rated at 13 amps will melt when a current in excess of 13 amps passes through the fuse thus stopping the flow of current. A **circuit breaker** throws a switch off when excess current passes and is similar in action to a fuse. Protection against overload is provided by fuses which detect a continuous marginal excess flow of current and energy. This protection is arranged to operate before damage occurs, either to the supply system or to the load which may be a motor or heater. When providing protection against overload, consideration needs to be made as to whether tripping the circuit could give rise to an even more dangerous situation, such as with fire-fighting equipment.

The prime objective of a fuse is to protect equipment or an installation from over-heating and becoming a fire hazard. It is not an effective protection against electric shock due to the time that it takes to cut the current flow.

The examination of fuses is a vital part of an inspection programme to ensure that the correct size or rating is fitted at all times.

Insulation

Insulation is used to protect both people from electric shock, the short circuiting of live conductors and the dangers associated with fire and explosions. Insulation is achieved by covering the conductor with an insulating material. Insulation is often accompanied by the enclosure of the live conductors so that they are out of reach to people. A break-down in insulation can cause electric shock, fire, explosion or instrument damage.

Isolation

The isolation of an electrical circuit involves more than 'switching off' the current in that the circuit is made dead and cannot be accidentally re-energized. It, therefore, creates a barrier between the equipment and the electrical supply which only an authorized person should be able to remove. When it is intended to carry out work, such as mechanical maintenance or a cleaning operation on plant or machinery, isolation of electrical equip-ment will be required to ensure safety during the work process. Isolators should always be locked off when work is to be done on electrical equipment.

Before earthing or working on an isolated circuit, checks must be made to ensure that the circuit is dead and that the isolation switch is 'locked off' and clearly labelled.

Reduced low voltage systems

When the working conditions are relatively severe either due to wet conditions or heavy and frequent usage of equipment, reduced voltage systems should be used.

All portable tools used on construction sites, vehicle washing stations or near swimming pools, should operate on 110 volts or less, preferably with a centre tapped to earth at 55 volts. This means that while the full 110 volts are available to power the tool, only 55 volts is available to shock the worker. At this level of voltage, the effect of any electric shock should not be severe. For lighting, even lower voltages can be used and are even safer. Another way to reduce the voltage is to use **battery (cordless) operated hand tools**.

Residual current devices

If electrical equipment must operate at mains voltage, the best form of protection against electric shock is the residual current device (RCD). RCDs, also known as earth leakage circuit breakers, monitor and compare the current flowing in the live and neutral conductors supplying the protected equipment. Such devices are very

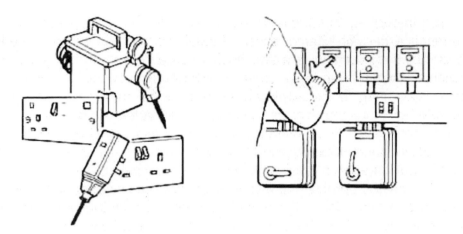

Figure 10.6 Transformer, RCD and isolators. Source HSE. Crown copyright material is reproduced with the permission of the Controller of HMSO and the Queen's Printer for Scotland.

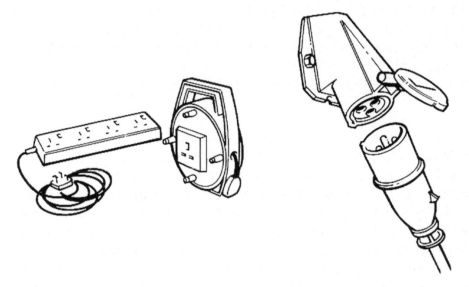

Figure 10.7 Multiplugs, extension lead and special plugs and sockets. Source HSE. Crown copyright material is reproduced with the permission of the Controller of HMSO and the Queen's Printer for Scotland.

sensitive to differences of current between the live and neutral power lines and will cut the supply to the equipment in a very short period of time when a difference of only a few milliamps occurs. It is the speed of the reaction which offers the protection against electric shock.

RCDs can be used to protect installations against fire since they will interrupt the electrical supply before sufficient energy to start a fire has been dissipated. For protection against electric shock, the RCD must have a rated residual current of 30 mA or less and an operating time of 40 milliseconds or less at a residual current of 250 mA. The protected equipment must be properly protected by insulation and enclosure in addition to the RCD. The RCD will not prevent shock or limit the current resulting from an accidental contact, but it will ensure that the duration of the shock is limited to the time taken for the RCD to operate. The RCD has a test button which should be tested frequently to ensure that it is working properly.

Double insulation

To remove the need for earthing on some portable power tools, double insulation is used. Double insulation employs two independent layers of insulation over the live conductors, each layer alone being adequate to insulate the electrical equipment safely. Since such tools are not protected by an earth, they must be inspected and maintained regularly and must be discarded if damaged.

Figure 10.8 shows the symbol which is marked on double insulated portable power tools.

Figure 10.8 Double insulation sign.

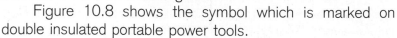

10.6 Inspection and maintenance strategies

10.6.1 Maintenance strategies

Regulation 4(2) of the Electricity at Work Regulations 1989 requires that 'as may be necessary to prevent danger, all systems shall be maintained so as to prevent so far as is reasonably practicable, such danger'. Regular maintenance is, therefore, required to ensure that a serious risk of injury or fire does not result from installed electrical equipment. Maintenance standards should be set as high as possible so that a more reliable and safe electrical system will result. Inspection and maintenance periods should be determined by reference to the recommendations of the manufacturer, consideration of the operating conditions and the environment in which equipment is located. The importance of equipment within the plant, from the plant safety and operational viewpoint, will also have a bearing on inspection and maintenance periods. The mechanical safety of driven machinery is vital and the electrical maintenance and isolation of the electrically powered drives is an essential part of that safety.

The particular areas of interest for inspection and maintenance are:

➤ the cleanliness of insulator and conductor surfaces
➤ the mechanical and electrical integrity of all joints and connections
➤ the integrity of mechanical mechanisms, such as switches and relays
➤ the calibration, condition and operation of all protection equipment, such as circuit breakers, RCDs and switches.

Safe operating procedures for the isolation of plant and machinery during both electrical and mechanical maintenance, must be prepared and followed. All electrical isolators must, wherever possible, be fitted with mechanisms which can be locked in the 'open/off' position and there must be a procedure to allow fuse withdrawal wherever isolators are not fitted.

Working on live equipment with voltages in excess of 110 volts must not be permitted except where fault finding or testing measurements cannot be done in any other way. Reasons, such as the inconvenience of halting production, are not acceptable.

Part of the maintenance process should include an appropriate system of visual inspection. By concentrating on a simple, inexpensive system of looking for visible signs of damage or faults, many of the electrical risks can be controlled, although more systematic testing may be necessary at a later stage.

All fixed electrical installations should be inspected and tested periodically by a competent person, such as a member of the National Inspection Council for Electrical Installation Contracting (NICEIC).

10.6.2 Inspection strategies

Regular inspection of electrical equipment is an essential component of any preventative maintenance programme and, therefore, regular inspection is required under Regulation 4(2) of the Electricity at Work Regulations 1989, which is quoted previously. Any strategy for the inspection of electrical equipment, particularly portable appliances, should involve the following considerations:

➤ a means of identifying the equipment to be tested
➤ the number and type of appliances to be tested
➤ the competence of those who will undertake the testing (whether in-house or bought-in)
➤ the legal requirements for portable appliance testing (PAT) and other electrical equipment testing and the guidance available
➤ organizational duties of those with responsibilities for PAT and other electrical equipment testing
➤ test equipment selection and re-calibration
➤ the development of a recording, monitoring and review system
➤ the development of any training requirements resulting from the test programme.

10.7 Portable electrical appliances testing

Portable appliances should be subject to three forms of checks – a user check, a formal visual inspection and a combined inspection and test.

10.7.1 User checks

When any portable electrical hand tool, appliance, extension lead or similar item of equipment is taken into use, and at least once each week or, in the case of heavy work, before each shift, the following visual check and associated questions should be asked:

➤ is there a recent portable appliance test (PAT) label attached to the equipment?
➤ are any bare wires visible?
➤ is the cable covering undamaged and free from cuts and abrasions (apart from light scuffing)?
➤ is the cable too long or too short? (Does it present a trip hazard?)
➤ is the plug in good condition, for example, the casing is not cracked and the pins are not bent (see Figure 10.9)?
➤ are there no taped or other non-standard joints in the cable?
➤ is the outer covering (sheath) of the cable gripped where it enters the plug or the equipment? (The coloured insulation of the internal wires should not be visible)
➤ is the outer case of the equipment undamaged or loose and are all screws in place?

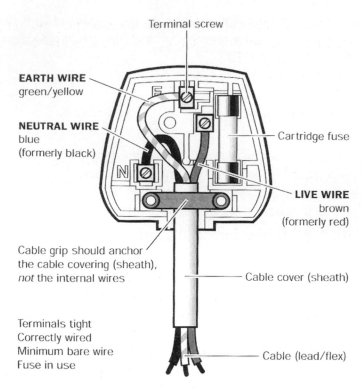

Terminal screw

EARTH WIRE
green/yellow

NEUTRAL WIRE
blue
(formerly black)

N

Cartridge fuse

LIVE WIRE
brown
(formerly red)

Cable grip should anchor
the cable covering (sheath),
not the internal wires

Cable cover (sheath)

Terminals tight
Correctly wired
Minimum bare wire
Fuse in use

Cable (lead/flex)

Figure 10.9 UK standard 3-pin plug wiring. Source HSE. Crown copyright material is reproduced with the permission of the Controller of HMSO and the Queen's Printer for Scotland.

➤ are there any overheating or burn marks on the plug, cable, sockets or the equipment?
➤ are the trip devices (RCDs) working effectively (by pressing the 'test' button)?

10.7.2 Formal visual inspections and tests

There should be a **formal visual inspection** routinely carried out on all portable electrical appliances. Faulty equipment should be taken out of service as soon as the damage is noticed. At this inspection the plug cover (if not moulded) should removed to check that the correct fuse is included, but the equipment itself should not be taken apart. This work can normally be carried out by a trained person who has sufficient information and knowledge.

Some faults, such as the loss of earth continuity due to wires breaking or loosening within the equipment, the breakdown of insulation and internal contamination (for example, dust containing metal particles may cause short circuiting if it gets inside the tool), will not be spotted by visual inspections. To identify these problems, a programme of testing and inspection will be necessary.

This formal **combined testing and inspection** should be carried out by a competent person either when there is reason to suspect the equipment may be faulty, damaged or contaminated, but this cannot be confirmed by visual inspection or after any repair, modification or similar work to the equipment, which could have affected its electrical safety. The competent person could be a person who has been specifically trained to carry out the testing of portable appliances using a simple 'pass/fail' type of tester. When more sophisticated tests are required, a competent person with the necessary technical electrical knowledge and experience would be needed. The inspection and testing should normally include the following checks:

- that the polarity is correct
- that the correct fuses are being used
- that all cables and cores are effectively terminated
- that the equipment is suitable for its environment.

Testing need not be expensive in many low-risk premises like shops and offices, if an employee is trained to perform the tests and appropriate equipment is purchased.

10.7.3 Frequency of inspection and testing

The frequency of inspection and testing should be based on a risk assessment which is related to the usage, type and operational environment of the equipment. The harsher the working environment is, the more frequent the period of inspection. Thus tools used on a construction site should be tested much more frequently than a visual display unit which is never moved from a desk. Manufacturers or suppliers may recommend a suitable testing period. Table 10.1 lists the suggested intervals for inspection and testing derived from HSE publications *Maintaining portable and transportable electrical equipment* (HSG(107) and INDG 236 and 237).

It is very important to stress that there is no 'correct' interval for testing – it depends on the frequency of usage, type of equipment, how and where it is used. A few years ago, a young trainee was badly scalded by a boiling kettle of water which exploded while in use. On investigation, an inspection report indicated that the kettle had been checked by a competent person and passed just a few weeks before the accident. Further investigation showed that this kettle was the only method of boiling water on the premises and was in use continuously for 24 hours each day. It was therefore unsuitable for the purpose and a plumbed-in continuous-use hot water heater would have been far more suitable.

10.7.4 Records of inspection and testing

Schedules which give details of the inspection and maintenance periods and the respective programmes, must be kept together with records of the inspection findings and the work done during maintenance. Records must include both individual items of equipment and a description of the complete system or section of the system. They should always be kept up-to-date and with an audit procedure in place to monitor the records and any required actions. The records do not have to be paper based but could be stored electronically on a computer. It is good practice to label the piece of equipment with the date of the last combined test and inspection.

The effectiveness of the equipment maintenance programme may be monitored and reviewed if a record of tests is kept. It can also be used as an inventory of portable appliances and help to regulate the use of unauthorized appliances. The record will enable any adverse trends to be monitored and to check that suitable equipment has been selected. It may also give an indication as to whether the equipment is being used correctly.

Table 10.1 Suggested intervals for portable appliance inspection and testing

Type of business/equipment	User checks	Formal visual inspection	Combined inspection and electrical tests
Equipment hire	Yes	Before issue and after return	Before issue
Construction	Yes	Before initial use and then every month	3 months
Industrial	Yes	Before initial use and then every 3 months	6–12 months
Hotels and offices low-risk environments			
Battery operated (less than 20 volts)	No	No	No
Extra low voltage: (less than 50 volts ac) e.g. telephone equipment, low voltage desk lights	No	No	No
Computers/photocopiers/fax machines	No	Yes 2–4 years	No if double insulated otherwise up to 5 years
Double insulated equipment: not hand-held. Moved occasionally, e.g. fans, table lamps, slide projectors	No	Yes 2–4 years	No
Double insulated equipment: hand-held, e.g. some floor cleaners, some kitchen equipment and irons	Yes	Yes 6 months–1 year	Yes 1–2 years
Earthed equipment (class 1): e.g. electric kettles, some floor cleaners, portable electric heaters	Yes	Yes 6 months–1 year	
Cables (leads) and plug connected to above. Extension leads (mains voltage)	Yes	Yes 6 months–4 years depending on the type of equipment it is connected to	Yes 1–5 years depending on the type of equipment it is connected to

Note: Operational experience may demonstrate that the above intervals can be reviewed. Derived from HSE

10.7.5 Advantages and limitations of portable appliance testing

The advantages of portable appliance testing include:

➤ an earlier recognition of potentially serious equipment faults, such as poor earthing, frayed and damaged cables and cracked plugs
➤ discovery of incorrect or inappropriate electrical supply and/or equipment
➤ discovery of incorrect fuses being used
➤ a reduction in the number of electrical accidents
➤ the misuse of portable appliances may be monitored
➤ equipment selection procedures can be checked
➤ an increased awareness of the hazards associated with electricity
➤ a more regular maintenance regime should result.

The limitations of portable appliance testing include:

➤ some fixed equipment is tested too often leading to excessive costs
➤ some unauthorized portable equipment, such as personal kettles, are never tested since there is no record of them
➤ equipment may be misused or over-used between tests due to a lack of understanding of the meaning of the test results
➤ all faults, including trivial ones, are included on the action list so that the list becomes very long and the more significant faults are forgotten or overlooked
➤ the level of competence of the tester can be too low
➤ the testing equipment has not been properly calibrated and/or checked before testing takes place.

Most of the limitations may be addressed and the reduction in electrical accidents and injuries enables the advantages of portable appliance testing to greatly outweigh the limitations.

10.8 Practice NEBOSH questions for Chapter 10

1. (a) Outline the dangers associated with electricity.
   (b) Outline the emergency action to take if a person suffers a severe electric shock. (March 2001)
2. Outline a range of checks that should be made to ensure electrical safety in an office environment. (December 2001)
3. State the items that should be included on a checklist for the routine inspection of portable electrical appliances. (June 2001)
4. In relation to the use of electrical cables and plugs in the workplace:
   (i) identify four examples of faults and bad practices that could contribute to electrical accidents
   (ii) outline the corresponding precautions that should be taken for each of the examples identified in (i). (March 2000)
5. (a) Outline the effects on the human body from a severe electric shock.
   (b) Describe how earthing can reduce the risk of receiving an electric shock. (June 1999)

Fire hazards and control

11.1 Introduction

Chapter 11 covers fire prevention in the workplace and how to ensure that people are properly protected if fire does occur. Each year UK fire brigades attend over 35 000 fires at work in which over 30 people are killed and about 2000 are injured. Fire and explosions at work account for about 2% of the major injuries reported under RIDDOR.

The financial costs associated with serious fires are very high including, in many cases, (believed to be over 40%) the failure to start up business again. Never underestimate the potential of any fire. What may appear to be a small fire in a waste bin, if not dealt with can quickly spread through a building or structure. The Bradford City Football ground in 1985 or King's Cross Underground station in 1987 are examples of where small fires quickly became raging infernos, resulting in many deaths and serious injuries.

Since the introduction of the Fire Services Act of 1947, the fire authorities have had the responsibility to fight fires in all types of premises. In 1971, the Fire Precautions Act gave the fire authorities control over certain fire procedures, means of escape and basic fire protection equipment through the drawing up and issuing of Fire Certificates in certain categories of building. The Fire Certification was mainly introduced to combat a number of serious industrial fires that had occurred, with a needless loss of life, where simple well-planned protection would have allowed people to escape unhurt.

11.2 Legal considerations

Action, regarding general fire precautions at work, by the occupier or the person in control of premises is covered in the Fire Precautions (Workplace) Regulations 1997, or for construction sites, the Construction (Health, Safety and Welfare) Regulations 1996 and the Management of Health and Safety at Work Regulations 1999. These regulations are summarized in Chapter 17 and are enforced by the fire authorities except in high-risk premises, such as chemical plants. Process-related fire precautions come under the HSE or local authority, and are covered by the general duties imposed by the HSW Act or other specific regulations.

11.2.1 Fire safety legislation and duty holders

Rules for different premises

The workplace is covered by the Fire Precautions (Workplace) Regulations 1997 (as amended by the Fire Precautions (Workplace) (Amendment) Regulations 1999). It may also be covered by a further provision, the main one being a valid fire certificate under the Fire Precautions Act 1971 (as amended by the Fire Safety and Safety of Places of Sport Act 1987).

Other alternative provisions cover premises such as sports grounds, underground railway stations, construction sites, docks, mines, offshore installations, means of transport, remote agricultural or forestry workplaces, or other special cases.

The local fire authority enforces fire safety legislation within its area. The authority may be asked whether or not a fire certificate is required for particular premises and will be pleased to offer guidance on which provisions cover the premises.

Integration of fire precautions and safety management rules

For premises covered by the 1997 regulations, the fire authority also enforces the Management of Health and Safety at Work Regulations 1999 as regards fire safety, in particular the rules on:

➤ appointment of competent person(s) by the employer to assist in complying with fire safety legislation
➤ assessment of fire risks to determine the measures needed both to prevent fire and to protect people in the event of fire
➤ fire safety arrangements, such as for managing the specific requirements of the Fire Precautions (Workplace) Regulations and for managing all other fire safety measures including fire prevention, practising fire drills, employee fire safety training, fire investigations, records and whatever else is determined from risk assessments as necessary in helping to control fire risks so far as is reasonably practicable
➤ procedures to be followed in the event of fire
➤ provision of information to employees and visiting workers as regards fire emergency procedures and the identity of fire marshals
➤ where employers share a workplace, cooperation and coordination over fire precautions.

Employee consultation

The 1997 Regulations have added fire safety as a further specific matter on which employees must be consulted, either directly or through their representatives. However, the regulations left fire safety consultation as a matter to be enforced by the Health and Safety at Work authorities and not by the fire authorities.

Duty holders

Employers and others, such as landlords, with any control over work premises all have duties to ensure compliance with the fire safety legislation applying at the workplace. The obligations of a particular duty holder relate to matters within their control. It is therefore advisable that arrangements between duty holders, such as contracts or tenancy agreements, should clarify the division of responsibilities to avoid unnecessary prosecutions by the fire authority where one duty holder simply assumed the other would satisfy a particular requirement.

In addition to fire safety legislation, health and safety at work legislation also covers the elimination or minimization of fire risks. As well as the particular and main general duties under the HSW Act, fire risks are also covered by specific rules, such as for work equipment, certain types of substance, electricity and other hazards. Thus, environmental health officers or HSE inspectors may enforce health and safety standards for the assessment and removal or control of fire risks, where it is necessary, for the protection of workers and others so far as is reasonably practicable.

Employees whose conduct leads to a breach of Health and Safety at Work legislation as regards fire risks can be prosecuted alongside or instead of a duty holder.

Fire Authorities officers have similar powers to those under the HSW Act and can issue improvement and prohibition notices. Cases can be tried on summary conviction in a magistrates court (fine not exceeding the statutory maximum) or on indictment (for which there is a fine or up to two years imprisonment or both).

11.3 Fire certificate

Figure 11.1 Fire certificate.

11.3.1 Application for fire certificate

A fire certificate must be applied for in the case of:

> offices, shops, factories or railway premises where more than 20 people are at work
> offices, shops, factories or railway premises where more than 10 are at work other than on the ground floor
> buildings with two or more individual offices, shops, factories or railway premises when the aggregate of people at work exceeds the above totals
> factories where, irrespective of the number of people at work, explosive or highly flammable materials are stored or used unless the fire authority has already confirmed there is no serious risk to employees

> hotels or boarding houses with sleeping accommodation for more than 6 guests or staff or, irrespective of numbers, with any sleeping accommodation above the first floor or below the ground floor.

In other cases, or in cases of doubt, it is advisable to check with the local fire authority whether a fire certificate is required since it is an offence to use premises for which one is required without applying for a certificate.

For premises already covered by a fire certificate, it is advisable to check with the local fire authority whether the certificate needs to be updated or has become unnecessary. This is because the premises may:

> have a certificate issued before 1977 (or before 1972 in the case of hotels) which is in need of revision
> no longer require a certificate (for example due to exemption by the fire authority) and are covered by the Fire Precautions (Workplace) Regulations 1997.

11.3.2 Duties while waiting for issue of certificate

Where a fire certificate is required, there are duties that apply during the interim period before the fire certificate is issued setting out the fire authority's precise requirements for the particular workplace. The interim duties are to maintain existing fire fighting equipment and means of escape; and provide fire safety training for personnel.

11.3.3 Contents of fire certificate

A fire certificate sets out the fire authority's precise requirements for the particular workplace. These requirements are minimum standards and may need to be exceeded to comply with health and safety duties. The employer's risk assessment as regards fire should therefore consider the adequacy of fire certificate requirements for eliminating or controlling fire risks.

A fire certificate specifies:

> the particular use or uses of premises which it covers
> the means of escape in the event of fire (shown in a plan)
> how to ensure that the means of escape can safely and effectively be used (such as by means of direction signs, emergency lighting, fire doors, smoke doors, and the like)
> fire fighting means for use by people at the premises
> how fire warnings must be given
> particulars as to any explosives or highly flammable materials stored or used on the premises.

In addition, a fire certificate may require:

> maintenance of the means of escape and their freedom from obstruction
> maintenance of other fire precautions set out in the certificate
> employee training on what to do in the event of fire and keeping of suitable records of such training
> limitation of number of people who may at any time be on the premises
> any other fire precautions.

In addition to complying with the requirements of the fire certificate, the employer must comply with the Management of Health and Safety Regulations as regards assessment of fire risks, fire arrangements and the like.

11.4 Basic principles of fire

11.4.1 Fire triangle

Fire cannot take place unless three things are present. These are shown in Figure 11.2.

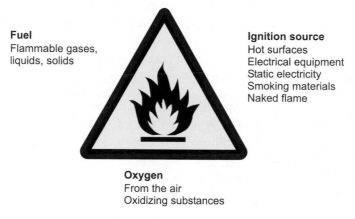

Fuel
Flammable gases, liquids, solids

Ignition source
Hot surfaces
Electrical equipment
Static electricity
Smoking materials
Naked flame

Oxygen
From the air
Oxidizing substances

Figure 11.2 Fire triangle.

The absence of any one of these elements will prevent a fire starting. Prevention depends on avoiding these three coming together. Fire extinguishing depends on removing one of the elements from an existing fire, and is particularly difficult if an oxidizing substance is present.

Once a fire starts it can spread very quickly from fuel to fuel as the heat increases.

11.4.2 Sources of ignition

Workplaces have numerous sources of ignition, some of which are obvious but others may be hidden inside machinery. Most of the sources may cause an accidental fire from sources inside but, in the case of arson (about 13% of industrial fires), the source of ignition may be brought from outside the workplace and will be deliberately used. The following are potential sources of ignition in the typical workplace:

➤ **naked flames** – from smoking materials, cooking appliances, heating appliances and process equipment
➤ **external sparks** – from grinding metals, welding, impact tools, electrical switch gear
➤ **internal sparking** – from electrical equipment (faulty and normal), machinery, lighting
➤ **hot surfaces** – from lighting, cooking, heating appliances, process equipment, poorly ventilated equipment, faulty and/or badly lubricated equipment, hot bearings and drive belts

> **static electricity** – causing significant high voltage sparks from the separation of materials such as unwinding plastic, pouring highly flammable liquids, walking across insulated floors, or removing synthetic overalls.

11.4.3 Sources of fuel

If it will burn it can be fuel for a fire. The things which will burn easily are the most likely to be the initial fuel, which then burns quickly and spreads the fire to other fuels. The most common things that will burn in a typical workplace are:

Solids – these include, wood, paper, cardboard, wrapping materials, plastics, rubber, foam (e.g. polystyrene tiles and furniture upholstery), textiles (e.g. furnishings and clothing), wall paper, hardboard and chipboard used as building materials, waste materials (e.g. wood shavings, dust, paper, etc.), hair.

Liquids – these include, paint, varnish, thinners, adhesives, petrol, white spirit, methylated spirits, paraffin, toluene, acetone and other chemicals. Most flammable liquids give off vapours which are heavier than air so they will fall to the lowest levels. A flash flame or an explosion can occur if the vapour catches fire in the correct concentrations of vapour and air.

Gases – flammable gases include LPG (liquefied petroleum gas in cylinders, usually butane or propane), acetylene (used for welding) and hydrogen. An explosion can occur if the air/gas mixture is within the explosive range.

11.4.4 Oxygen

Oxygen is of course provided by the air all around but this can be enhanced by wind, or by natural or powered ventilation systems which will provide additional oxygen to continue burning.

Cylinders providing oxygen for medical purposes or welding can also provide an additional very rich source of oxygen.

In addition some chemicals such as nitrates, chlorates, chromates and peroxides can release oxygen as they burn and therefore need no external source of air.

11.5 Methods of extinction

There are four main methods of extinguishing fires, which are explained as follows:

> **cooling** – reducing the ignition temperature by taking the heat out of the fire – using water to limit or reduce the temperature

➤ **smothering** – limiting the oxygen available by smothering and preventing the mixture of oxygen and flammable vapour – by the use of foam or a fire blanket

➤ **starving** – limiting the fuel supply – by removing the source of fuel by switching off electrical power, isolating the flow of flammable liquids or removing wood and textiles etc.

➤ **chemical reaction** – by interrupting the chain of combustion and combining the hydrogen atoms with chlorine atoms in the hydrocarbon chain for example with Halon extinguishers. (Halons have generally been withdrawn because of their detrimental effect on the environment, as ozone depleting agents).

11.6 Classification of fire

Fires are classified in accordance with British Standard *EN 2: Classification of Fires.* There are four main categories based on fuel and the means of extinguishing as follows:

Class A – fires which involve solid materials such as wood, paper, cardboard, textiles, furniture and plastics where there are normally glowing embers during combustion. Such fires are extinguished by cooling which is achieved using water.

Class B – fires which involve liquids or liquefied solids such as paints, oils or fats. These can be further sub-divided into:

➤ **Class B1** – fires which involve liquids that are soluble in water such as methanol. They can be extinguished by carbon dioxide, dry powder, water spray, light water and vaporizing liquids.

➤ **Class B2** – fires which involve liquids not soluble in water, such as petrol and oil. They can be extinguished by using foam, carbon dioxide, dry powder, light water and vaporizing liquid.

Class C – fires which involve gases such as natural gas or liquefied gases such as butane or propane. They can be extinguished using foam or dry powder in conjunction with water to cool any containers involved or nearby.

Class D – fires which involve metals such as aluminium or magnesium. Special dry powder extinguishers are required to extinguish these fires, which may contain powdered graphite or talc.

In addition there are:

Class F – fires which involve high temperature cooking oils or fats in large catering establishments or restaurants. This is not yet a recognized class within the British Standard but may be added when it is revised.

Electrical fires – fires involving electrical equipment or circuitry do not constitute a fire class on their own, as electricity is a source of ignition that will feed a fire until switched off or isolated. But there are some pieces of equipment that can store, within capacitors, lethal voltages even when isolated. Extinguishers specifically designed for electrical use like carbon dioxide or dry powder should always be used for this type of fire hazard.

Fire extinguishers are usually designed to tackle one or more class of fire. This is discussed later.

11.7 Principles of heat transmission and fire spread

Fire transmits heat in several ways, which need to be understood in order to prevent, plan escape from, and fight fires. Heat can be transmitted by convection, conduction, radiation and direct burning (Figure 11.3).

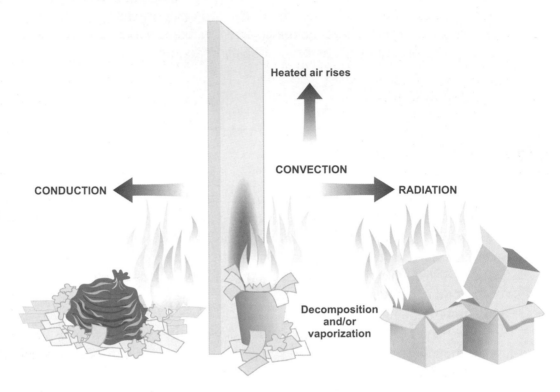

Figure 11.3   Principles of heat transmission.

11.7.1 Convection

Hot air becomes less dense and rises drawing in cold new air to fuel the fire with more oxygen. The heat is transmitted upwards at sufficient intensity to ignite combustible materials in the path of the very hot products of combustion and flames. This is particularly important inside buildings or other structures where the shape may effectively form a chimney for the fire.

11.7.2 Conduction

This is the transmission of heat through a material with sufficient intensity to melt or destroy the material and ignite combustible materials which come into contact or close to a hot section. Metals like copper, steel and aluminium are very effective or good conductors of heat. Other materials like concrete, brickwork and insulation materials are very ineffective or poor conductors of heat.

Poor conductors or good insulators are used in fire protection arrangements. When a poor conductor is also incombustible it is ideal for fire protection. Care is necessary to ensure that there are no other hazards like a health problem with such materials.

Asbestos is a very poor conductor of heat and is incombustible. Unfortunately, it has, of course, very severe health problems which now far outweigh its value as a fire protection material and it is banned in the UK, although still found in many buildings where it was used extensively for fire protection.

11.7.3 Radiation

Often in a fire the direct transmission of heat through the emission of heat waves from a surface can be so intense that adjacent materials are heated sufficiently to ignite. A metal surface glowing red-hot would be typical of a severe radiation hazard in a fire.

11.7.4 Direct burning

This is the effect of combustible materials catching fire through direct contact with flames which causes fire to spread, in the same way that lighting an open fire, with a range of readily combustible fuels is spread within a grate.

11.7.5 Smoke spread in buildings

Where fire is not contained and people can move away to a safe location there is little immediate risk to those people. However, where fire is confined inside buildings the fire behaves differently.

The smoke rising from the fire gets trapped inside the space by the immediate ceiling then spreads horizontally across the space deepening all the time until the entire space is filled. The smoke will also pass through any holes or gaps in the walls, ceiling or floor and get into other parts of the building. It moves rapidly up staircases or lift wells and into any areas that are left open, or rooms which have open doors connecting to the staircase corridors. The heat from the building gets trapped inside, raising the temperature very rapidly. The toxic smoke and gases are an added danger to people inside the building who must be able to escape quickly to a safe location.

Figure 11.4   Fire spread in buildings.

11.8 Common causes of fire and consequences

11.8.1 Causes

The Home Office statistics show that the causes of fires in buildings, excluding dwellings, in 1999 was as shown in Figure 11.5. The total was 43 600, a rise of 4% on the previous year but below the ten-year peak of 47 900 in 1995.

The sources of ignition are shown in Figure 11.6. Out of the 25 000 accidental fires in 1999, it shows that cooking appliances and electrical equipment account for 25% of the total.

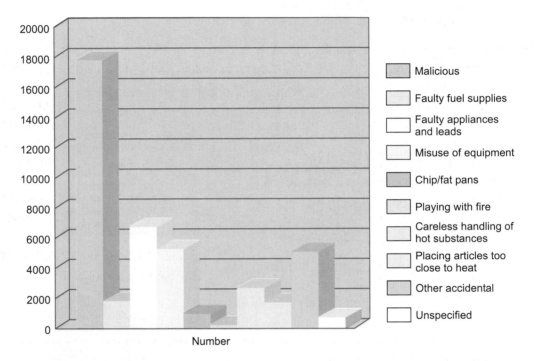

Figure 11.5   Causes of fire, 1999.

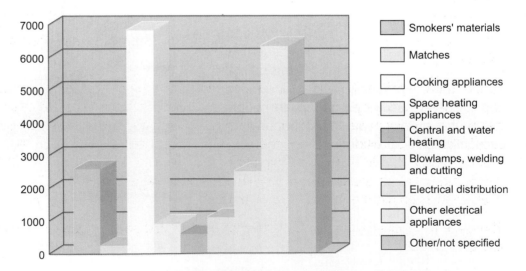

Figure 11.6   Accidental fires – sources of ignition.

11.8.2 Consequences

The main consequences of fire are:

> death – although this is a very real risk, relatively few people die in building fires that are not dwellings. In 1999, 38 (6%) people died out of a total of 663 in all fires.

The main causes of all deaths were:

> ➤ Overcome by gas or smoke   46%
> ➤ Burns   27%
> ➤ Burns and overcome by gas or smoke   20%
> ➤ Other   7%

Clearly gas and smoke are the main risks

> personal injury – some 1900 people were injured (11%) of total injuries in all fires
> building damage – can be very significant, particularly if the building materials have poor resistance to fire and there is little or no built in fire protection
> flora and fauna damage – can be significant, particularly in a hot draught or forest fire
> loss of business and jobs – it is estimated that about 40% of businesses do not start up again after a significant fire. Many are under- or not insured and small companies often cannot afford the time and expense of setting up again when they probably still have the old debts to service
> transport disruption – rail routes, roads and even airports are sometimes closed because of a serious fire. The worst case was of course 11 September 2001 when airports around the world were disrupted
> environmental damage from the fire and/or fighting the fire – fire-fighting water, the products of combustion and exploding building materials, such as asbestos cement roofs, can contaminate significant areas around the fire site.

11.9 Fire risk assessment

11.9.1 General

What fire precautions are needed depends on the risks. Managers need to take the following action:

> assess the fire risks in the workplace (either as part of the general review of health and safety risks which they already carry out or, if they wish, as a separate exercise;
> check that a fire can be detected in a reasonable time and that people can be warned;
> check that people who may be in the building can get out safely;
> provide reasonable fire-fighting equipment;
> check that those in the building know what to do if there is a fire;
> check and maintain the fire safety equipment.

There are five main hazards produced by fire that should be considered when assessing the level of risk:

> oxygen depletion

➤ flames and heat
➤ smoke
➤ gaseous combustion products
➤ structural failure of buildings.

11.9.2 Will anyone be hurt if there is a fire?

There may be places where people are at more serious risk than others. People may come into the workplace from outside, such as visitors, the public or other workers. The assessor must decide whether the current arrangements are satisfactory or if changes are needed.

If more than four people are employed, a formal record of the significant findings and any measures proposed to deal with them must be recorded. (For more detail on general risk assessment requirements see Chapter 5.)

If a fire occurs in the workplace, there is a risk that people will be trapped by the fire or injured as they attempt to escape. The purpose of the risk assessment is to identify where fires may start in the workplace and anyone who may be put at risk from that fire.

Fires occur when combustible materials come into contact with an ignition source. Fires can be started accidentally, through carelessness or by arsonists. When carrying out a fire risk assessment the following is required:

➤ any sources of ignition that may cause a fire should be identified. Where possible, take steps to reduce the risk of fire occurring;
➤ combustible materials in the workplace need to be identified. Steps should be taken to store them away from sources of ignition;
➤ identify those people who are at significant risk from fire and take steps to reduce that risk;
➤ structural features that could promote the spread of fire should be identified. Where possible, steps should be taken to reduce the potential for rapid fire growth;
➤ during periods of maintenance or refurbishment it will be necessary to take steps to monitor the introduction of sources of heat or combustible materials.

11.9.3 Sources of ignition

Possible sources of ignition may be heaters, boilers, engines, smoking materials or heat from processes or electrical apparatus, whether in normal use or through carelessness or accidental failure. The potential for an arson attack must be considered.

Where possible, sources of ignition should be removed from the workplace or replaced with safer forms. Where this cannot be done the ignition source should be kept well away from combustible materials or made the subject of management controls.

Particular care should be taken in areas where portable heaters are used or where smoking is permitted. Where heat is used as part of a process, it should be used carefully to reduce the chance of a fire as much as possible. Good security both inside and outside the workplace will help to combat the risk of arson.

11.9.4 Combustible materials

Most workplaces contain combustible materials. Usually, the presence of normal stock in trade should not cause concern, provided the materials are used safely and stored away from sources of ignition. Good standards of housekeeping are essential to minimize the risk of a fire starting or spreading quickly.

The amount of combustible material in a workplace should be kept as low as is reasonably practicable. Materials should not be stored in gangways, corridors or stairways or where they may obstruct exit doors and routes. Fires often start and are assisted to spread by combustible waste in the workplace. Such waste should be collected frequently and removed from the workplace, particularly where processes create large quantities of it.

Some combustible materials, such as flammable liquids, gases or plastic foams, ignite more readily than others and quickly produce large quantities of heat and/or dense toxic smoke. Ideally, such materials should be stored away from the workplace or in fire-resisting stores. The quantity of these materials kept or used in the workplace should be as small as possible, normally no more than one-half day's supply.

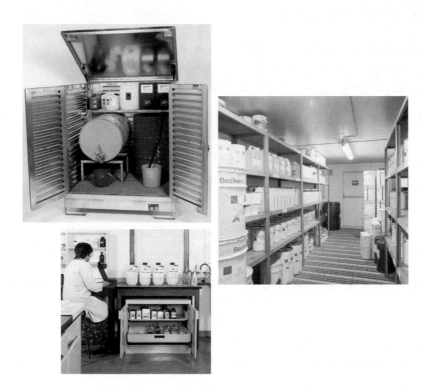

Figure 11.7 Various storage arrangements for highly flammable liquids. (Courtesy of Armaguard.)

11.9.5 Highly flammable liquids and LPG

Highly flammable liquids should be kept in a safe place in a separate building or the open air. If highly flammable liquids have to be kept inside the workplace no more than 50 litres should be allowed and this should be held in a special metal cupboard or container. Any larger amounts should be kept in special fire-resisting store, which should be:

231

➤ properly ventilated
➤ provided with spillage retaining arrangements such as sills
➤ free of sources of ignition, such as unprotected electrical equipment, sources of static electrical sparks, naked flames or smoking materials
➤ arranged so that incompatible chemicals do not become mixed together either in normal use or in a fire situation
➤ of fire-resisting construction
➤ used for empty as well as full containers – all containers must be kept closed
➤ kept clear of combustible materials such as cardboard or foam plastic packaging materials.

The Petroleum Consolidation Act & Regulations of 1928 no longer apply to workplaces except petrol filling stations.

The Dangerous Substances and Explosive Atmospheres Regulations, DSEAR 2002 (see Chapter 17, Section 17.30) replace the Highly Flammable Liquids and Liquified Petroleum Gases Regulations 1972. The requirements are good practice and should be adopted in all workplaces as appropriate. They include:

➤ reducing the amount kept at the workplace to a minimum
➤ choosing a safe location with good ventilation either through natural or mechanical air movement for the dispensing and use of materials
➤ using safety containers with self-closing lids or caps
➤ making sure that spillages are contained in a tray with supplies of absorbent material to hand
➤ controlling sources of ignition, such as naked flames and sparks from electrical equipment
➤ making sure that no-smoking rules are enforced
➤ storing contaminated rags and the like in a metal bin with a lid
➤ disposing of waste safely.

LPG and other highly flammable gas cylinders also need to be stored and used safely. The following guidance should be adopted:

➤ both full and empty cylinders should be stored outside. They should be kept in a separate secure compound at ground level with sufficient ventilation. Open mesh is preferable;
➤ valves should be uppermost during storage to retain them in the vapour phase of the LPG;
➤ cylinders must be protected from mechanical damage. Unstable cylinders should be together, for example and make sure cylinders are protected from the heat of the summer sun;
➤ the correct fittings must be used. These include hose, couplers, clamps and regulators;
➤ gas valves must be turned off after use at the end of the shift;
➤ take precautions against welding flame 'flash back' into the hoses or cylinders. People need training in the proper lighting up and safe systems of work procedures; non-return valves and flame arrestors also need to be fitted;
➤ change cylinders in a well-ventilated area remote from any sources of ignition;
➤ test joints for gas leaks using soapy/detergent water – never use a flame;
➤ flammable material must be removed or protected before welding or similar work;

➤ cylinders should be positioned outside buildings with gas piped through in fixed metal piping;

➤ make sure that both high and low ventilation is maintained where LPG appliances are being used;

➤ flame failure devices are necessary to shut off the gas supply in the event of flame failure.

11.9.6 People at risk

Because fire is a dynamic event, which, if unchecked, will spread throughout the workplace, all people present will eventually be at risk if fire occurs. Where people are at risk, adequate means of escape from fire should be provided together with arrangements for detecting and giving warning of fire. Fire-fighting equipment suitable for the hazards in the workplace should be provided.

Some people may be at significant risk because they work in areas where fire is more likely or where rapid fire growth can be anticipated. Where possible, the hazards creating the high level of risk should be reduced. Specific steps should be taken to ensure that people affected are made aware of the danger and the action they should take to ensure their safety and the safety of others.

11.9.7 Structural features

The workplace may contain features that could promote the rapid spread of fire, heat or smoke and affect escape routes. These features may include ducts or flues, openings in floors or walls, or combustible wall or ceiling linings. Where people are put at risk from these features, appropriate steps should be taken to reduce the potential for rapid fire spread by, for example, non-combustible automatic dampers fitted in ducts or to provide an early warning of fire so that people can leave the workplace before their escape routes become unusable.

Combustible wall or ceiling linings should not be used on escape routes and large areas should be removed wherever they are found. Other holes in fire-resisting floors, walls or ceilings should be filled in with fire-resisting material to prevent the passage of smoke, heat and flames.

11.9.8 Maintenance and refurbishment

Sources of heat or combustible materials may be introduced into the workplace during periods of maintenance or refurbishment. Where the work involves the introduction of heat, such as welding, this should be carefully controlled by a safe system of work, for example, a Fire Permit (see Chapter 6 for details). All materials brought into the workplace in connection with the work being carried out should be stored away from sources of heat and not obstruct exit routes.

11.9.9 Fire plans

Fire plans should be produced and attached to the risk assessment. A copy should be posted in the workplace. A single line plan of the area or floor should be produced or an existing plan should be used which needs to show:

> escape routes, numbers of exits, number of stairs, fire-resisting doors, fire-resisting walls and partitions, places of safety, and the like
> fire safety signs and notices including pictorial fire exit signs and fire action notices
> the location of fire warning call points and sounders or rotary gongs
> the location of emergency lights
> the location and type of fire-fighting equipment.

11.10 Fire detection and warning

In the event of fire it is vital that everyone in the workplace is alerted as soon as possible. The earlier the fire is discovered, the more likely it is that people will be able to escape before the fire takes hold and before it blocks escape routes or makes escape difficult.

Every workplace should have detection and warning arrangements. Usually the people who work there will detect the fire and in many workplaces nothing further will be needed.

It is important to consider how long a fire is likely to burn before it is discovered. Fires are likely to be discovered quickly if they occur in places that are frequently visited by employees, or in occupied areas of a building. For example, employees are likely to smell burning or see smoke if a fire breaks out in an office.

Where there is concern that fire may break out in an unoccupied part of the premises, for example, in a basement, some form of automatic fire detection should be fitted. Commercially available heat or smoke detection systems can be used. In small premises a series of interlinked domestic smoke alarms that can be heard by everyone present will be sufficient. In most cases, staff can be relied upon to detect a fire. In small workplaces where occupancy is low, a shouted warning should be all that is needed, so long as the warning can be heard and understood everywhere on the premises.

If the size or occupancy of a workplace means that a shouted warning is insufficient, hand-operated devices such as bells, gongs or sirens can be used. They should be installed on exit routes and should be clearly audible throughout the workplace. In all other places an electrically operated fire alarm system should

Figure 11.8 Typical fire point in offices with extinguishers, fire notice and alarm break-glass call point. (Courtesy of NEBOSH.)

234

be fitted. This will have call points adjacent to exit doors and enough bells or sounders to be clearly audible throughout the premises.

If it is thought that there might be some delay in fire being detected, automatic fire detection should be considered, linked into an electrical fire alarm system. Where a workplace provides sleeping accommodation or where fires may develop undetected, automatic detection must be provided. If a workplace provides sleeping accommodation for fewer than six people, interlinked domestic smoke alarms (wired to the mains electricity supply) can be used provided that they are audible throughout the workplace while people are present.

11.11 Means of escape in case of fire

11.11.1 General

It is essential to ensure that people can escape quickly from a workplace if there is a fire. Normally the entrances and exits to the workplace will provide escape routes, particularly if staff have been trained in what to do in case of fire and if it is certain that an early warning will be given.

In modern buildings which have had Building Regulation approval, and where there have not been significant changes to the building or where the workplace has recently been inspected by the fire authorities and found to be satisfactory, it is likely that the means of escape will be adequate. It may occasionally be necessary to improve the fire protection on existing escapes routes, or to provide additional exits. In making a decision about the adequacy of means of escape, the following points should be considered:

➤ people need to be able to turn away from a fire as they escape or be able to pass a fire when it is very small;
➤ if a single-direction escape route is in a corridor, the corridor may need to be protected from fire by fire-resisting partitions and self-closing fire doors;
➤ stair openings can act as natural chimneys in fires. This makes escape from the upper parts of some workplaces difficult. Most stairways, therefore, need to be separated from the workplace by fire-resisting partitions and self-closing fire doors. Where stairways serve no more than two open areas, in shops for example, which people may need to use as escape routes, there may be no need to use this type of protection.

11.11.2 Doors

Some doors may need to open in the direction of travel, such as:

➤ doors from a high-risk area, such as a paint spraying room or large kitchen
➤ doors that may be used by more than 50 persons
➤ doors at the foot of stairways where there may be a danger of people being crushed
➤ some sliding doors may be suitable for escape purposes provided that they do not put people using them at additional risk, slide easily and are marked with the direction of opening
➤ doors which only revolve and do not have hinged segments are not suitable as escape doors.

11.11.3 Escape routes

Escape routes should meet the following criteria:

➤ where two or more escape routes are needed they should lead in different directions to places of safety;

➤ escape routes need to be short and to lead people directly to a place of safety, such as the open air or an area of the workplace where there is no immediate danger;

➤ it should be possible for people to reach the open air without returning to the area of the fire. They should then be able to move well away from the building;

➤ ensure that the escape routes are wide enough for the volume of people using them. A 750 mm door will allow up to 40 people to escape in one minute, so most doors and corridors will be wide enough. If the routes are likely to be used by people in wheelchairs, the minimum width will need to be 800 mm.

While the workplace is in use, it must be possible to open all doors easily and immediately from the inside, without using a key or similar device, and doors must readily be opened in the direction of escape. Make sure that there are no trip hazards on the floor. Fire doors should be self-closing (fire doors to cupboards or lockers can be simply latched or locked).

Make sure that there are no obstructions on escape routes, especially on corridors and stairways where people who are escaping could dislodge stored items or be caused to trip. Any fire hazards must be removed from exit routes as a fire on an exit route could have very serious consequences.

Escape routes need regular checks to make sure that they are not obstructed and that exit doors are not locked. Self-closing fire-resisting doors should be checked to ensure doors close fully, including those fitted with automatic release mechanisms.

11.11.4 Lighting

Escape routes must be well lit. If the route has only artificial lighting or if it is used during the hours of darkness, alternative sources of lighting should be considered in case the power fails during a fire. Check the routes when it is dark as, for example, there may be street lighting outside that provides sufficient illumination. In small workplaces it may be enough to provide the staff with torches that they can use if the power fails. However, it may be necessary to provide battery operated emergency lights so that if the mains lighting fails the lights will operate automatically. Candles, matches and cigarette lighters are not adequate forms of emergency lighting.

11.11.5 Signs

Exit signs on doors or indicating exit routes should be provided where they will help people to find a safe escape route. Signs on exit routes should have directional arrows, 'up' for straight on and 'left, right or down' according to the route to be taken. Advice on the use of all signs including exit signs can be found in an HSE publication, *Safety Signs and Signals*, available from all good bookshops.

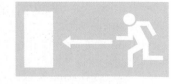

11.11.6 Escape times

Everyone in the building should be able to get to the nearest place of safety in between 2 and 3 minutes. This means that escape routes should be kept short. Where there is only one means of escape, or where the risk of fire is high, people should be able to reach a place of safety, or a place where there is more than one route available, in one minute.

The way to check this is to pace out the routes, walking slowly and noting the time. Start from where people work and walk to the nearest place of safety. Remember that the more people there are using the route, the longer they will take. People take longer to negotiate stairs and they are also likely to take longer if they have a disability.

Where fire drills are held, check how long it takes to evacuate each floor in the workplace. This can be used as a basis for assessment. If escape times are too long, it may be worth re-arranging the workplace so that people are closer to the nearest place of safety, rather than undertake expensive alterations to provide additional escape routes.

Reaction time needs to be considered. This is the amount of time people will need for preparation before they escape. It may involve, for example, closing down machinery, issues of security or helping visitors or members of the public out of the premises. Reaction time needs to be as short as possible to reduce risk to staff. Assessment of escape routes should include this. If reaction times are too long, additional routes may need to be provided. It is important that people know what to do in case of fire as this can speed up the time needed to evacuate the premises.

11.12 Principles of fire protection in buildings

11.12.1 General

The design of all new buildings and the design of extensions or modifications to existing buildings must be approved by the local planning authority.

Design data for new and modified buildings must be retained throughout the life of the structure.

Building Legislation Standards are concerned mainly with safety to life and therefore consider the early stage of fire and how it affects the means of escape, but also aim to prevent eventual spread to other buildings.

Asset protection requires extra precautions that will have effect at both early and later stages of the fire growth by controlling fire spread through and between buildings and preventing structural collapse. However, this extra fire protection will also improve life safety not only for those escaping at the early stages of the fire but also for fire fighters who will subsequently enter.

If a building is carefully designed and suitable materials are used to build it and maintain it, then the risk of injury or damage from fire can be substantially reduced. Three objectives must be met:

➤ it must be possible for everyone to leave the building quickly and safely
➤ the building must remain standing for as long as possible
➤ the spread of fire and smoke must be reduced.

These objectives can be met through the selection of materials and design of buildings.

11.12.2 Fire loading

The fire load of a building is used to classify types of building use. It may be calculated simply by multiplying the weight of all combustible materials by their energy values and dividing by the floor area under consideration. The higher the fire load the more effort needs to be made to offset this by building to higher standards of fire resistance.

11.12.3 Surface spread of fire

Combustible materials, when present in a building as large continuous areas, such as for lining walls and ceilings, readily ignite and contribute to spread of fire over their surfaces. This can represent a risk to life in buildings, particularly where walls of fire escape routes and stairways are lined with materials of this nature.

Materials are tested by insurance bodies and fire research establishments. The purpose of the tests is to classify materials according to the tendency for flame to spread over their surfaces. As with all standardized test methods, care must be take when applying test results to real applications.

In the UK, a material is classified as having a surface in one of the following categories:

➤ Class 1 Surface of very low flame spread
➤ Class 2 Surface of low flame spread
➤ Class 3 Surface of medium flame spread
➤ Class 4 Surface of rapid flame spread.

The test shows how a material would behave in the initial stages of a fire.

As all materials tested are combustible, in a serious fire they would burn or be consumed. Therefore, there is an additional Class 0 of materials which are non-combustible throughout or, under specified conditions, non-combustible on one face and combustible on the other. The spread of flame rating of the combined Class 0 product must not be worse than Class 1.

Internal partitions of walls and ceilings should be Class 0 materials wherever possible and must not exceed Class 1.

11.12.4 Fire resistance of structural elements

If elements of construction such as walls, floors, beams, columns and doors are to provide effective barriers to fire spread and to contribute to the stability of a building, they should be of a required standard of fire resistance.

In the UK, tests for fire resistance are made on elements of structure, full size if possible, or on a representative portion having minimum dimensions of 3 m long for columns and beams and 1 m^2 for walls and floors. All elements are exposed to the same standard fire provided by furnaces in which the temperature increases with time at a set rate. The conditions of exposure are appropriate to the element tested. Freestanding columns are subjected to heat all round, and walls and floors are exposed

Figure 11.9 Steel structures will collapse in the heat of a fire.

to heat on one side only. Elements of structure are graded by the length of time they continue to meet three criteria:

➤ the element must not collapse
➤ the element must not develop cracks through which flames or hot gases can pass
➤ the element must have enough resistance to the passage of heat so the temperature of the unexposed face does not rise by more than a prescribed amount.

The term fire resistance has a precise meaning. It should not be applied to such properties of materials as resistance to ignition or resistance to flame propagation. For example, steel has a high resistance to ignition and flame propagation but will distort quickly in a fire and allow the structure to collapse – it therefore has poor 'Fire Resistance'. It must be insulated to provide good fire protection. This is normally done by encasing steel frames in concrete.

In the past, asbestos has been made into a paste and plastered onto steel frames, giving excellent fire protection, but it has caused major health problems and its use in new work is banned.

Building materials with high fire resistance are, for example, brick, stone, concrete, very heavy timbers (the outside chars and insulates the inside of the timber), and some specially made composite materials used for fire doors.

11.12.5 Insulating materials

Building materials used for thermal or sound insulation could contribute to the spread of fire. Only approved fire resisting materials should be used.

11.12.6 Fire compartmentation

A compartment is a part of a building that is separated from all other parts by walls and floors, and is designed to contain a fire for a specified time. The principal object is to limit the effect of both direct fire damage and consequential business interruption caused by not only fire spread but also smoke and water damage in the same floor and other storeys.

Buildings are classified in purpose groups, according to their size. To control the spread of fire, any building whose size exceeds that specified for its purpose group must be divided into compartments that do not exceed the prescribed limits of volume and floor area. Otherwise, they must be provided with special fire protection. In the UK, the normal limit for the size of a compartment is 7000 m³. Compartments must be separated by walls and floors of sufficient fire resistance. Any openings needed in these walls or floors must be protected by fire-resisting doors to ensure proper fire-tight separation.

Ventilation and heating ducts must be fitted with fire dampers where they pass through compartment walls and floors. Firebreak walls must extend completely across a building from outside wall to outside wall. They must be stable; they must be able to stand even when the part of the building on one side or the other is destroyed.

No portion of the wall should be supported on unprotected steelwork nor should it have the ends of unprotected steel members embedded in it. The wall must extend up to the underside of a non-combustible roof surface, and sometimes above it. Any openings must be protected to the required minimum grade of fire resistance. If an external wall joins a firebreak wall and has an opening near the join, the firebreak wall may need to extend beyond the external wall.

An important function of external walls is to contain a fire within a building, or to prevent fire spreading from outside.

The fire resistance of external walls should be related to the:

➤ purpose for which the building is used
➤ height, floor area and volume of the building
➤ distance of the building from relevant boundaries and other buildings
➤ extent of doors, windows and other openings in the wall.

A wall, which separates properties from each other, should have no doors or other openings in it.

11.13 Provision of fire-fighting equipment

If fire breaks out in the workplace and trained staff can safely extinguish it using suitable fire-fighting equipment, the risk to others will be removed. Therefore, all workplaces where people are at risk from fire should be provided with suitable fire-fighting equipment.

The most useful form of fire-fighting equipment for general fire risks is the water-type extinguisher or suitable alternative. One such extinguisher should be provided for around each 200 m² of floor space with a minimum of one per floor. If each floor has a hose reel, which is known to be in working order and of sufficient length for the floor it serves, there may be no need for water-type extinguishers to be provided.

Areas of special risks involving the use of oil, fats or electrical equipment may need carbon dioxide, dry powder or other types of extinguisher.

Fire extinguishers should be sited on exit routes, preferably near to exit doors or where they are provided for specific risks, near to the hazards they protect. Notices indicating the location of fire-fighting equipment should be displayed where the location of the equipment is not obvious or in areas of high fire risk where the notice will assist in reducing the risk to people in the workplace.

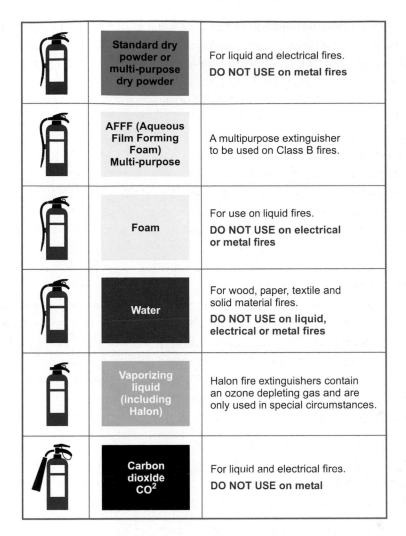

Figure 11.10 Types of fire extinguishers and labels.

11.14 Maintenance and testing of fire equipment

It is important that equipment is fit for its purpose and is properly maintained and tested. One way in which this can be achieved, is through companies that specialize in the test and maintenance of fire equipment.

All equipment provided to assist escape from the premises, such as fire detection and warning systems and emergency lighting, and all equipment provided to assist with fighting fire, should be regularly checked and maintained by a suitably competent person in accordance with the manufacturer's recommendations. Table 11.1 gives guidance on the frequency of test and maintenance and provides a simple guide to good practice.

11.15 Planning for an emergency and training staff

Each workplace should have an emergency plan. The plan should include the action to be taken by staff in the event of fire, the evacuation procedure and the arrangements for calling the fire brigade.

Table 11.1 Maintenance and testing of fire equipment

Equipment	Period	Action
Fire-detection and fire-warning systems including self-contained smoke alarms and manually operated devices	**Weekly**	• Check all systems for state of repair and operation • Repair or replace defective units • Test operation of systems, self-contained alarms and manually operated devices
	Annually	• Full check and test of system by competent service engineer • Clean self-contained smoke alarms and change batteries
Emergency lighting including self-contained units and torches	**Weekly**	• Operate torches and replace batteries as required • Repair or replace any defective unit
	Monthly	• Check all systems, units and torches for state of repair and apparent function
	Annually	• Full check and test of systems and units by competent service engineer • Replace batteries in torches
Fire-fighting equipment including hose-reels	**Weekly**	• Check all extinguishers including hose-reels for correct installation and apparent function
	Annually	• Full check and test by competent service engineer

For small workplaces this could take the form of a simple fire action notice posted in positions where staff can read it and become familiar with it.

High-fire-risk or larger workplaces will need more detailed plans, which take account of the findings of the risk assessment, for example, the staff significantly at risk and their location. For large workplaces, notices giving clear and concise instructions of the routine to be followed in case of fire should be prominently displayed. The notice should include the method of raising an alarm in the case of fire and the location of an assembly point to which staff escaping from the workplace should report.

More information is given in Chapter 6.

11.16 Fire procedures and people with a disability

People with a disability (including members of the public) need special consideration, when planning for emergencies. But the problems this raises are seldom great. Employers should:

➤ identify everyone who may need special help to get out;
➤ allocate responsibility to specific staff to help people with a disability in emergency situations;
➤ consider possible escape routes;
➤ enable the safe use of lifts;
➤ enable people with a disability to summon help in emergencies;
➤ train staff to be able to help their colleagues.

Advice on the needs of people with a disability, including sensory impairment, is available from the organizations which represent various groups. (Names and addresses can be found in the telephone directory.)

People with impaired vision must be encouraged to familiarize themselves with escape routes, particularly those not in regular use. A 'buddy' system would be helpful. But, to take account of absences, more than one employee working near anyone with impaired vision should be taught how to help them.

Staff with impaired hearing may not hear alarms in the same way as those with normal hearing but may still be able to recognize the sound. This may be tested during the weekly alarm audibility test. There are alternative means of signalling, such as lights or other visual signs, vibrating devices or specially selected sound signals. The Royal National Institute for Deaf People, (19–23 Featherstone St, London EC1), can advise. Ask the fire brigade before installing alternative signals.

Wheelchair users or others with impaired mobility may need help to negotiate stairs, etc. Anyone selected to provide this help should be trained in the correct methods. Advice on the lifting and carrying of people can be obtained from the Fire Service, Ambulance Service, British Red Cross Society, St John Ambulance Brigade or certain disability organizations.

Lifts should not be used as a means of escape in the event of a fire. If the power fails, the lift could stop between floors, trapping occupants in what may become a chimney of fire and smoke. But BS 5588: Part 8 provides advice on specially designed lifts for use by the disabled in the event of a fire.

Employees with a mental handicap may also require special provision. Management should ensure that the colleagues of any employee with a mental handicap know how to reassure them and lead them to safety.

11.17 Practice NEBOSH questions for Chapter 11

1. (a) Explain, using a suitable sketch, the significance of the 'fire triangle'.
   (b) List the types of ignition source that may cause a fire to occur, and give an example of each type. (March 2001)
2. Identify the four methods of heat transfer and explain how each can cause the spread of fire. (December 2001)
3. List eight ways of reducing the risk of a fire starting in a workplace. (June 2001)
4. Outline the measures that should be taken to minimize the risk of fire from electrical equipment. (December 2000)
5. Outline the main requirements for a safe means of escape from a building in the event of a fire. (December 2000)
6. (a) With reference to the fire triangle, outline two methods of extinguishing fires.

(b) State the ways in which persons could be harmed by a fire in work premises. (June 2000)

7. List eight features of a safe means of escape from a building in the event of fire. (March 2000)

8. Outline reasons for undertaking regular fire drills in the workplace. (June 1999)

9. (a) Outline the main factors to be considered in the siting of fire extinguishers.
   (b) Outline the inspection and maintenance requirements for fire extinguishers in the workplace. (June 1998)

10. (a) Identify four different types of ignition source that may cause a fire.
    (b) For each type of ignition source identified in (a), outline the precautions that could be taken to prevent a fire starting. (June 1997)

Chemical and biological health hazards and control

12.1 Introduction

Occupational health is as important as occupational safety but generally receives less attention from managers. Every year twice as many people suffer ill-health caused or exacerbated by the workplace than suffer workplace injury. Although these illnesses do not usually kill people, they can lead to many years of discomfort and pain. Such illnesses include respiratory disease, hearing problems, asthmatic conditions and back pain. Furthermore, it has been estimated in the UK that 30% of all cancers probably have an occupational link – that linkage is known for certain in 8% of cancer cases.

Work in the field of occupational health has been taking place for the last four centuries and possibly longer. The main reason for the relatively low profile for occupational health over the years has been the difficulty in linking the ill-health effect with the workplace cause. Many illnesses, such as asthma or back pain, can have a workplace cause but can also have other causes. Many of the advances in occupational health have been as a result of statistical and epidemiological studies (one well-known such study linked the incidence of lung cancer to cigarette smoking). While such studies are invaluable in the assessment of health risk, there is always an element of doubt when trying to link cause and effect. The measurement of gas and dust concentrations is also subject to doubt when a correlation is made between a measured sample and the workplace environment from which it was taken. Occupational health, unlike occupational safety, is generally more concerned with probabilities than certainties.

In this chapter, chemical and biological health hazards will be considered – other forms of health hazard will be covered in Chapter 13.

The chemical and biological health hazards described in this chapter are covered by the following health and safety regulations:

➤ Control of Substances Hazardous to Health Regulations 2002
➤ Control of Lead at Work Regulations 2002
➤ Control of Asbestos at Work Regulations 2002.

12.2 Forms of chemical agent

Chemicals can be transported by a variety of agents and in a variety of forms. They are normally defined in the following ways.

Dusts are solid particles slightly heavier than air but often suspended in it for a period of time. The size of the particles ranges from about 0.4 μm (fine) to 10 μm (coarse). Dusts are created either by mechanical processes (e.g. grinding or pulverizing) or construction processes (e.g. concrete laying, demolition or sanding), or by specific tasks (e.g. furnace ash removal). The fine dust is much more hazardous because it penetrates deep into the lungs and remains there – known as **respirable dust**. Examples of such fine dust are cement, granulated plastic materials and silica dust produced from stone or concrete dust. Repeated exposure may lead to permanent lung disease.

Gases are any substances at a temperature above their boiling point. Steam is the gaseous form of water. Common gases include carbon monoxide, carbon dioxide, nitrogen and oxygen. Gases are absorbed into the bloodstream where they may be beneficial (oxygen) or harmful (carbon monoxide).

Vapours are substances which are at or very close to their boiling temperatures. They are gaseous in form. Many solvents, such as cleaning fluids, fall into this category. The vapours, if inhaled, enter the bloodstream and some can cause short-term effects (dizziness) and long-term effects (brain damage).

Liquids are substances which normally exist at a temperature between freezing (solid) and boiling (vapours and gases). They are sometimes referred to as fluids in health and safety regulations.

Mists are similar to vapours in that they exist at or near their boiling temperature but are closer to the liquid phase. This means that there are suspended very small liquid droplets present in the vapour. A mist is produced during a spraying process (such as paint spraying). Many industrially produced mists can be very damaging if inhaled producing similar effects to vapours. It is possible for some mists to enter the body through the skin or by ingestion with food.

Fume is a collection of very small metallic particles (less than 1 μm) which have condensed from the gaseous state. They are most commonly generated by the welding process. The particles tend to be within the respirable range (approximately 0.4–1.0 μm) and can lead to long-term permanent lung damage. The exact nature of any harm depends on the metals used in the welding process and the duration of the exposure.

12.3 Forms of biological agent

As with chemicals, biological hazards may be transported by any of the following forms of agent.

Fungi are very small organisms, sometimes consisting of a single cell, and can appear as plants (e.g. mushrooms and yeast). Unlike other plants, they cannot produce their own food but either live on dead organic matter or on living animals or plants as parasites. Fungi reproduce by producing spores which can cause allergic reactions when inhaled. The infections produced in man by fungi may be mild, such as athlete's foot, or severe such as ringworm. Many fungal infections can be treated with antibiotics.

Moulds are a particular group of very small fungi which, under damp conditions, will grow on things such as walls, bread, cheese, leather and canvas. They can be beneficial (penicillin) or cause allergic reactions (asthma). Asthma attacks, athlete's foot and farmer's lung are all examples of fungal infections.

Bacteria are very small single–celled organisms which are much smaller than cells within the human body. They can live outside the body and be controlled and destroyed by antibiotic drugs. There is evidence that bacteria are developing which are becoming resistant to most antibiotics. This has been caused by the widespread misuse of anti-biotics. It is important to note that not all bacteria are harmful to humans. Bacteria aid the digestion of food and babies would not survive without their aid to break down the milk in their digestive systems. Legionella, tuberculosis and tetanus are all bacterial diseases.

Viruses are minute non–cellular organisms which can only reproduce within a host cell. They are very much smaller than bacteria and cannot be controlled by antibiotics. They appear in various shapes and are continually developing new strains. They are usually only defeated by the defence and healing mechanisms of the body. Drugs can be used to relieve the symptoms of a viral attack but cannot cure it. The common cold is a viral infection as are hepatitis, AIDS (HIV) and influenza.

12.4 Classification of hazardous substances and their associated health risks

A hazardous substance is one which can cause ill-health to people at work. Such substances may include those used directly in the work processes (glues and paints), those produced by work activities (welding fumes) or those which occur naturally (dust). Hazardous substances are classified according to the severity and type of hazard which they may present to people who may come into contact with them. The contact may occur while working or transporting the substances or might occur during a fire or accidental spillage. There are several classifications but here only the five most common will be described.

Irritant is a non-corrosive substance which can cause skin (dermatitic) or lung (bron-chial) inflammation after repeated contact. People who react in this way to a particular substance are **sensitized or allergic** to that substance. In most cases, it is likely that the concentration of the irritant may be more significant than the exposure time. Many household substances, such as wood preservatives, bleaches and glues are irritants. Many chemicals used as solvents are also irritants (white spirit, toluene and acetone). Formaldehyde and ozone are other examples of irritants.

Corrosive substances are ones which will attack, normally by burning, living tissue. Usually a strong acid or alkali, examples include sulphuric acid and caustic soda. Many tough cleaning substances, such as kitchen oven cleaners, are corrosives as are many dishwasher crystals.

Harmful is the most commonly used classification and describes a substance which, if swallowed, inhaled or penetrates the skin, **may** pose limited health risks. These risks can usually be minimized or removed by following the instruction provided with the substance (e.g. by using personal protective equipment). There are many household substances which fall into this category including bitumen-based paints and paint brush restorers. Many chemical cleansers, such as trichloroethylene, are categorized as harm-ful. It is very common for substances labelled harmful also to be categorized as irritant.

Toxic substances are ones which impede or prevent the function of one or more organs within the body, such as the kidneys, liver and heart. A toxic substance is, therefore, a poisonous one. Lead, mercury, pesticides and the gas carbon monoxide are toxic substances.

Carcinogenic substances are ones which are known or suspected of promoting abnormal development of body cells to become cancers. Asbestos, hard wood dust, creosote and some mineral oils are carcinogenic. It is very important that the health and safety rules accompanying the substance are strictly followed.

Each of the classifications may be identified by a symbol and a symbolic letter – these are shown in Figure 12.1.

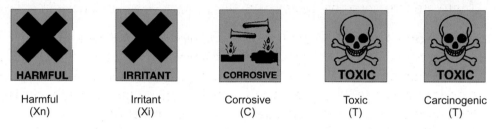

| Harmful (Xn) | Irritant (Xi) | Corrosive (C) | Toxic (T) | Carcinogenic (T) |

Figure 12.1 Classification symbols.

The effects on health of hazardous substances may be either acute or chronic.
Acute effects are of short duration and appear fairly rapidly, usually during or after a single or short-term exposure to a hazardous substance. Such effects may be severe and require hospital treatment but are usually reversible. Examples include asthma-type attacks, nausea and fainting.

Chronic effects develop over a period of time which may extend to many years. The word 'chronic' means 'with time' and should not be confused with 'severe' as its use in everyday speech often implies. Chronic health effects are produced from prolonged or repeated exposures to hazardous substances resulting in a gradual, latent and often irreversible illness, which may remain undiagnosed for many years. Many cancers and mental diseases fall into the chronic category. During the development stage of a chronic disease, the individual may experience no symptoms.

12.4.1 The role of COSHH

The Control of Substances Hazardous to Health Regulations 1988 (COSHH) were the most comprehensive and significant piece of Health and Safety legislation to be introduced since the Health and Safety at Work Act 1974. They were enlarged to cover biological agents in 1994 and further amended in 1997, 1999 and 2002. A detailed summary of these Regulations appears in Chapter 17. The Regulations impose duties on employers to protect employees and others who may be exposed to substances hazardous to their health and requires employers to control exposure to such substances.

The COSHH Regulations offer a framework for employers to build a management system to assess health risks and to implement and monitor effective controls. Adherence to these Regulations will provide the following benefits to the employer and employee:

> improved productivity due to lower levels of ill-health and more effective use of materials
> improved employee morale
> lower numbers of civil court claims
> better understanding of health and safety legal requirements.

Organizations which ignore COSHH requirements, will be liable for enforcement action, including prosecution, under the Regulations.

12.5 Routes of entry to the human body

There are three principal routes of entry of hazardous substances into the human body:

> **inhalation** – breathing in the substance with normal air intake. This is the main route of contaminants into the body. These contaminants may be chemical (e.g. solvents or welding fume) or biological (e.g. bacteria or fungi) and become airborne by a variety of modes, such as sweeping, spraying, grinding and bagging. They enter the lungs where they have access to the bloodstream and many other organs
> **absorption through the skin** – the substance comes into contact with the skin and enters either through the pores or a wound. Tetanus can enter in this way as can toluene, benzene and various phenols
> **ingestion** – through the mouth and swallowed into the stomach and the digestive system. This is not a significant route of entry to the body. The most common occurrences are due to airborne dust or poor personal hygiene (not washing hands before eating food).

Another very rare entry route is by **injection**. The abuse of compressed air lines by shooting high pressure air at the skin can lead to air bubbles entering the bloodstream. Accidents involving hypodermic syringes in a health or veterinary service setting are rare but illustrate this form of entry route.

The most effective control measures which can reduce the risk of infection from biological organisms are disinfection, proper disposal of clinical waste (including syringes), good personal hygiene and, where appropriate, personal protective equipment. Other measures include vermin control, water treatment and immunization.

There are five major functional systems within the human body – respiratory, nervous, cardiovascular (blood), urinary and the skin.

12.5.1 The respiratory system

This comprises the lungs and associated organs (e.g. the nose). Air is breathed in through the nose, passes through the trachea (windpipe) and the bronchi into the two lungs. Within the lungs, the air enters many smaller passageways (bronchioli) and thence to one of 300 000 terminal sacs called alveoli. The alveoli are approximately 0.1 mm across, although the entrance is much smaller. On arrival in the alveoli, there is a diffusion of oxygen into the bloodstream through blood capillaries and an effusion of carbon dioxide from the bloodstream. While soluble dust which enters the alveoli will be absorbed into the bloodstream, insoluble dust (respirable dust) will remain permanently, leading to possible chronic illness.

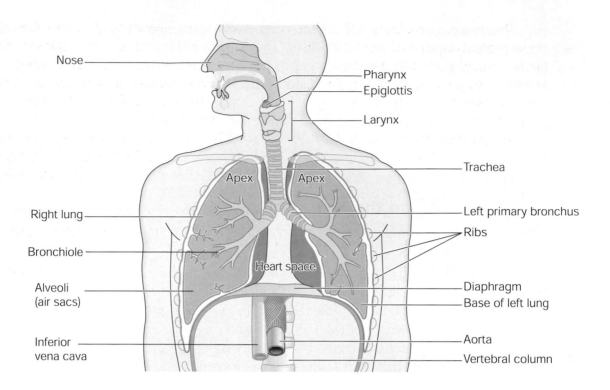

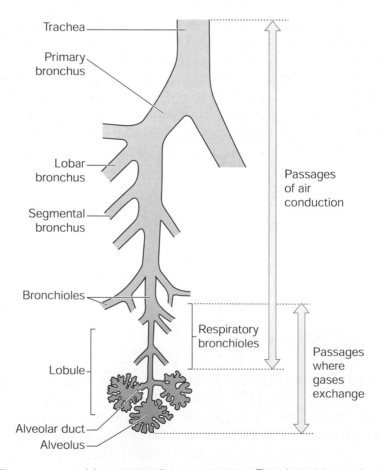

Figure 12.2 The upper and lower respiratory system. Reprinted from *Anatomy and Physiology in Health and Illness* Ninth edition, Waugh and Grant, pages 240 and 248, 2002 by permission of the publisher Churchill Livingstone.

The whole of the bronchial system is lined with hairs, known as cilia. The cilia offer some protection against insoluble dusts. These hairs will arrest all non-respirable dust (above 7 μm) and, with the aid of mucus, pass the dust from one hair to a higher one and thus bring the dust back to the throat. (This is known as the ciliary escalator). It has been shown that smoking damages this action. The nose will normally trap large particles (greater than 20 μm) before they enter the trachea.

Respirable dust tends to be long thin particles with sharp edges which puncture the alveoli walls. The puncture heals producing scar tissues which are less flexible than the original walls – this can lead to fibrosis. Such dusts include asbestos, coal, silica, some plastics and talc.

Acute effects on the respiratory system include bronchitis and asthma and chronic effects include fibrosis and cancer. Hardwood dust, for example, can produce asthma attacks and nasal cancer.

Finally, asphyxiation, due to a lack of oxygen, is a problem in confined spaces particularly when MIG (metal inert gas) welding is taking place.

12.5.2 The nervous system

The nervous system consists primarily of the brain, the spinal cord and nerves extending throughout the body. Any muscle movement or sensation (e.g. hot and cold) is controlled or sensed by the brain through small electrical impulses transmitted through the spinal cord and nervous system. The effectiveness of the nervous system can be reduced by **neurotoxins** and lead to changes in mental ability (loss of memory and anxiety), epilepsy and narcosis (dizziness and loss of consciousness). Organic solvents (trichloroethylene) and heavy metals (mercury) are well-known neurotoxins. The expression 'mad hatters' originated from the mental deterioration of top hat polishers in the 19th century who used mercury to produce a shiny finish on the top hats.

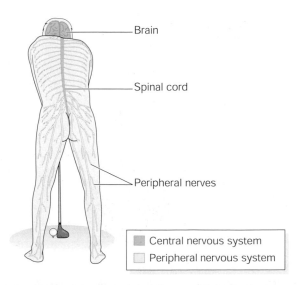

Figure 12.3   The nervous system. Reprinted from *Anatomy and Physiology in Health and Illness* Ninth edition, Waugh and Grant, page 9, 2002 by permission of the publisher Churchill Livingstone.

12.5.3 The cardiovascular system

The blood system uses the heart to pump blood around the body through arteries, veins and capillaries. Blood is produced in the bone marrow and consists of a plasma within which are red cells, white cells and platelets. The system has three basic objectives:

> to transport oxygen to vital organs, tissues and the brain and carbon dioxide back to the lungs (red cell function);
> to attack foreign organisms and build up a defence system (white cell function);
> to aid the healing of damaged tissue and prevent excessive bleeding by clotting (platelets).

There are several ways in which hazardous substances can interfere with the cardiovascular system. Benzene can affect the bone marrow by reducing the number of blood cells produced. Carbon monoxide prevents the red cells from absorbing sufficient oxygen and the effects depend on its concentration. Symptoms begin with headaches and end with unconsciousness and possibly death.

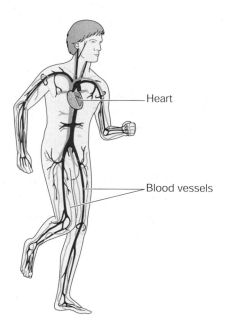

Heart

Blood vessels

Figure 12.4   The cardiovascular system. Reprinted from *Anatomy and Physiology in Health and Illness* Ninth edition, Waugh and Grant, page 8, 2002 by permission of the publisher Churchill Livingstone.

12.5.4 The urinary system

The urinary system extracts waste and other products from the blood. The two most important organs are the liver (normally considered part of digestive system) and the kidneys, both of which can be affected by hazardous substances within the bloodstream.

The liver removes toxins from the blood, maintains the levels of blood sugars and produces protein for the blood plasma. Hazardous substances can cause the liver to be too active or inactive (e.g. xylene), lead to liver enlargement (e.g. cirrhosis caused by alcohol) or liver cancer (e.g. vinyl chloride).

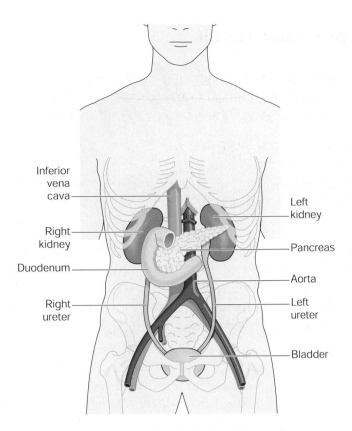

Figure 12.5 Parts of the urinary system. Reprinted from *Anatomy and Physiology in Health and Illness* Ninth edition, Waugh and Grant, page 340, 2002 by permission of the publisher Churchill Livingstone.

The kidneys filter waste products from the blood as urine, regulate blood pressure and liquid volume in the body and produce hormones for making red blood cells. Heavy metals (e.g. cadmium and lead) and organic solvents (e.g. glycol ethers used in screen printing) can restrict the operation of the kidneys possibly leading to failure.

12.5.5 The skin

The skin holds the body together and is the first line of defence against infection. It regulates body temperature, is a sensing mechanism, provides an emergency food store (in the form of fat) and helps to conserve water. There are two layers – an outer layer called the epidermis (0.2 mm) and an inner layer called the dermis (4 mm). The epidermis is a tough protective layer and the dermis contains the sweat glands, nerve endings and hairs.

The most common industrial disease of the skin is **dermatitis** (non-infective dermatitis). It begins with a mild irritation on the skin and develops into blisters which can peel and weep becoming septic. It can be caused by various chemicals, mineral oils and solvents. There are two types:

➢ **irritant contact dermatitis** occurs soon after contact with the substance and the condition reverses after contact ceases (detergents and weak acids)
➢ **allergic contact dermatitis** is caused by a sensitizer such as turpentine, epoxy resin, solder flux and formaldehyde.

253

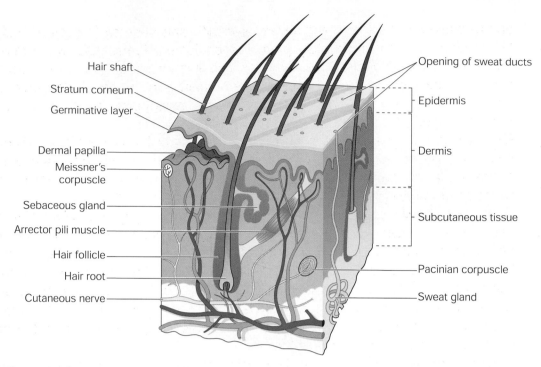

Figure 12.6 The skin – main structures in the dermis. Reprinted from *Anatomy and Physiology in Health and Illness* Ninth edition, Waugh and Grant, page 363, 2002 by permission of the publisher Churchill Livingstone.

For many years, dermatitis was seen as a 'nervous' disease which was psychological in nature. Nowadays, it is recognized as an industrial disease which can be controlled by good personal hygiene, personal protective equipment, use of barrier creams and health screening of employees.

The risks of dermatitis occurring increases with the presence of skin cuts or abrasions, which allow chemicals to be more easily absorbed, and the type, sensitivity and existing condition of the skin.

12.6 Health hazards of specific agents

The health hazards associated with hazardous substances can vary from very mild (momentary dizziness or a skin irritation) to very serious, such as a cancer.
Cancer is a serious body cell disorder in which the cells develop into tumours. There are two types of tumour – benign and malignant. Benign tumours do not spread but remain localized within the body and grow slowly. Malignant tumours are called cancers and often grow rapidly, spreading to other organs using the bloodstream and lymphatic glands. Survival rates have improved dramatically in recent years as detection methods have improved and the tumours are found in their early stages of development. A minority of cancers are believed to be occupational in origin.

The following common agents of health hazards will be described together with the circumstances in which they may be found:
Ammonia is a colourless gas with a distinctive odour which, even in small concentrations, causes the eyes to smart and run and a tightening of the chest. It is a corrosive substance which can burn the skin, burn and seriously damage the eye, cause soreness

and ulceration of the throat and severe bronchitis and oedema (excess of fluid) of the lungs. Good eye and respiratory protective equipment is essential when maintaining equipment containing ammonia. Any such equipment should be tested regularly for leaks and repaired promptly if required. Ammonia is also used in the production of fertilizers and synthetic fibres. Most work on ammonia plant should require a permit-to-work procedure.

Chlorine is a greenish, toxic gas with a pungent smell which is highly irritant to the respiratory system, producing severe bronchitis and oedema of the lungs and may also cause abdominal pain, nausea and vomiting. It is used as a disinfectant for drinking water and swimming pool water and in the manufacture of chemicals.

Organic solvents are used widely in industry as cleansing and degreasing agents. There are two main groups – the hydrocarbons (includes the aromatic and aliphatic hydrocarbons) and the non-hydrocarbons (such as toluene, white spirit, trichloroethylene and carbon tetrachloride). All organic solvents are heavier than air and most are sensitizers and irritants. Some are narcotics, while others can cause dermatitis and after long exposure periods liver and kidney failure. It is very important that the hazard data sheet accompanying the particular solvent is read and the recommended personal protective equipment is worn at all times.

Carbon dioxide is a colourless and odourless gas which is heavier than air. It represses the respiratory system, eventually causing death by asphyxiation. At low concentrations it will cause headaches and sweating followed by a loss of consciousness. The greatest hazard occurs in confined spaces, particularly where the gas is produced as a by-product.

Carbon monoxide is a colourless, tasteless and odourless gas which makes it impossible to detect without special measuring equipment. As explained earlier, carbon monoxide enters the blood (red cells) more readily than oxygen and restricts the supply of oxygen to vital organs. At low concentrations (less than 5%), headaches and breathlessness will occur, while at higher concentrations unconsciousness and death will result. The most common occurrence of carbon monoxide is as an exhaust gas either from a vehicle or a heating system. In either case, it results from inefficient combustion and, possibly, poor maintenance.

Isocyanates are volatile organic compounds widely used in industry for products such as printing inks, adhesives, two-pack paints (particularly in vehicle body shops) and in the manufacture of plastics (polyurethane products). They are irritants and sensitizers. Inflammation of the nasal passages, the throat and bronchitis are typical reactions to many isocyanates. When a person becomes sensitized to an isocyanate, very small amounts of the substance often provoke a serious reaction similar to an extreme asthma attack. Isocyanates also present a health hazard to fire fighters. They are subject to a maximum exposure limit (MEL) and respiratory protective equipment should normally be worn.

Asbestos appears in three main forms – crocidolite (blue), amosite (brown) and chrysotile (white). The blue and brown asbestos are considered to be the most dangerous and may be found in older buildings where they were used as heat insulators around boilers and hot water pipes and as fire protection of structure. White asbestos has been used in asbestos cement products and brake linings. It is difficult to identify an asbestos product by its colour alone – laboratory identification is usually required. Asbestos produces a fine fibrous dust of respirable dust size which can become lodged in the lungs. The fibres can be very sharp and hard causing damage to the lining of the lungs over a period of many years. This can lead to one of the following diseases:

> asbestosis or fibrosis (scarring) of the lungs
> lung cancer
> mesothelioma – cancer of the lining of the lung or, in rarer cases, the abdominal cavity.

If asbestos is discovered during the performance of a contract, work should cease immediately and the employer be informed. Typical sites of asbestos include ceiling tiles, asbestos cement roof and wall sheets, sprayed asbestos coatings on structural members, loft insulation and asbestos gaskets. Asbestos has its own set of Regulations (Control of Asbestos at Work Regulations 2002 updated 21 November 2002) and a summary of these is given in Chapter 17. These cover the need for a risk assessment, a method statement covering the removal and disposal, air monitoring procedures and the control measures (including personal protective equipment and training) to be used.

Lead is a heavy, soft and easily worked metal. It is used in many industries but is most commonly associated with plumbing and roofing work. Lead enters the body normally by inhalation but can also enter by ingestion and skin contact. The main targets for lead are the central nervous system (and the brain) and the blood (and blood production). The effects are normally chronic and develop as the quantity of leads builds up. Headaches and nausea are the early symptoms followed by anaemia, muscle weakening and (eventually) coma. Regular blood tests are a legal and sensible requirement as are good ventilation and the use of appropriate personal protective equipment. High personal hygiene standards and adequate welfare (washing) facilities are essential and must be used before smoking or food is consumed. The reduction in the use of leaded petrol was an acknowledgement of the health hazard represented by lead in the air. Lead is covered by its own set of regulations – the Control of Lead at Work Regulations 2002 (summarized in Chapter 17). These regulations require risk assessments to be undertaken and engineering controls to be in place. They also recognize that lead can be transferred to an unborn child through the placenta and, therefore, offer additional protection to women of reproductive capacity. Medical surveillance, in the form of a blood test, of all employees, who come into contact with lead operations, is required by the regulations. Such tests should take place at least once a year.

Silica is the main component of most rocks and is a crystalline substance made of silicon and oxygen. It occurs in quartz (found in granite), sand and flint. Harm is caused by the inhalation of silica dust which can lead to silicosis (acute and chronic), fibrosis and pneumoconiosis. The dust which causes the most harm is respirable dust which becomes trapped in the alveoli. This type of dust is sharp and very hard and, probably, causes wounding and scarring of lung tissue. As silicosis develops, breathing becomes more and more difficult and eventually as it reaches its advanced stage, lung and heart failure occur. It has also been noted that silicosis can result in the development of tuberculosis as a further complication. Hard rock miners, quarrymen, stone and pottery workers are most at risk. Health surveillance is recommended for workers in these occupation at initial employment and at subsequent regular intervals. Prevention is best achieved by the use of good dust extraction systems and respiratory personal protective equipment.

Leptospirosis or Weil's disease is caused by a bacterium found in the urine of rats. In humans, the kidneys and liver are attacked causing high temperatures and headaches followed by jaundice and, in up to 20% of cases, it can be fatal. It enters the body either through the skin or by ingestion. The most common source is contaminated water in a river, sewer or ditch and workers, such as canal or sewer workers, are most at risk.

Leptospirosis is always a risk where rats are present, particularly if the associated environment is damp. Good impervious protective clothing, particularly wellington boots, is essential in these situations and the covering of any skin wounds. For workers who are frequently in high-risk environments (sewer workers), immunization with a vaccine may be the best protection.

Legionella is an airborne bacterium and is found in a variety of water sources. It produces a form of pneumonia caused by the bacteria penetrating to the alveoli in the lungs. This disease is known as Legionnaires' disease, named after the first documented outbreak at a State Convention of the American Legion held at Pennsylvania in 1976. During this outbreak, 200 men were affected, of whom 29 died. That outbreak and many subsequent ones were attributed to air-conditioning systems. The legionella bacterium cannot survive at temperatures above 60°C but grows between 20°C and 45°C, being most virulent at 37°C. It also requires food in the form of algae and other bacteria. Control of the bacteria involves the avoidance of water temperatures between 20°C and 45°C, avoidance of water stagnation and the build-up of algae and sediments and the use of suitable water treatment chemicals. This work is often done by a specialist contractor.

The most common systems at risk from the bacterium are:

➤ water systems incorporating a cooling tower
➤ water systems incorporating an evaporative condenser
➤ hot and cold water systems and other plant where the water temperature may exceed 20°C.

An approved code of practice (*Legionnaires' disease – The control of legionella bacteria in water systems* – L8) was produced by the Health and Safety Executive in 2000. Where plant at risk of the development of legionella exists, the following is required:

➤ a written 'suitable and sufficient' risk assessment
➤ the preparation and implementation of a written control scheme involving the treatment, cleaning and maintenance of the system
➤ appointment of a named person with responsibility for the management of the control scheme
➤ the monitoring of the system by a competent person
➤ record keeping and the review of procedures developed within the control scheme.

The code of practice also covers the design and construction of hot and cold water systems and cleaning and disinfection guidance. There have been several cases of members of the public becoming infected from a contaminated cooling tower situated on the roof of a building. It is required that all cooling towers are registered with the local authority. People are more susceptible to the disease if they are older or weakened by some other illness. It is, therefore, important that residential and nursing homes and hospitals are particularly vigilant.

Hepatitis is a disease of the liver and can cause high temperatures, nausea and jaundice. It can be caused by hazardous substances (some organic solvents) or by a virus. The virus can be transmitted from infected faeces (hepatitis A) or by infected blood (hepatitis B and C). The normal precautions include good personal hygiene particularly when handling food and in the use of blood products. Hospital workers who come into contact with blood products are at risk of hepatitis as are drug addicts who share needles. It is also important that workers at risk regularly wash their hands and wear protective disposable gloves.

12.7 Requirements of the COSHH Regulations

When hazardous substances are considered for use (or in use) at a place of work, the COSHH Regulations impose certain duties on employers and requires employees to cooperate with the employer by following any measures taken to fulfil those duties. The principal requirements are as follows:

1 employers must undertake a suitable and sufficient assessment of the health risks created by work which is liable to expose their employees to substances hazardous to health and of the steps that need to be taken by employers to meet the requirements of these Regulations (Regulation 6)

2 employers must prevent, or where this is not reasonably practicable, adequately control the exposure of their employees to substances hazardous to health. Maximum exposure limits (MEL), which should not be exceeded, are specified by the Health and Safety Executive for certain substances. Occupational exposure standards (OES) are approved by the Health and Safety Executive and are levels above which controls are necessary to reduce concentrations to or below the OES as soon as is reasonably practicable. As far as inhalation is concerned, control should be achieved by means other than personal protective equipment. If, however, respiratory equipment, for example, is used, then the equipment must conform to HSE standards (Regulation 7)

3 employers and employees must make proper use of any control measures provided (Regulation 8)

4 employers must maintain any installed control measures on a regular basis and keep suitable records (Regulation 9)

5 monitoring must be undertaken of any employee exposed to items listed in schedule 5 of the Regulations or in any other case where monitoring is required for the maintenance of adequate control or the protection of employees. Records of this monitoring must be kept for at least 5 years, or 40 years where employees can be identified (Regulation 10)

6 health surveillance must be provided to any employees who are exposed to any substances listed in schedule 6. Records of such surveillance must be kept for at least 40 years after the last entry (Regulation 11)

7 employees who may be exposed to substances hazardous to their health must be given information, instruction and training sufficient for him to know the health risks created by the exposure and the precautions which should be taken (Regulation 12).

12.8 Details of a COSHH assessment

Not all hazardous substances are covered by the COSHH Regulations. If there is no warning symbol on the substance container or it is a biological agent which is not directly used in the workplace (such as an influenza virus), then no COSHH assessment is required. The COSHH Regulations do not apply to those hazardous substances which are subject to their own individual regulations (asbestos, lead or radioactive substances). The COSHH Regulations do apply to the following substances:

➤ substances having occupational exposure limits as listed in the HSE publication EH40 (Occupational Exposure Limits)

➤ substances or combinations of substances listed in the Chemicals (Hazard, Information and Packaging for Supply) Regulations 2002 better known as the CHIP Regulations

➤ biological agents connected with the workplace

➤ substantial quantities of airborne dust (more the 10 mg/m^3 of total inhalable dust or 4 mg/m^3 of respirable dust, both 8 hour time-weighted average, when there is no indication of a lower value)

➤ any substance creating a comparable hazard but for technical reasons may not be covered by CHIP.

12.8.1 Assessment requirements

A COSHH assessment is very similar to a risk assessment but is applied specifically to hazardous substances. There are six stages to a COSHH assessment:

1 identify the hazardous substances present in the workplace and those who could be affected by them;

2 gather information about the hazardous substances;

3 evaluate the risks to health;

4 decide on the control measures, if any, required including information, instruction and training;

5 record the assessment;

6 review the assessment.

It is important that the assessment is conducted by somebody who is competent to undertake it. Such competence will require some training, the extent of which will depend on the complexities of the workplace. For large organizations with many high-risk operations, a team of competent assessors will be needed. If the assessment is simple and easily repeated, a written record is not necessary. In other cases, a concise and dated record of the assessment together with recommended control measures should be made available to all those likely to be affected by the hazardous substances. The assessment should be reviewed on a regular basis, particularly when there are changes in work process or substances or when adverse ill-health is reported.

12.8.2 Occupational exposure limits

One of the main purposes of a COSHH assessment is to control adequately the exposure of employees and others to hazardous substances. This means that such substances should be reduced to levels which do not pose a health threat to those exposed to them day after day at work. The HSC has assigned occupational exposure limits (OEL) to a large number of hazardous substances and the HSE publishes an annual update in a publication called *Occupational Exposure Limits* EH40. Two types of exposure limit are published:

1 Maximum Exposure Limit (MEL) – this limit is set for substances which may cause serious health problems and must not be exceeded and should be reduced to as low as is reasonably practicable

259

Table 12.1 Examples of Occupational Exposure Limits (OEL)

Maximum exposure limit (MEL)	LTEL (8-hTWA)	STEL (15 min)
All isocyanates	0.02 mg/m^3	0.07 mg/m^3
Styrene	430 mg/m^3	1080
Occupational exposure standards (OES)		
Ammonia	18 mg/m^3	25 mg/m^3
Toluene	191 mg/m^3	574 mg/m^3

2 **Occupational Exposure Standard (OES)** – is the level at or below which little health risk is likely to occur to persons exposed to the hazardous substance. If the level is exceeded, then steps should be taken to reduce the exposure. Work need not cease when the OES is exceeded provided that reasons are given and that the exposure levels will be reduced as soon as is reasonably practicable.

The two exposure limits are subject to time-weighted averaging. There are two such time-weighted averages (TWA) – the long-term exposure limit (LTEL) or 8-hour reference period and the short-term exposure limit (STEL) or 15-minute reference period. The 8-hour TWA is the maximum exposure allowed over an 8-hour period, so that if the exposure period was less than 8 hours the maximum exposure limit is increased accordingly with the proviso that exposure above the LTEL value continues for no longer than 1 hour.

For example, if a person was exposed to a hazardous substance with a OES of 100 mg/m^3 (8-hour TWA) for 4 hours, no action would be required until an exposure level of 200 mg/m^3 was reached. (Exposure at levels between 100 and 200 mg/m^3 should be restricted to 1 hour.)

If, however, the substance has an STEL of 150 mg/m^3, then action would be required when the exposure level rose above 150 mg/m^3 for more than 15 minutes.

The STEL always takes precedence over the LTEL. When a STEL is not given, it should be assumed that it is three times the LTEL value.

The publication EH40 is a valuable document for the Health and Safety Professional since it contains much additional advice on hazardous substances for use during the assessment of health risks, particularly where new medical information has been made public.

It is important to stress that if an MEL is exceeded, the process and use of the substance should cease immediately. In the longer term, the process, the control and monitoring measures should be reviewed and health surveillance considered. The over-riding requirement for any hazardous substance, which has an MEL, is to reduce exposure to as low as is reasonably practicable.

Finally, there are certain limitations on the use of the published exposure limits:

> they are specifically quoted for an 8-hour period (with an additional STEL for many hazardous substances). Adjustments must be made when exposure occurs over a continuous period longer than 8 hours;

➤ they can only be used for exposure in a workplace and not to evaluate or control non-occupational exposure (e.g. to evaluate exposure levels in a neighbourhood close to the workplace, such as a playground);

➤ OESs are only approved where the atmospheric pressure varies from 900 to 1100 millibars. This could exclude their use in mining and tunnelling operations;

➤ they should also not be used when there is a rapid build-up of a hazardous substance due to a serious accident or other emergency. Emergency arrangements should cover these eventualities.

12.8.3 Sources of information

There are other important sources of information available for a COSHH assessment in addition to the HSE Guidance Note EH40.

Product labels include details of the hazards associated with the substances contained in the product and any precautions recommended. It may also bear one or more of the CHIP hazard classification symbols.

Product safety data sheets are another very useful source of information for hazard identification and associated advice. Manufacturers of hazardous substances are obliged to supply such sheets to users giving details of the name, chemical composition and properties of the substance. Information on the nature of the hazards and any

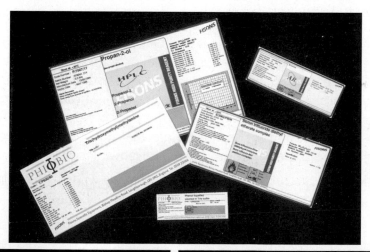

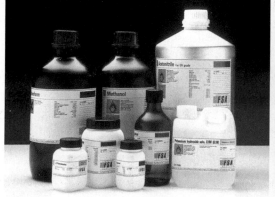

Figure 12.7 Typical product labels and data sheets.

relevant standards (OEL) should also be given. The sheets contain useful additional information on first aid and fire-fighting measures and handling, storage, transport and disposal information. The data sheets should be stored in a readily accessible and known place for use in the event of an emergency.

Other sources of information include trade association publications, industrial codes of practice and specialist reference manuals.

12.8.4 Survey techniques for health risks

An essential part of the COSHH assessment is the measurement of the quantity of the hazardous substance in the atmosphere surrounding the workplace. This is known as air sampling. There are four common types of air sampling techniques used for the measurement of air quality:

1 **Stain tube detectors** use dire-ct reading glass indicator tubes filled with chemical crystals which change colour when a particular hazardous substance passes through them. The method of operation is very similar to the breathalyser used by the police to check alcohol levels in motorists. The glass tube is opened at each end and fitted into a pumping device (either hand or electrically operated). A specific quantity of contaminated air, containing the hazardous substance, is drawn through the tube and the crystals in the tube change colour in the direction of the air flow. The tube is calibrated such that the extent of the colour change along the tube indicates the concentration of the hazardous substance within the air sample.

This method can only be effective if there are no leakages within the instrument and the correct volume of sampled air is used. The instrument should be held within 30 cm of the nose of the person whose atmosphere is being tested. A large range of different tubes are available. This technique of sampling is known as **grab** or **spot** sampling since it is taken at one point.

The advantages of the technique are that it is quick, relatively simple to use and inexpensive. There are, however, several disadvantages:

> the instrument cannot be used to measure concentrations of dust or fume

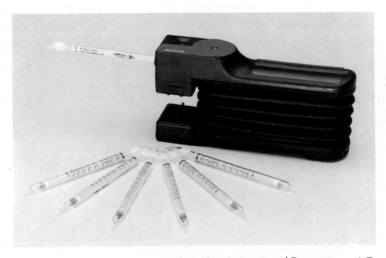

Figure 12.8 Hand pump and stain detector. (Courtesy of Draeger.)

> the accuracy of the reading is approximately ± 25%
> it will yield false reading if other contaminants present react with the crystals
> it can only give an instantaneous reading not an average reading over the working period (TWA)
> the tubes are very fragile with a limited shelf life.

2 **Passive sampling** is measured over a full working period by the worker wearing a badge containing absorbent material. The material will absorb the contaminant gas and at the end of the measuring period, the sample is sent to a laboratory for analysis. The advantages of this method over the stain tube are that there is less possibility of instrument errors and it gives a time-weighted average reading (TWA).

3 **Sampling pumps and heads** can be used to measure gases and dusts. The worker, whose breathing zone is being monitored, wears a collection head as a badge and a battery operated pump on his back at waist level. The pump draws air continuously through a filter, fitted in the head, which will either absorb the contaminant gas or trap hazardous dust particles. After the designated testing period, it is sent to a laboratory for analysis. This system is more accurate than stain tubes and gives a TWA result but can be uncomfortable to wear over long periods. Such equipment can only be used by trained personnel.

4 **Direct reading instruments** are available in the form of sophisticated analysers which can only be used by trained and experienced operatives. Infra-red gas analysers are the most common but other types of analyser are also available. They are very accurate and give continuous or TWA readings. They tend to be very expensive and are normally hired or used by specialist consultants.

Other common monitoring instruments include **vane anemometers** used for measuring air flow speeds and **hygrometers** which are used for measuring air humidity.

Qualitative monitoring techniques include smoke tubes and the dust observation lamp. **Smoke tubes** generate a white smoke which may be used to indicate the direction of flow of air – this is particularly useful when the air speed is very low or when testing the effectiveness of ventilation ducting. **A dust observation lamp** enables dust particles which are normally invisible to the human eye to be observed in the light beam. This dust is usually in the respirable range and, although the lamp does not enable any measurements of the dust to be made, it will illustrate the operation of a ventilation system and the presence of such dust.

Regulation 10 requires routine sampling or monitoring of exposure where there could be serious health effects if the controls failed, exposure limits might be exceeded or the control measures may not be working properly. Air monitoring should also be undertaken for any hazardous substances listed in Schedule 5 of the COSHH Regulations. Records of this monitoring should be kept for 5 years unless an employee is identifiable in the records in which case they should be kept for 40 years.

12.9 The control measures required under the COSHH Regulations

12.9.1 Hierarchy of control measures

The COSHH Regulations require the prevention or adequate control of exposure by measures other than personal protective equipment, so far as is reasonably practicable,

taking into account the degree of exposure and current knowledge of the health risks and associated technical remedies. The hierarchy of control measures are as follows:

- elimination
- substitution
- provide engineering controls
- provide supervisory (people) controls
- personal protective equipment.

Examples where engineering controls are not reasonably practicable include emergency and maintenance work, short-term and infrequent exposure and where such controls are not technically feasible.

Measures for preventing or controlling exposure to hazardous substances include one or a combination of the following:

- elimination
- substitution
- total or partial enclosure of the process
- local exhaust ventilation
- dilute or general ventilation
- reduction of the number of employees exposed to a strict minimum
- reduced time exposure
- housekeeping
- training
- personal protective equipment
- welfare (including first aid)
- medical records
- health surveillance.

12.9.2 Preventative control measures

Prevention is the safest and most effective of the control measures and is achieved either by changing the process completely or by substituting for a less hazardous substance (the change from oil-based to water-based paints is an example of this). It may be possible to use a substance in a safer form, such as a brush paint rather than a spray.

12.9.3 Engineering controls

The simplest and most efficient engineering control is the segregation of people from the process and a chemical fume cupboard is an example of this as is the handling of toxic substances in a glove box. Modification of the process is another effective control to reduce human contact with hazardous substances.

More common methods, however, involve the use of forced ventilation – local exhaust ventilation and dilute ventilation.

Local exhaust ventilation

Local exhaust ventilation removes the hazardous gas, vapour or fume at its source before it can contaminate the surrounding atmosphere and harm people working in the vicinity. Such systems are commonly used for the extraction of welding fumes and

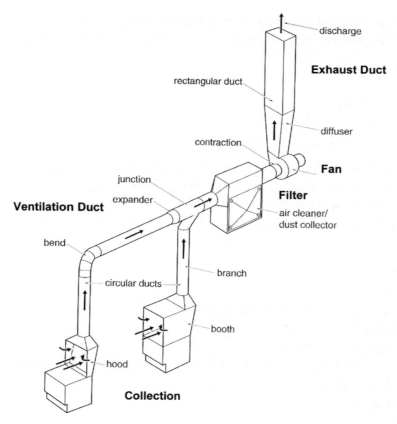

Figure 12.9 Typical local exhaust and ventilation system. Source HSE. Crown copyright material is reproduced with the permission of the Controller of HMSO and the Queen's Printer for Scotland.

dust from woodworking machines. All exhaust ventilation systems have the following five basic components:

1 **a collection hood and intake** – sometimes this is a nozzle-shaped point which is nearest to the work-piece, while at other times it is simply a hood placed over the workstation. The speed of the air entering the intake nozzle is important, if it is too low then hazardous fume may not be removed (air speeds of up to 1 m/s are normally required)

2 **ventilation ducting** – this normally acts as a conduit for the contaminated air and transports it to a filter and settling section. It is very important that this section is inspected regularly and any dust deposits removed. It has been known for ventilation ducting attached to a workshop ceiling to collapse under the added weight of metal dust deposits

3 **filter or other air cleaning device** – normally located between the hood and the fan, the filter removes the contaminant from the air stream. The filter requires regular attention to remove contaminant and to ensure that it continues to work effectively

4 **fan** – this moves the air through the system. It is crucial that the correct type and size of fan is fitted to a given system and should only be selected by a competent person. It should also be positioned so that it can be easily maintained but does not create a noise hazard to nearby workers

5 **exhaust duct** – this exhausts the air to the outside of the building. It should be checked regularly to ensure that the correct volume of air is leaving the system and that there are no leakages. The exhaust duct should also be checked to ensure that there is no corrosion due to adverse weather conditions.

An Introduction to Local Exhaust Ventilation, HSG(37), HSE Books, is a very useful document on ventilation systems.

The COSHH Regulations require that such ventilation systems must be inspected at least every 14 months by a competent person to ensure that they are still operating effectively.

The effectiveness of a ventilation system will be reduced by damaged ducting, blocked or defective filters and poor fan performance. More common problems include the unauthorized extension of the system, poor initial design, poor maintenance, incorrect adjustments and a lack of inspection or testing.

Routine maintenance should include repair of any damaged ducting, checking filters, examination of the fan blades to ensure that there has been no dust accumulation, tightening all drive belts and a general lubrication of moving parts.

Dilution (or general) ventilation

Dilution (or general) ventilation uses either natural ventilation (doors and windows) or fan-assisted forced ventilation system to ventilate the whole working room by inducing a flow of clean air by using extraction fans fitted into the walls and the roof, sometimes assisted by inlet fans. It operates by either removing the contaminant or reducing its concentration to an acceptable level. It is used either when airborne contaminants are of low toxicity, low concentration and low vapour density or contamination occurs uniformly across the workroom. Paint spraying operations often use this form of ventilation as does glass reinforced plastics (GRP) boat-building industry – these being instances where there are not discrete points of release of the hazardous substances. It is also widely used in kitchens and bathrooms. It is not suitable for dust extraction and where it is reasonably practicable to reduce levels by other means.

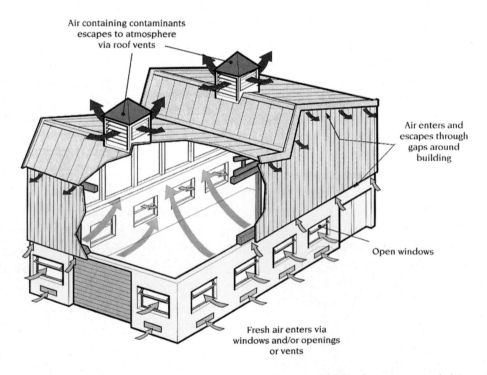

Air containing contaminants escapes to atmosphere via roof vents

Air enters and escapes through gaps around building

Open windows

Fresh air enters via windows and/or openings or vents

Figure 12.10 Natural ventilation in a building. Source HSE. Crown copyright material is reproduced with the permission of the Controller of HMSO and the Queen's Printer for Scotland.

There are limitations to the use of dilution ventilation. Certain areas of the work room (e.g. corners and beside cupboards) will not receive the ventilated air and a build-up of hazardous substances occurs. These areas are known as 'dead areas'. The flow patterns are also significantly affected by doors and windows being opened or the rearrangement of furniture or equipment.

12.9.4 Supervisory or people controls

Many of the supervisory controls required for COSHH purposes are part of a good safety culture and were discussed in detail in Chapter 4. These include items such as systems of work, arrangements and procedures, effective communications and training. Additional controls when hazardous substances are involved are as follows:

> reduced time exposure – thus ensuring that workers have breaks in their exposure periods. The use of this method of control depends very much on the nature of the hazardous substance and its short-term exposure limit (STEL);
> reduced number of workers exposed – only persons essential to the process should be allowed in the vicinity of the hazardous substance. Walkways and other traffic routes should avoid any area where hazardous substances are in use;
> eating, drinking and smoking must be prohibited in areas where hazardous substances are in use;
> any special rules, such as the use of personal protective equipment, must be strictly enforced.

12.9.5 Personal protective equipment

Personal protective equipment is to be used as a control measure only as a last resort. It does not eliminate the hazard and will present the wearer with the maximum health risk if the equipment fails. Successful use of personal protective equipment relies on good user training, the availability of the correct equipment at all times and good supervision and enforcement.

The 'last resort' rule applies in particular to respiratory protective equipment (RPE) within the context of hazardous substances. There are some working conditions when RPE may be necessary:

> during maintenance operations
> as a result of a new assessment, perhaps following the introduction of a new substance
> during emergency situations, such as fire or plant breakdown
> where alternatives are not technically feasible.

All personal protective equipment, except respiratory protection equipment (which is covered by specific Regulations, such as COSHH, lead, etc.) is controlled by its own set of regulations – the Personal Protective Equipment at Work Regulations 1992. A detailed summary of these Regulations is given in Chapter 17 and need to be studied in detail with this section. The principal requirements of these regulations are as follows:

> personal protective equipment which is suitable for the wearer and the task
> compatibility and effectiveness of the use of multiple personal protective equipment

Figure 12.11   Personal protective equipment. (Courtesy of Draper.)

➢ a risk assessment to determine the need and suitability of proposed personal protective equipment
➢ a suitable maintenance programme for the personal protective equipment
➢ suitable accommodation for the storage of the personal protective equipment when not in use
➢ information, instruction and training for the user of personal protective equipment
➢ the supervision of the use of personal protective equipment by employees and a reporting system for defects.

Types of personal protective equipment

There are several types of personal protective equipment, such as footwear, hearing protectors and hard hats, which are not primarily concerned with protection from hazardous substances; those which are used for such protection, include:

➢ respiratory protection
➢ hand and skin protection
➢ eye protection
➢ protective clothing.

For all types of personal protective equipment, there are some basic standards that should be reached. The personal protective equipment should fit well, be comfortable to wear and not interfere either with other equipment being worn or present the user with additional hazards (e.g. impaired vision due to scratched eye goggles). Training in the use of particular personal protective equipment is essential, so that it is not only used correctly, but the user knows when to change an air filter or to change a type of glove. Supervision is essential with disciplinary procedures invoked for non-compliance with personal protective equipment rules.

Respiratory protection equipment

Respiratory protection equipment can be sub-divided into two categories – respirators (or face masks), which filter and clean the air, and breathing apparatus which supplies breathable air.

Respirators should not be worn in air which is dangerous to health, including oxygen deficient atmospheres. They are available in several different forms but the common ones are:

> a **filtering half mask** often called disposable respirator – made of the filtering material. It covers the nose and mouth and removes respirable size dust particles. It is normally replaced after 8/10 hours of use. They offer protection against some vapours and gases;

> a **half mask respirator** – made of rubber or plastic and covers the nose and mouth. Air is drawn through a replaceable filter cartridge. These can be used for vapours, gases or dusts but it is very important that the correct filter is used (a dust filter will not filter vapours);

> a **full face mask respirator** – similar to the half mask type but covers the eyes with a visor;

> a **powered respirator** – a battery-operated fan delivers air through a filter to the face mask, hood, helmet or visor.

Breathing apparatus is used in one of three forms:

> **self-contained breathing apparatus** – where air is supplied from compressed air in a cylinder and forms a completely sealed system;

> **fresh air hose apparatus** – fresh air is delivered through a hose to a sealed face mask from an uncontaminated source. The air may be delivered by the wearer, by natural breathing or mechanically by a fan;

> **compressed air line apparatus** – air is delivered through a hose from a compressed air line. This can be either continuous flow or on demand. The air must be properly filtered to remove oil, excess water and other contaminants and the air pressure must be reduced. Special compressors are normally used.

(a)   **Filtering half masks to protect against particles**

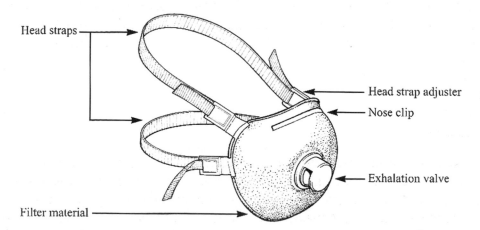

Head straps

Head strap adjuster

Nose clip

Exhalation valve

Filter material

Figure 12.12 Types of respiratory protective equipment. (a) Filtering half mask; (b) half mask reusable with filters; (c) compressed air line breathing apparatus with full face mask fitted with demand valve. Source HSE. Crown copyright material is reproduced with the permission of the Controller of HMSO and the Queen's Printer for Scotland.

(b) **Half masks reusable with filters**

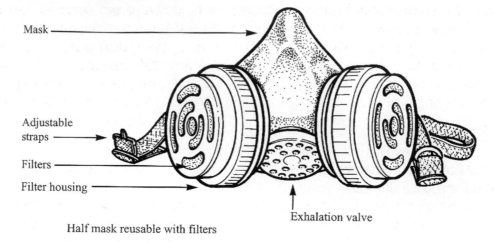

Mask

Adjustable straps

Filters

Filter housing

Exhalation valve

Half mask reusable with filters

(c) Compressed airline BA with full-face mask fitted with demand valve

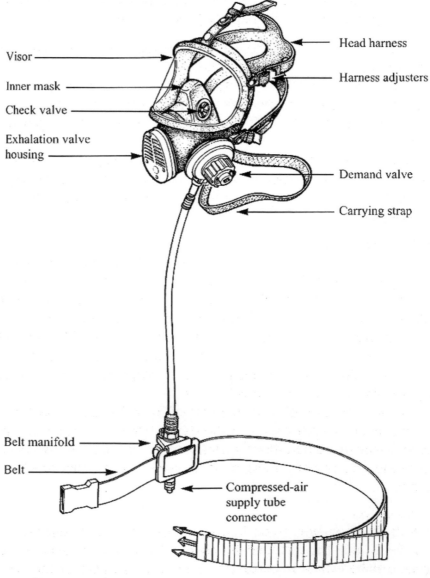

Visor

Inner mask

Check valve

Exhalation valve housing

Head harness

Harness adjusters

Demand valve

Carrying strap

Belt manifold

Belt

Compressed-air supply tube connector

Figure 12.12 (Continued.)

270

The selection of appropriate respiratory protection equipment and correct filters for particular hazardous substances is best done by a competent specialist person.

There are several important technical standards which must be considered during the selection process. Respiratory protection equipment must either be CE marked or HSE approved (HSE approval ceased on 30 June 1995 but such approved equipment may still be used). Other standards include the minimum protection required (MPR) and the assigned protection factor (APF). The CE mark does not indicate that the equipment is suitable for a particular hazard. The following information will be needed before a selection can be made:

➤ details of the hazardous substances in particular whether it is a gas, vapour or dust or a combination of all three;
➤ presence of a beard or other facial hair which could prevent a good leak-free fit (a simple test to see whether the fit is tight or not is to close off the air supply, breathe in and hold the breath. The respirator should collapse onto the face. It should then be possible to check to see if there is a leak);
➤ the size and shape of the face of the wearer and physical fitness;
➤ compatibility with other personal protective equipment, such as ear defenders;
➤ the nature of the work and agility and mobility required.

Filters and masks should be replaced at the intervals recommended by the supplier or when taste or smell is detected by the wearer.

All respiratory protection equipment should be examined at least once a month except for disposable respirators. A record of the inspection should be kept for at least 5 years. There should be a routine cleaning system in place and proper storage arrangements.

Respiratory protective equipment – a practical guide for users, HSG(53), HSE Books, contains comprehensive advice and guidance on respiratory protective equipment selection, use, storage, maintenance and training and should be consulted for more information.

Hand and skin protection
Hand and skin protection is mainly provided by gloves (arm shields are also available). A wide range of safety gloves are available for protection from chemicals, sharp objects, rough working and temperature extremes. Many health and safety catalogues give helpful guidance for the selection of gloves. For protection from chemicals, including paints and solvents, impervious gloves are recommended. These may be made of PVC, nitrile or neoprene. For sharp objects, such as trimming knives, a Kevlar based glove is the most effective. Gloves should be regularly inspected for tears or holes since this will obviously allow skin contact to take place.

Another effective form of skin protection is the use of barrier creams and these come in two forms – pre-work and after-work. Pre-work creams are designed to provide a barrier between the hazardous substance and the skin. After-work creams are general purpose moisteners which replace the natural skin oils removed either by solvents or by washing.

Eye protection
Eye protection comes in three forms – spectacles (safety glasses), goggles and face visors. Eyes may be damaged by chemical and solvent splashes or vapours, flying particles, molten metals or plastics, non-ionizing radiation (arc welding and lasers) and dust. Spectacles are suitable for low-risk hazards (low speed particles such as

Figure 12.13 Variety of eye protection goggles. (Courtesy of Draper.)

machine swarf). Some protection against scratching of the lenses can be provided but this is the common reason for replacement. Prescription lenses are also available for people who normally wear spectacles.

Goggles are best to protect the eyes from dust or solvent vapours because they fit tightly around the eyes. Visors offer protection to the face as well as the eyes and do not steam up so readily in hot and humid environments. For protection against very bright lights, special light filtering lenses are used (e.g. in arc welding). Maintenance and regular cleaning are essential for the efficient operation of eye protection.

When selecting eye protection, several factors need to be considered. These include the nature of the hazard (the severity of the hazard and its associated risk will determine the quality of protection required), comfort and user acceptability, compatibility with other personal protective equipment, training and maintenance requirements and costs.

Protective clothing
Protective clothing includes aprons, boots and headgear (hard hats and bump caps). Aprons are normally made of PVC and protect against spillages but can become uncomfortable to wear in hot environments. Other lighter fabrics are available for use in these circumstances. Safety footwear protects against falling objects, collision with hard or sharp objects, hot or molten materials, slippery surfaces and against chemical spills. They have a metal toe-cap and come in the form of shoes, ankle boots or knee-length boots and are made of a variety of materials dependent on the particular hazard (e.g. thermally insulated against cold environments). They must be used with care near live unprotected electricity. Specialist advice is needed for use with flammable liquids.

It is important to note that appropriate personal protective equipment should be made available to visitors and other members of the public when visiting workplaces where hazardous substances are being used. It is also important to stress that managers and supervisors must lead by example, particularly if there is a legal requirement to wear particular personal protective equipment. Refusals by employees to wear mandatory personal protective equipment must lead to some form of disciplinary action.

12.10 Health surveillance and personal hygiene

Health surveillance enables the identification of those employees most at risk from occupational ill-health. It should not be confused with health monitoring procedures such as per-employment health checks or drugs and alcohol testing, but it covers a wide range of situations, from a responsible person looking for skin damage on hands to medical surveillance by a medical doctor. Health surveillance detects the start of an ill-health problem and collects data on ill-health occurrences. It also gives an indication of the effectiveness of the control procedures.

It is required when there appears to be a reasonable chance that ill-health effects are occurring in a particular workplace as a result of reviewing sickness records or when a substance listed in Schedule 6 under Regulation 11 of the COSHH Regulations is being used. Schedule 6 lists the substances and the processes in which they are used. There are a limited number of such substances. The health surveillance includes medical surveillance by an employment medical adviser or appointed doctor at intervals not exceeding 12 months. Records of the health surveillance must contain approved particulars and be kept for 40 years. The need for health surveillance is not that common and further advice on the necessary procedures is available from the Employment Advisory Medical Advisory Service.

Personal hygiene has already been covered under supervisory controls. It is very important for workers exposed to hazardous substances to wash their hands thoroughly before eating, drinking or smoking. Protection against biological hazards can be increased significantly by vaccination (e.g. tetanus). Finally, contaminated clothing and overalls need to be removed and cleaned on a regular basis.

To provide an effective occupational health programme.

Figure 12.14   Health commitment.

12.11 Maintenance and emergency controls

Engineering control measures will only remain effective if there is a programme of preventative maintenance available. Indeed the COSHH Regulations require that systems of adequate control are in place and the term 'adequate control' spans normal operations, emergencies and maintenance. Maintenance will involve the cleaning, testing and, possibly, the dismantling of equipment. It could involve the changing of filters in extraction plant or entering confined spaces. It will almost certainly require hazardous substances to be handled and waste material to be safely disposed. It may also require a permit-to-work procedure to be in place since the control equipment will be inoperative during the maintenance operations. Records of maintenance should be kept for at least 5 years.

Emergencies can range from fairly trivial spillages to major fires involving serious air pollution incidents. The following points should be considered when emergency procedures are being developed:

> the possible results of a loss of control (e.g. lack of ventilation)
> dealing with spillages and leakages (availability of effective absorbent materials)
> raising the alarm for more serious emergencies
> evacuation procedures, including the alerting of neighbours
> fire-fighting procedures and organization
> availability of respiratory protection equipment
> information and training.

The Emergency Services should be informed of the final emergency procedures and, in the case of the Fire Service, consulted for advice during the planning of the procedures. (See Chapter 6 for more details on emergency procedures.)

12.12 Environmental considerations

Organizations must also be concerned with aspects of the environment. There will be an interaction between the health and safety policy and the environmental policy which many organizations are now developing. Many of these interactions will be concerned with good practice, the reputation of the organization within the wider community and the establishment of a good health and safety culture. The health and safety data sheet, used for a COSHH assessment, also contains information of an environmental nature covering ecological information and disposal considerations.

There are three environmental issues which place statutory duties on employers and are directly related to the health and safety function. These are:

> air pollution
> water pollution
> waste disposal.

The statutory duties are contained in the Environmental Protection Act 1990 (EPA) and several of its subsequent regulations. The Act is enforced by various state agencies (the Environmental Agency, Local Authorities and the National Rivers Authority) and these agencies have very similar powers to the Health and Safety Executive (e.g. enforcement and prohibition notices and prosecution). There is a summary of Parts 1 and 2 of the EPA in Chapter 17.

12.12.1 Air pollution

The most common airborne pollutants are carbon monoxide, benzene, 1,3-butadiene, sulphur dioxide, nitrogen dioxide and lead. Air pollution is currently monitored by

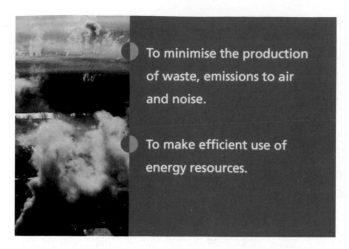

- To minimise the production of waste, emissions to air and noise.

- To make efficient use of energy resources.

Figure 12.15   Environmental protection commitment.

Integrated Pollution Control (IPC). This is a system established by Part 1 of the EPA which introduced two tiers of pollution control:

➤ **Part A processes**, which are certain large-scale manufacturing processes with a potential to cause serious environmental damage to the air, water or the land. In England and Wales this is enforced by the Environment Agency. In Scotland there is a parallel system enforced by the Scottish Environment Protection Agency

➤ **Part B processes**, which may be classified as those from less polluting industries with only emissions released to air being subject to regulatory control. For such processes local authorities are the enforcing body through the Environmental Health Officers and the system is known as Local Air Pollution Control (LAPC).

This division has led to some anomalies in that some Part A processes create less pollution than some Part B processes. However, the grouping of three pollution destinations under one arrangement tends to a more holistic approach. The aim of IPC is to control pollution of the whole environment under a single enforcement system and to achieve the 'Best Practicable Environmental Option' (BPEO) by using BATNEEC (Best Available Techniques Not Entailing Excessive Cost) to minimize the overall environmental impact of a process. Part B processes only need satisfy the BATNEEC requirement.

IPC prescribes certain listed substances from being released to air, water or land. All prescribed processes must have authorization. The EPA extends BATNEEC beyond pollution control technology to include employee training and competence and building design and maintenance.

An operator of a prescribed process (such as a vehicle spray booth) must apply to the Environmental Agency for prior authorization to operate the process. If the application is granted, the operator must monitor emissions and report them to the Environmental Agency on a yearly basis. The Agency has the power to revoke the authorization, enforce the terms of the authorization or prohibit the operation of the process.

Further information on the authorization process is given in Chapter 17.

The Pollution Prevention and Control Act 1999 will, by 2007, replace the Integrated Pollution Control regulations made under Part 1 of the EPA by extending those powers to cover waste minimization, energy efficiency, noise and site restoration.

12.12.2 Water pollution

Pollution of rivers and other water courses can produce very serious effects on the health of plants and animals which rely on that water supply. The National Rivers Authority is responsible for coastal waters, inland fresh water and ground waters (known as 'controlled waters'). The EC Groundwater Directive seeks to protect groundwater from pollution since this is a source of drinking water. Such sources can become polluted by leakage from industrial soakaways. Discharges to a sewer are controlled by the Water Industry Act which defines trade effluent and those substances which are prohibited from discharge (e.g. petroleum spirit). It is an offence to pollute any controlled waters or sewage system.

The local Water Company has a right to sample discharges into its sewers because it is required to keep a public trade effluent register. There are two lists of prescribed substances which can only be discharged into a public sewer with the permission of the Water Company.

12.12.3 Waste disposal

The statutory duty of care for the management of waste derives from Part 2 of the EPA. The principal requirements are as follows:

➤ to handle waste so as to prevent any unauthorized escape into the environment;
➤ to pass waste only to an authorized person as defined by EPA;
➤ to ensure that a written description accompanies all waste. The Environmental Protection (Duty of Care) Regulations 1991 requires holders or producers of waste to complete a 'Transfer Note' giving full details of the type and quantity of waste for collection and disposal. Copies of the note should be kept for at least 2 years;
➤ to ensure that no person commits an offence under the Act.

The EPA is concerned with controlled waste. Controlled waste comprises household, industrial or commercial waste. It is a criminal offence to deposit controlled waste without a waste management licence and/or in a manner likely to cause environmental pollution or harm to human health.

The EPA also covers 'special wastes' which are so dangerous that they can only be disposed using special arrangements. These are usually substances which are life threatening (toxic, corrosive or carcinogenic) or highly flammable. Clinical waste falls within this category. A consignment note system accompanies this waste at all the stages to its final destination. Before special waste is removed from the originating premises, a contract should be in place with a licensed carrier. Special waste should be stored securely prior to collection to ensure that the environment is protected.

In 1998, land disposal accounted for approximately 58% of waste disposal, 26% was recycled and the remainder was incinerated with some of the energy recovered as heat. The Producer Responsibility Obligations (Packaging Waste) Regulations 1997 placed legal obligations on employers to reduce their packaging waste by either recycling or recovery as energy (normally as heat from an incinerator attached to a district heating system). A series of targets have been stipulated which will reduce the amount of waste progressively over the years. These regulations are enforced by the Environment Agency who have powers of prosecution in the event of non-compliance.

12.13 Practice NEBOSH questions for Chapter 12

1. For each of the following types of hazardous substance, give a typical example and state its primary effect on the body:
   (i) toxic
   (ii) corrosive
   (iii) carcinogenic
   (iv) irritant. (March 2001)

2. An employee is engaged in general cleaning activities in a large veterinary practice.
   (i) Identify four specific types of hazard that the cleaner might face when undertaking the cleaning.
   (ii) Outline the precautions that could be taken to minimize the risk of harm from these hazards. (March 2001)

3. Identify four different types of hazard that may necessitate the use of special footwear, explaining in each case how the footwear affords protection. (March 2001)

4. A furniture factory uses solvent-based adhesives in its manufacturing process.
   (i) Identify the possible effects on the health of employees using the adhesives.
   (ii) State four control measures to minimize such health effects. (December 2001)

5. (a) Identify possible routes of entry of biological organisms into the body.
   (b) Outline control measures that could be used to reduce the risk of infection from biological organisms. (December 2001)

6. (a) Explain the health and safety benefits of restricting smoking in the workplace.
   (b) Outline the ways in which an organization could effectively implement a no-smoking policy. (December 2001)

7. In relation to the spillage of a toxic substance from a ruptured drum stored in a warehouse:
   (i) identify three ways in which persons working in close vicinity to the spillage might be harmed
   (ii) outline a procedure to be adopted in the event of such a spillage. (June 2001)

8. (a) List four respiratory diseases that could be caused by exposure to dust at work.
   (b) Identify the possible indications of a dust problem in a workplace. (June 2001)

9. (a) Identify the types of hazard against which gloves could offer protection.
   (b) Outline the practical limitations of using gloves as a means of protection. (June 2001)

10. Outline the health and safety risks associated with welding operations. (June 2001)

11. (a) Explain the meaning of the term 'dilution ventilation'.
    (b) Outline the circumstances in which the use of dilution ventilation may be appropriate. (December 2000)

12. An employee is required to install glass-fibre insulation in a loft.
    (i) Identify two hazards connected with this activity.
    (ii) Outline the precautions that might be taken to minimize harm to the employee carrying out this operation. (December 2000)

13. (a) Identify three types of hazard for which personal eye protection would be required.
    (b) Outline the range of issues that should be addressed when training employees in the use of personal eye protection. (December 2000)

14. (a) Describe, by means of a labelled sketch, a chemical indicator (stain detector) tube suitable for atmospheric monitoring.
    (b) List the main limitations of chemical indicator (stain detector) tubes. (June 2000)

15. The manager of a company is concerned about a substance to be introduced into one of its manufacturing processes. Outline four sources of information that might be consulted when assessing the risk from this substance. (March 2000)

16. (a) Describe the typical symptoms of occupational dermatitis.
    (b) State the factors that could affect the likelihood of dermatitis occurring in workers handling dermatitic substances. (March 2000)

17. An essential raw material for a process is delivered in powdered form and poured by hand from bags into a mixing vessel. Outline the control measures that might be considered in this situation in order to reduce employee exposure to the substance. (March 2000)

18. Local exhaust ventilation (LEV) systems must be thoroughly examined at least every 14 months. Outline the routine maintenance that should be carried out between statutory examinations in order to ensure the continuing efficiency of an LEV system. (March 2000)

19. (a) State two respiratory diseases that may be caused by exposure to asbestos.
    (b) Identify where asbestos is likely to be encountered in a building during renovation work. (June 1998)

20. (a) Explain the meaning of the terms:
    (i) 'occupational exposure standard' (OES)
    (ii) 'maximum exposure limit' (MEL)
    (b) Outline four actions management could take when an MEL has been exceeded. (June 1998)

Physical and psychological health hazards and control

13.1 Introduction

Occupational health is concerned with physical and psychological hazards as well as chemical and biological hazards. The physical occupational hazards have been well-known for many years and the recent emphasis has been on the development of lower risk workplace environments. Physical hazards include topics such as electricity and manual handling which were covered in earlier chapters and noise, display screen equipment and radiation which are discussed in this chapter.

However, it is only really in the last twenty years that psychological hazards have been included among the occupational health hazards faced by many workers. This is now the most rapidly expanding area of occupational health and includes topics such as mental health and workplace stress, violence to staff, passive smoking, drugs and alcohol.

The physical and psychological hazards discussed in this chapter are covered by the following health and safety regulations:

➤ Workplace (Health, Safety and Welfare) Regulations 1992
➤ Health and Safety (Display screen Equipment) Regulations 1992
➤ Manual Handling Operations Regulations 1992
➤ Noise at Work Regulations 1989
➤ Ionising Radiations Regulations 1985.

13.2 Task and workstation design

13.2.1 The principles and scope of ergonomics

Ergonomics is the study of the interaction between the worker and his work in the broadest sense in that it encompasses the whole system surrounding the work process. It is, therefore, as concerned with the work organization, process and design

Before   After

Figure 13.1 Workstation ergonomic design improvements. Source HSE. Crown copyright material is reproduced with the permission of the Controller of HMSO and the Queen's Printer for Scotland.

of the workplace and work methods as it is with work equipment. The common definitions of ergonomics, the 'man–machine interface' or 'fitting the man to the machine rather than vice versa' are far too narrow. It is concerned about the physical and mental capabilities of an individual as well as their understanding of the job under consideration. Ergonomics includes the limitations of the worker in terms of skill level, perception and other personal factors in the overall design of the job and the system supporting and surrounding it. It is the study of the relationship between the worker, the machine and the environment in which it operates and attempts to optimize the whole worksystem, including the job, to the capabilities of the worker so that maximum output is achieved for minimum effort and discomfort by the worker. Cars, buses and lorries are all ergonomically designed so that all the important controls, such as the steering wheel, brakes, gear stick and instrument panel are easily accessed by most drivers within a wide range of sizes. Ergonomics is sometimes described as human engineering and as working practices become more and more automated, the need for good ergonomic design becomes essential.

The scope of ergonomics and an ergonomic assessment is very wide incorporating the following areas of study:

➤ personal factors of the worker, in particular physical, mental and intellectual abilities, body dimensions and competence in the task required
➤ the machine and associated equipment under examination
➤ the interface between the worker and the machine – controls, instrument panel or gauges and any aids including seating arrangements and hand tools
➤ environmental issues affecting the work process such as lighting, temperature, humidity, noise and atmospheric pollutants
➤ the interaction between the worker and the task, such as the production rate, posture and system of working
➤ the task or job itself – the design of a safe system of work, checking that the job is not too strenuous or repetitive and the development of suitable training packages
➤ the organization of the work, such as shift work, breaks and supervision.

The reduction of the possibility of human error is one of the major aims of ergonomics and an ergonomic assessment. An important part of an ergonomic study is to design the workstation or equipment to fit the worker. For this to be successful, the physical measurement of the human body and an understanding of the variations in these measurements between people is essential. Such a study is known as **anthropometry**, which is defined as the scientific measurement of the human body and its movement. Since there are considerable variations in, for example, the heights of people, it is common for some part of the workstation to be variable (e.g. an adjustable seat).

13.2.2 The ill-health effects of poor ergonomics

Ergonomic hazards are those hazards to health resulting from poor ergonomic design. They generally fall within the physical hazard category and include the manual handling and lifting of loads, pulling and pushing loads, prolonged periods of repetitive activities and work with vibrating tools. The condition of the working environment, such as low lighting levels, can present health hazards to the eyes. It is also possible for psychological conditions, such as occupational stress, to result from ergonomic hazards.

The common ill-health effects of ergonomic hazards are musculoskeletal disorders (back injuries, covered in Chapter 8, and work-related upper limb disorders (WRULDs) including repetitive strain injury being the main disorders) and deteriorating eyesight.

Work Related Upper Limb Disorders (WRULDs)

WRULDs describe a group of illnesses which can affect the neck, shoulders, arms, elbows, wrists, hands and fingers. **Tenosynovitis** (affecting the tendons), **carpal tunnel syndrome** (affecting the tendons which pass through the carpal bone in the hand) and **frozen shoulder** are all examples of WRULDs which differ in the manifestation and site of the illness. The term **repetitive strain injury (RSI)** is commonly used to describe WRULDs.

WRULDs are caused by repetitive movements of the fingers, hands or arms which involve pulling, pushing, reaching, twisting, lifting, squeezing or hammering. These disorders can occur to workers in offices as well as in factories or on construction sites. Typical occupational groups at risk include painters and decorators, riveters and pneumatic drill operators and desk top computer users.

The main symptoms of WRULDs are aching pain to the back, neck and shoulders, swollen joints and muscle fatigue accompanied by tingling, soft tissue swelling, similar

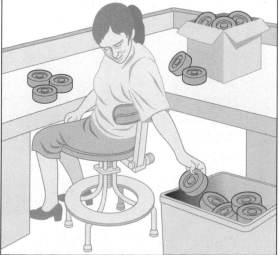

Figure 13.2   Poor workstation layout may cause WRULD.

to bruising, and a restriction in joint movement. The sense of touch and movement of fingers may be affected. The condition is normally a chronic one in that it gets worse with time and may lead eventually to permanent damage. The injury occurs to muscle, tendons and/or nerves. If the injury is allowed to heal before being exposed to the repetitive work again, no long-term damage should result. However, if the work is repeated again and again, healing cannot take place and permanent damage can result leading to a restricted blood flow to the arms, hands and fingers.

The risk factors, which can lead to the onset of WRULDs, are repetitive actions of lengthy duration, the application of significant force and unnatural postures, possibly involving twisting and over-reaching and the use of vibrating tools. Cold working environments, work organization and worker perception of the work organization have all been shown in studies to be risk factors, as is the involvement of vulnerable workers such as those with pre-existing ill-health conditions and pregnant women.

Ill-health due to vibrations

Hand-held vibrating machinery (such as pneumatic drills, sanders and grinders, powered lawn mowers and strimmers and chainsaws) can produce health risks from hand–arm or whole body vibrations.

Hand–arm vibration syndrome (HAVS)

HAVS describes a group of diseases caused by the exposure of the hand and arm to external vibration. Some of these have been described under WRULDs, such as carpal tunnel syndrome.

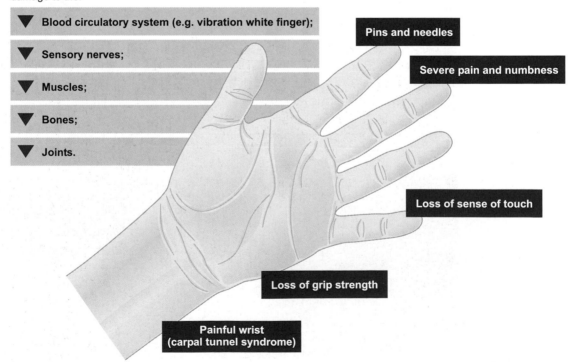

Regular exposure to HAV can cause a range of permanent injuries to hands and arms, collectively known as hand-arm vibration syndrome (HAVS). The injuries can include damage to the:

▼ Blood circulatory system (e.g. vibration white finger);

▼ Sensory nerves;

▼ Muscles;

▼ Bones;

▼ Joints.

Pins and needles

Severe pain and numbness

Loss of sense of touch

Loss of grip strength

Painful wrist (carpal tunnel syndrome)

Figure 13.3 Injuries which can be caused by hand–arm vibration. Source HSE. Crown copyright material is reproduced with the permission of the Controller of HMSO and the Queen's Printer for Scotland.

However, the best known disease is **vibration white finger (VWF)** in which the circulation of the blood, particularly in the hands, is adversely affected by the vibrations. The early symptoms are tingling and numbness felt in the fingers, usually some time after the end of the working shift. As exposure continues, the tips of the fingers go white and then the whole hand may become affected. This results in a loss of grip strength and manual dexterity. Attacks can be triggered by damp and/or cold conditions and, on warming, 'pins and needles' are experienced. If the condition is allowed to persist, more serious symptoms become apparent including discoloration and enlargement of the fingers. In very advanced cases, gangrene can develop leading to the amputation of the affected hand or finger. VWF was first detailed as an industrial disease in 1911.

The risk of developing HAVS depends on the frequency of vibration and the length of exposure.

Whole body vibration (WBV)

WBV is caused by vibrations from machinery passing into the body, either through the feet of standing workers or the buttocks of sitting workers. The most common ill-health effect is severe back pain which, in severe cases, may result in permanent injury. The two most common occupations which are affected by WBV, are pneumatic drillers and agricultural or horticultural machinery operatives. There is growing concern throughout the European Community about this problem. Control measures include the proper use of the equipment, including correct adjustments of air or hydraulic pressures, seating and, in the case of vehicles, correct suspension, tyre pressures and appropriate speeds to suit the terrain. Other control measures include the selection of suitable equipment with low vibration characteristics, work rotation, good maintenance and fault reporting procedures.

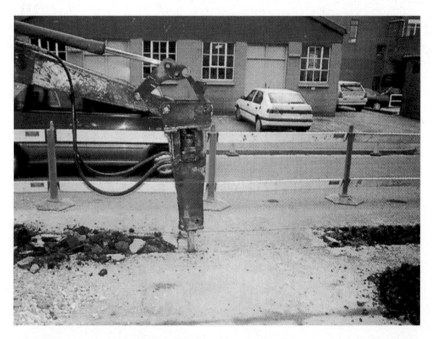

Figure 13.4 Mounted breaker to reduce vibrations. Source HSE. Crown copyright material is reproduced with the permission of the Controller of HMSO and the Queen's Printer for Scotland.

Preventative and precautionary measures

The control strategy outlined in Chapter 6 can certainly be applied to ergonomic risks. The common measures used to control ergonomic ill-health effects are:

> the elimination of vibrating or hazardous tasks by performing the job in a different way;
> ensure that the correct equipment (properly adjusted) is always used;
> introduce job rotation so that workers have a reduced time exposure to the hazard;
> during the design of the job ensure that poor posture is avoided;
> undertake a risk assessment;
> introduce a programme of health surveillance;
> ensure that employees are given adequate information on the hazards and develop a suitable training programme;
> ensure that a programme of preventative maintenance is introduced and include the regular inspection of items such as vibration isolation mountings;
> keep up to date with advice from equipment manufacturers, trade associations and health and safety sources (more and more low vibration equipment is becoming available).

A very useful and extensive checklist for the identification and reduction of work-related upper limb disorders is given in Appendix 2 of the HSE *Guide to Work related upper limb disorders* (HSG 60).

13.2.3 Display screen equipment (DSE)

Display screen equipment, which includes visual display units, is a good example of a common work activity which relies on an understanding of ergonomics and the ill-health conditions which can be associated with poor ergonomic design.

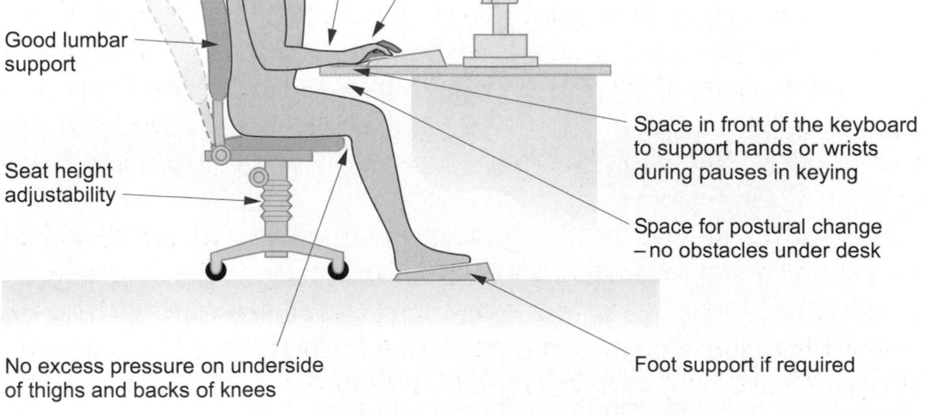

Seating and posture for typical office tasks

Figure 13.5 Workstation design.

Legislation governing DSE is covered by the Health and Safety (Display Screen Equipment) Regulations 1992 and a detailed summary of them is given in Chapter 17. The regulations apply to a user or operator of DSE which define a user or operator. The definition is not as tight as many employers would like it to be, but usage in excess of approximately one hour continuously each day would define a user. The definition is important since users are entitled to free eye tests and, if required, a pair of spectacles. The basic requirements of the regulations are:

➤ a suitable and sufficient risk assessment of the workstation, including the software in use and the surrounding environment
➤ workstation compliance with the minimum specifications laid down in the schedules appended to the regulations
➤ a plan of the work programme to ensure that there are adequate breaks in the work pattern of workers
➤ the provision of free eye sight tests and, if required, spectacles to users of DSE
➤ a suitable programme of training and sufficient information given to all users.

There are three basic ill-health hazards associated with DSE. These are:

➤ musculoskeletal problems
➤ visual problems
➤ psychological problems.

A fourth hazard, of radiation, has been shown from several studies to be very small and is now no longer normally considered in the risk assessment.

Similarly, in the past, there have been suggestions that DSE could cause epilepsy and there were concerns about adverse health effects on pregnant women and their unborn children. All these particular risks have been shown in various studies to be very low.

Musculoskeletal problems

Tenosynovitis is the most common and well-known problem which affects the wrist of the user. The symptoms and effects of this condition have already been covered. Suffice it to say that if the condition is ignored, then the tendon and tendon sheath around the wrist will become permanently injured. Tenosynovitis, better known as RSI, is caused by the continual use of a keyboard and can be relieved by the use of wrist supports. Other WRULDs are caused by poor posture and can produce pains in the back, shoulders, neck or arms. Less commonly, pain may also be experienced in the thighs, calves and ankles. These problems can be mitigated by the application of ergonomic principles in the selection of working desks, chairs, foot rests and document holders. It is also important to ensure that the desk is at the correct height and the computer screen is tilted at the correct angle to avoid putting too much strain on the neck. (Ideally the user should look down on the screen at a slight angle.)

The keyboard should be detachable so that it can be positioned anywhere on the desktop and a correct posture adopted while working at the keyboard. The chair should be adjustable in height, stable and have an adjustable backrest. If the knees of the user are lower than the hips when seated, then a footrest should be provided. The surface of the desk should be non-reflecting and uncluttered.

Visual problems

There does not appear to be much medical evidence that DSE causes deterioration in eye sight, but many users suffer from visual fatigue which results in eye strain, sore eyes and headaches. Less common ailments are skin rashes and nausea.

The use of DSE may indicate that reading spectacles are needed and the Regulations make provision for this. It is possible that any prescribed lenses may only be suitable for DSE work since they will be designed to give optimum clarity at the normal distance at which screens are viewed (50–60 cm).

The screen should be adjustable in tilt angle and screen brightness and contrast. Finally, the lighting around the workstation is important. It should be bright enough to allow documents to be read easily but not too bright so that either headaches are caused or there are reflective glares on the computer screen.

Psychological problems

These are generally stress-related problems. They may have environmental causes, such as noise, heat, humidity or poor lighting, but they are usually due to high speed working, lack of breaks, poor training and poor workstation design. One of the most common problems is the lack of understanding of all or some of the software packages being used.

There are several other processes and activities where ergonomic considerations are important. These include the assembly of small components (microelectronics assembly lines) and continually moving assembly lines (car assembly plants).

13.3 Welfare and work environment issues

Welfare and work environment issues are covered by the Workplace (Health, Safety and Welfare) Regulations 1992 together with an Approved Code of Practice and additional guidance.

13.3.1 Welfare

Welfare arrangements include the provision of sanitary conveniences and washing facilities, drinking water, accommodation for clothing, facilities for changing clothing and facilities for rest and eating meals. First-aid provision is also a welfare issue, but is covered in Chapter 6.

Sanitary conveniences and washing facilities must be provided together and in a proportion to the size of the workforce. The Approved Code of Practice provides two tables offering guidance on the requisite number of water closets, wash stations and urinals for varying sizes of workforce (approximately one of each for every 25 employees). Special provision should be made for disabled workers and there should normally be separate facilities for men and women. There should be adequate protection from the weather and only as a last resort should public conveniences be used. A good supply of warm water, soap and towels must be provided as close to the sanitary facilities as possible. Hand dryers are permitted but there are concerns about their effectiveness in drying hands completely and thus removing all bacteria. In the case of temporary or remote worksites, sufficient chemical closets and sufficient washing water in containers must be provided.

All such facilities should be well ventilated and lit and cleaned regularly.

Drinking water must be readily accessible to all the workforce. The supply of drinking water must be adequate and wholesome. Normally mains water is used and should be marked as 'drinking water' if water not fit for drinking is also available.

Accommodation for clothing and facilities for changing clothing must be provided which is clean, warm, dry, well-ventilated and secure. Where workers are required to wear special or protective clothing, arrangements should be such that the workers' own clothing is not contaminated by any hazardous substances.

Facilities for rest and eating meals must be provided so that workers may sit down during break times in areas where they do not need to wear personal protective equipment. Separate rooms should be provided for smokers and non-smokers. Facilities should also be provided for pregnant women and nursing mothers to rest. Arrangements must be in place to ensure that food is not contaminated by hazardous substances.

13.3.2 Workplace environment

The issues governing the workplace environment are ventilation, heating and temperature, lighting, workstations and seating.

Ventilation

Ventilation of the workplace should be effective and sufficient and free of any impurity and air inlets should be sited clear of any potential contaminant (e.g. a chimney flue). Care needs to be taken to ensure that workers are not subject to uncomfortable draughts. The ventilation plant should have an effective visual or audible warning device fitted to indicate any failure of the plant. The plant should be properly maintained and records kept. The supply of fresh air should not normally fall below 5 to 8 litres per second per occupant.

Heating and temperature

During working hours, the temperature in all workplaces inside buildings shall be reasonable (not uncomfortably high or low). 'Reasonable' is defined in the Approved Code of Practice as at least 16°C, unless much of the work involves severe physical effort in which case the temperature should be at least 13°C. These temperatures refer to readings taken close to the workstation at working height and away from windows. The Approved Code of Practice recognizes that these minimum temperatures cannot be maintained where rooms open to the outside or where food or other products have to be kept cold. A heating or cooling method must not be used in the workplace which produces fumes, injurious or offensive to any person. Such equipment needs to be regularly maintained.

A sufficient number of thermometers should be provided and maintained to enable workers to determine the temperature in any workplace inside a building (but need not be provided in every workroom).

Where, despite the provision of local heating or cooling, the temperatures are still unreasonable, suitable protective clothing and rest facilities should be provided.

Lighting

Every workplace shall have suitable and sufficient lighting and this shall be natural lighting so far as is reasonably practicable. Suitable and sufficient emergency lighting must also be provided and maintained in any room where workers are particularly exposed to danger in the event of a failure of artificial lighting (normally due to a power cut and/or a fire). Windows and skylights should be kept clean and free from obstruction so far as is reasonably practicable unless it would prevent the shading of windows or skylights or prevent excessive heat or glare.

When deciding on the suitability of a lighting system, the general lighting requirements will be affected by the following factors:

> the availability of natural light
> the specific areas and processes, in particular any colour rendition aspects or concerns over stroboscopic effects (associated with fluorescent lights)
> the type of equipment to be used and the need for specific local lighting
> the location of visual display units and any problems of glare
> structural aspects of the workroom and the reduction of shadows
> the presence of atmospheric dust
> the heating effects of the lighting
> lamp and window cleaning and repair (and disposal issues)
> the need and required quantity of emergency lighting.

Light levels are measured in illuminance, having units of lux (lx), using a light meter and a general guide to lighting level in different workplaces is given in Table 13.1.

Poor lighting levels will increase the risk of accidents such as slips, trips and falls. More information is available on lighting from *Lighting at work,* HSG38, HSE Books.

Workstations and seating

Workstations should be arranged so that work may be done safely and comfortably. The worker should be at a suitable height relative to the work surface and there should be no need for undue bending and stretching. Workers must not be expected to stand for long periods of time particularly on solid floors. A suitable seat should be provided when a substantial part of the task can or must be done sitting. The seat should, where possible, provide adequate support for the lower back and a footrest provided for any worker whose feet cannot be placed flat on the floor. It should be made of materials suitable for the environment, be stable and, possibly, have arm rests.

Table 13.1   Typical workplace lighting levels

Workplace or type of work	Illuminance (lx)
Warehouses and stores	150
General factories or workshops	300
Offices	500
Drawing offices (detailed work)	700
Fine working (ceramics or textiles)	1000
Very fine work (watch repairs or engraving)	1400

It is also worth noting that sitting for prolonged periods can present health risks, such as blood circulation and pressure problems, vertebral and muscular damage.

Seating at work, HSG57, HSE Books, provides useful guidance on how to ensure that seating in the workplace is safe and suitable.

13.4 Noise

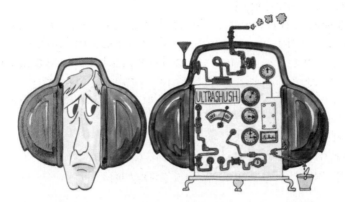

Figure 13.6   Better to control noise at source rather than wear ear protection

There was considerable concern for many years over the increasing cases of occupational deafness and this led to the introduction of Noise at Work Regulations in 1989. These regulations which are summarized in Chapter 17, require the employer:

➤ to assess noise levels and keep records
➤ reduce the risks from noise exposure by using engineering controls in the first instance and the provision and maintenance of hearing protection as a last resort
➤ provide employees with information and training
➤ if a manufacturer or supplier of equipment, to provide relevant noise data on that equipment (particularly if any of the three action levels is likely to be reached).

Sound is transmitted through the air by sound waves which are produced by vibrating objects. The vibrations cause a pressure wave which can be detected by a receiver, such as a microphone or the human ear. The ear may detect vibrations which vary from 20 to 20 000 (typically 50–16 000) cycles each second (or Hertz (Hz)). Sound travels through air at a finite speed (342 m/s at 20°C and sea level). The existence of this speed is shown by the time lag between lightning and thunder during a thunderstorm. Noise normally describes loud, sudden, harsh or irritating sounds.

Noise may be transmitted directly through the air, by reflection from surrounding walls or buildings or through the structure of a floor or building.

13.4.1 Health effects of noise

The human ear
There are three sections of the ear – the outer (or external) ear, the middle ear and the inner (or internal) ear. The sound pressure wave passes into and through the outer ear and strikes the eardrum causing it to vibrate. The eardrum is situated approximately 25 mm inside the head. The vibration of the eardrum causes the proportional movement of

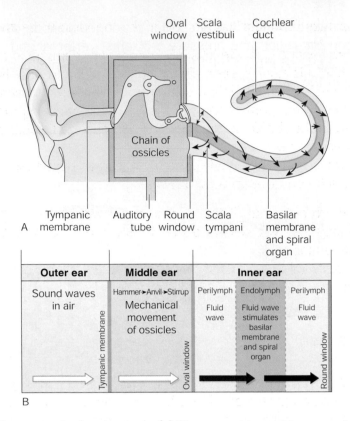

Figure 13.7 Passage of sound waves. (a) The ear with cochlea uncoiled; (b) summary of transmission. Reprinted from *Anatomy and Physiology in Health and Illness* Ninth edition, Waugh and Grant, page 195, 2002 by permission of the publisher Churchill Livingstone.

three interconnected small bones in the middle ear, thus passing the sound to the cochlea situated in the inner ear.

Within the cochlea the sound is transmitted to a fluid causing it to vibrate. The motion of the fluid induces a membrane to vibrate which, in turn, causes hair cells attached to the membrane to bend. The movement of the hair cells causes a minute electrical impulse to be transmitted to the brain along the auditory nerve. Those hairs nearest to the middle ear respond to high frequency, while those at the tip of the cochlea respond to lower frequencies.

There are about 30 000 hair cells within the ear and noise-induced hearing loss causes irreversible damage to these hair cells.

Ill-health effects of noise

Noise can lead to ear damage on a temporary (acute) or permanent (chronic) basis.

There are three principal **acute** effects:

➤ temporary threshold shift – caused by short excessive noise exposures and affects the cochlea by reducing the flow of nerve impulses to the brain. The result is a slight deafness, which is reversible when the noise is removed;

➤ tinnitus – is a ringing in the ears caused by an intense and sustained high noise level. It is caused by the over-stimulation of the hair cells. The ringing sensation continues for up to 24 hours after the noise has ceased;

➤ acute acoustic trauma – caused by a very loud noise such as an explosion. It effects either the eardrum or the bones in the middle ear and is usually reversible. Severe explosive sounds can permanently damage the eardrum.

Occupational noise can also lead to one of the following three **chronic** hearing effects:

➤ noise-induced hearing loss – results from permanent damage to the cochlea hair cells. It affects the ability to hear speech clearly but the ability to hear is not lost completely;

➤ permanent threshold shift – this results from prolonged exposure to loud noise and is irreversible due to the permanent reduction in nerve impulses to the brain. This shift is most marked at the 4000 Hz frequency, which can lead to difficulty in hearing certain consonants and some female voices;

➤ tinnitus – is the same as the acute form but becomes permanent. It is a very unpleasant condition, which can develop without warning.

It is important to note that, if the level of noise exposure remains unchanged, noise-induced hearing loss will lead to a permanent threshold shift affecting an increasing number of frequencies.

Presbycusis is the term used for hearing loss in older people which may have been exacerbated by occupational noise earlier in their lives.

13.4.2 Noise assessments

The Noise at Work Regulations 1989 specify action levels at which the hearing of employees must be protected. The conclusion as to whether any of those levels has been breached, is reached after an assessment of noise levels has been made. However, before noise assessment can be discussed, noise measurement and the statutory action levels must be described.

Noise measurement
Sound intensity is measured by a unit known as a pascal (Pa – N/m^2), which is a unit of pressure similar to that used when inflating a tyre. If noise was measured in this way, a large scale of numbers would be required ranging from 1 at one end to 1 million at the other. The sound pressure level (SPL) is a more convenient scale because:

➤ it compresses the size of the scale by using a logarithmic scale to the base 10
➤ it measures the ratio of the measured pressure, p, to a reference standard pressure, p_0, which is the pressure at the threshold of hearing (2×10^{-5} Pa).

The unit is called a decibel (dB) and is defined as:

$$SPL = 20 \log_{10} (p/p_0) \, dB$$

It is important to note that since a logarithmic scale to the base 10 is used, each increase of 3 dB is a doubling in the sound intensity. Thus, if a sound reading changes from 75 dB to 81 dB, the sound intensity or loudness has increased by four times.

Finally, since the human ear tends to distort its sensitivity to the sound it receives by being less sensitive to lower frequencies, the scale used by sound meters is weighted so that readings mimic the ear. This scale is known as the A scale and the readings known as dB(A). There are also three other scales known as B, C and D.

Table 13.2 gives some typical dB readings for common activities.

Noise is measured using a sound level meter which reads sound pressure levels in dB(A) and the **peak sound pressure** in pascals (Pa), which is the highest noise level reached by the sound. There are two basic types of sound meter – integrated and direct reading meters. Meters which integrate the reading provide an average over a particular

Table 13.2 Some typical sound
pressure levels (SPL)
dB(A) values

Activity or environment	SPL (dBA)
Threshold of pain	140
Pneumatic drill	125
Pop group or disco	110
Heavy lorry	93
Street traffic	85
Conversational speech	65
Business office	60
Living room	40
Bedroom	25
Threshold of hearing	0

time period which is an essential technique when there are large variations in sound levels. This value is known as the **continuous equivalent noise level (L_{Eq})** which is normally measured over an 8-hour period.

Direct reading devices, which tend to be much cheaper, can be used successfully when the noise levels are continuous at a near constant value.

Another important noise measurement is the **daily personal exposure level** of the worker, $L_{EP,d}$, which is measured over an 8-hour working day. Hence, if a person was exposed to 87 dB(A) over 4 hours, this would equate to a $L_{EP,d}$ of 84 dB(A) since a reduction of 3 dB(A) represents one half of the noise dose.

The Noise at Work Regulations include some very useful guides which help with the interpretation of the Regulations. Noise Guide No 3 covers 'equipment and procedures for noise surveys' and contains a nomogram which can be used to evaluate $L_{EP,d}$ when the noise occurs during several short intervals and/or at several different levels during the 8-hour period.

Noise action levels
Regulation 6, of the Noise at Work Regulations 1989, places a duty on employers to reduce the risk of damage to the hearing of employees from exposure to noise to the lowest level reasonably practicable.

The regulations further define three action levels at and above which actions to protect the hearing of employees must be taken. These levels are:

➤ first action level – daily exposure noise level, $L_{EP,d}$, of 85 dB(A)
➤ second action level – daily exposure noise level, $L_{EP,d}$, of 90 dB(A)

➤ peak action level – a peak sound pressure of 200 pascals (or 140 dB) (For practical purposes this is 140 dB(C).)

The peak action level is given to deal with loud impact or explosive noises, such as those emanating from guns or cartridge tools.

It is important to note that the first and second action levels are currently under discussion with a view to their reduction, although no decisions have yet been made.

If the daily noise exposure exceeds the first action level, then a noise assessment should be carried out and recorded by a competent person. There is a very simple test which can be done in any workplace to determine the need for an assessment. If people have to shout or have difficulty in being heard clearly by somebody who is 2 m away, then an assessment is needed.

The *Reducing Noise at Work Guide*, which accompanies the Regulations (L108, HSE Books), gives detailed guidance on noise assessments and surveys. The most important point is that the measurements should be taken at the working stations of the employees closest to the source of the noise. Other points to be included in a noise assessment are:

➤ details of the noise meter used and the date of its last calibration
➤ the number of employees using the machine, time period of usage and other work activities
➤ an indication of the condition of the machine and its maintenance schedule
➤ the work being done on the machine at the time of the assessment
➤ a schematic plan of the workplace showing the position of the machine being assessed
➤ other noise sources, such as ventilation systems, should be considered in the assessment (the control of these sources may help to reduce overall noise levels)
➤ recommendations for future actions, if any.

Other actions which the employer must undertake when the first action level is exceeded are:

➤ inform, instruct and train employees on the hearing risks;
➤ supply hearing protection to those employees requesting it;
➤ ensure that any equipment or arrangements provided under the regulations are correctly used or implemented.

The additional measures which the employer must take if the second action level is reached are:

➤ reduce and control exposure to noise by means other than hearing protection;
➤ establish hearing protection zones, marked by notices and ensure that anybody entering the zone is wearing hearing protection;
➤ supply hearing protection and ensure that it is worn.

If exposure continues over a long period, health surveillance of employees is recommended using audiometric testing. This will indicate whether there has been any deterioration in hearing ability.

Figure 13.8 Typical ear protection zone sign.

The Regulations also place statutory duties on employees:

> for noise levels above the first action level, they must use any control equipment (other than hearing protection) supplied by the employer and report any defects
> for noise levels above the second action level, they fulfil the obligations given above and wear the hearing protection provided.

13.4.3 Noise control techniques

In addition to **reduced time exposure** of employees to the noise source, there is a simple hierarchy of control techniques:

> reduction of noise at source
> reduction of noise levels received by the employee (known as attenuation)
> personal protective equipment, which should only be used when the above two remedies are insufficient.

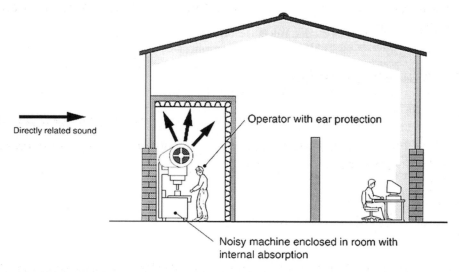

Directly related sound

Operator with ear protection

Noisy machine enclosed in room with internal absorption

Figure 13.9 Segregation of a noisy operation to benefit the whole workplace. Source HSE. Crown copyright material is reproduced with the permission of the Controller of HMSO and the Queen's Printer for Scotland.

Reduction of noise at source
There are several means by which noise could be reduced at source:

> change the process or equipment (e.g. replace solid tyres with rubber tyres or replace diesel engines with electric motors);
> change the speed of the machine;
> improve the maintenance regime by regular lubrication of bearings, tightening of belt drives.

Attenuation of noise levels
There are many methods of attenuating or reducing noise levels and these are covered in detail in the guide to the Regulations. The more common ones will be summarized here.

Orientation or re-location of the equipment – turn the noisy equipment away from the workforce or locate it away in separate and isolated areas.

Enclosure – surrounding the equipment with a good sound insulating material can reduce sound levels by up to 30 dB(A). Care will need to be taken to ensure that the machine does not become overheated.

Screens or absorption walls – can be used effectively in areas where the sound is reflected from walls. The walls of the rooms housing the noisy equipment are lined with sound absorbent material, such as foam or mineral wool, or sound absorbent (acoustic) screens are placed around the equipment.

Damping – the use of insulating floor mountings to remove or reduce the transmission of noise and vibrations through the structure of the building such as girders, wall panels and flooring.

Lagging – the insulation of pipes and other fluid containers to reduce sound transmission (and, incidentally, heat loss).

Silencers – normally fitted to engines which are exhausting gases to atmosphere. Silencers consist of absorbent material or baffles.

Isolation of the workers – the provision of sound-proofed workrooms or enclosures isolated away from noisy equipment (a power station control room is an example of worker isolation).

13.4.4 Personal ear protection

The following factors should be considered when selecting personal ear protection:

➤ suitability for the range of sound spectrum of frequencies to be encountered
➤ acceptability and comfort of the wearer
➤ durability
➤ hygiene considerations
➤ maintenance and storage arrangements
➤ cost.

There are two main types of ear protection – earplugs and ear defenders (earmuffs).
Earplugs are made of sound absorbent material and fit into the ear. They can be reusable or disposable and are able to fit most people. Their effectiveness depends on the quality of the fit in the ear which, in turn, depends on the level of training given to the wearer. Permanent earplugs come in a range of sizes so that a good fit is obtained. The effectiveness of earplugs decreases with age and should be replaced at the intervals specified by the supplier. The main disadvantage is that they do not reduce the sound transmitted through the bone structure which surrounds the ear.
Ear defenders (earmuffs) offer a far better reduction of all sound frequencies. They are generally more acceptable to workers because they are more comfortable to wear. They also reduce the sound intensity transmitted through the bone structure surrounding the ear. A communication system can be built into earmuffs. However, they may be less effective if the user has long hair or is wearing spectacles or large ear rings. They may also be less effective if worn with helmets or face shields. Maintenance is an important factor with earmuffs and should include checks for wear and tear and general cleanliness.
Selection of suitable ear protection is very important since they should not just reduce sound intensities below the statutory action levels but also reduce those intensities at

particular frequencies. Normally advice should be sought from a competent supplier who will be able to advise ear protection to suit a given spectrum of noise using 'octave band analysis'.

Finally, it is important to stress that the use of ear protection must be well supervised to ensure that, not only is it being worn correctly, but that it is, in fact, actually being worn.

13.5 Heat and radiation hazards

13.5.1 Extremes of temperature

The human body is very sensitive to relatively small changes in external temperatures. Food not only provides energy and the build-up of fat reserves, but also generates heat which needs to be dissipated to the surrounding environment. The body also receives heat from its surroundings. The body temperature is normally around 37°C and will attempt to maintain this temperature irrespective of the temperature of the surroundings. Therefore, if the surroundings are hot, sweating will allow heat loss to take place by evaporation caused by air movement over the skin. On the other hand, if the surroundings are cold, shivering causes internal muscular activity, which generates body heat.

At high temperatures, the body has more and more difficulty in maintaining its natural temperature unless sweating can take place and therefore water must be replaced by drinking. If the surrounding air has high humidity, evaporation of the sweat cannot take place and the body begins to overheat. This leads to heart strain and, in extreme cases, heat stroke. It follows that when working is required at high temperatures, a good supply of drinking water should be available and, further, if the humidity is high, a good supply of ventilation air is also needed. Heat exhaustion is a particular hazard in confined spaces.

At low temperatures, the body will lose heat too rapidly and the extremities of the body will become very cold leading to frostbite and possibly the loss of limbs. Under these conditions, thick warm clothing and external heating will be required.

In summary, extremes of temperature require special measures, particularly if it is accompanied by extremes of humidity. Frequent rest periods will be necessary to allow the body to acclimatize to the conditions.

13.5.2 Ionising radiation

Ionising radiation is emitted from radioactive materials, either in the form of directly ionising alpha and beta particles or indirectly ionising X- and gamma rays or neutrons. It has a high energy potential and an ability to penetrate, ionise and damage body tissue and organs.

All matter consists of atoms within which is a nucleus, containing protons and neutrons, and orbiting electrons. The number of electrons within the atom

Figure 13.10 Typical ionising sign.

defines the element – hydrogen has 1 electron and lead has 82 electrons. Some atoms are unstable and will change into atoms of another element, emitting ionising radiation in the process. The change is called radioactive decay and the ionising radiations most commonly emitted are alpha and beta particles and gamma rays. X-rays are produced by bombarding a metal target with electrons at very high speeds using very high voltage electrical discharge equipment. Neutrons are released by nuclear fission and are not normally found in manufacturing processes.

Alpha particles consist of two protons and two neutrons and have a positive charge. They have little power to penetrate the skin and can be stopped using very flimsy material, such as paper. Their main route into the body is by ingestion.

Beta particles are high speed electrons whose power of penetration depends on their speed, but penetration is usually restricted to 2 cm of skin tissue. They can be stopped using aluminium foil. There are normally two routes of entry into the body – inhalation and ingestion.

Gamma rays, which are similar to **X-rays**, are electromagnetic radiations and have far greater penetrating power than alpha or beta particles. They are produced from nuclear reactions and can pass through the body.

There are two principal units of radiation – the becquerel (1 Bq) which measures the amount of radiation in a given environment and the millisievert (1 mSv) which measures the ionising radiation dose received by a person.

Ionising radiation occurs naturally as well as from man-made processes and about 87% of all radiation exposure is from natural sources. The Ionising Radiations Regulations 1999 specify a range of dose limits, some of which are given in Table 13.3.

Harmful effects of ionising radiation

Ionising radiation attacks the cells of the body by producing chemical changes in the cell DNA by ionising it (thus producing free radicals) which leads to abnormal cell growth. The effects of these ionising attacks depend on the following factors:

> the size of the dose – the higher the dose then the more serious will be the effect
> the area or extent of the exposure of the body – the effects may be far less severe if only a part of the body (e.g. an arm) receives the dose

Table 13.3 Typical radiation dose limits

	Dose (mSv)	Area of body
Employees aged 18 years +	20	Whole body per year
Trainees 16–18 years	6	Whole body per year
Any other person	1	Whole body per year
Women employees of child-bearing age	13	The abdomen in any consecutive 3-month period
Pregnant employees	1	During the declared term of the pregnancy

➤ the duration of the exposure – a long exposure to a low dose is likely to be less harmful than a short exposure to the same quantity of radiation.

Acute exposure can cause, dependent on the size of the dose, blood cell changes, nausea and vomiting, skin burns and blistering, collapse and death. Chronic exposure can lead to anaemia, leukaemia and other forms of cancer. It is also known that ionising radiation can have an adverse effect on the function of human reproductive organs and processes. Increases in the cases of sterility, stillbirths and malformed fetuses have also been observed.

The health effects of ionising radiation may be summarized into two groups – **somatic effects** which refers to cell damage in the person exposed to the radiation dose and **genetic effects** which refers to the damage done to the children of the irradiated person.

Sources of ionising radiation

The principal workplaces which could have ionising radiation present are the nuclear industry, medical centres (hospitals and research centres) and educational centres. Radioactive processes are used for the treatment of cancers and radioactive isotopes are used for many different types of scientific research. X-rays are used extensively in hospitals, but they are also used in industry for non-destructive testing (e.g. crack detection in welds). Smoke detectors, used in most workplaces, also use ionising radiations.

Ionising radiations can also occur naturally – the best example being radon, which is a radioactive gas that occurs mainly at or near granite outcrops where there is a presence of uranium. It is particularly prevalent in Devon and Cornwall. The gas enters buildings normally from the substructure through cracks in flooring or around service inlets. The Ionising Regulations have set two action levels above which remedial action, such as sumps and extraction fans, have to be fitted to lower the radon level in the building. The first action level is $400\,Bq/m^3$ in workplaces and $200\,Bq/m^3$ in domestic properties. At levels above $1000\,Bq/m^3$, remedial action should be taken within one year.

Personal radiation exposure can be measured using a film badge, which is worn by the employee over a fixed time interval. The badge contains a photographic film which, after the time interval, is developed and an estimate of radiation exposure is made. A similar device, known as a radiation dose meter or detector, can be positioned on a shelf in the workplace for three months, so that a mean value of radiation levels may be measured. Instantaneous radiation values can be obtained from portable hand-held instruments, known as geiger counters, which continuously sample the air for radiation levels. Similar devices are available to measure radon levels.

13.5.3 Non-ionising radiation

Non-ionising radiation includes ultraviolet, visible light (this includes lasers which focus or concentrate visible light), infra-red and microwave radiations. Since the wavelength is relatively long, the energy present is too low to ionise atoms which make up matter. The action of non-ionising radiation is to heat cells rather than change their chemical composition.

Other than the Health and Safety at Work Act 1974, there is no specific set of regulations governing non-ionising radiation. However, the Personal Protective

Equipment at Work Regulations 1992 are particularly relevant since the greatest hazard is tissue burning of the skin or the eyes.

Ultraviolet radiation occurs with sunlight and with electric arc welding. In both cases, the skin and the eyes are at risk from the effect of burning. The skin will burn (as in sunburn) and repeated exposure can lead to skin cancer. Skin which is exposed to strong sun light, should be protected either by clothing or barrier creams. This problem has become more common with the reduction in the ozone layer (which filters out much ultraviolet light). The eyes can be affected by a form of conjunctivitis which feels like grit in the eye, and is called by a variety of names dependent on the activity causing the problem. Arc welders call it 'arc eye' or 'welder's eye' and skiers 'snow blindness'. Cataracts caused by the action of ultraviolet radiation on the eye lens is another possible outcome of exposure.

Lasers use visible light and light from the invisible wavelength spectrum (infra-red and ultraviolet). As the word laser implies, they produce 'light **a**mplification by **s**timulated **e**mission of **r**adiation'. This light is highly concentrated and does not diverge or weaken with distance and the output is directly related to the chemical composition of the medium used within the particular laser. The output beam may be pulsating or continuous – the choice being dependent on the task of the laser. Lasers have a large range of applications including bar code reading at a supermarket checkout, the cutting and welding of metals and accurate measurement of distances and elevations required in land and mine surveying. They are also extensively used in surgery for cataract treatment and the sealing of blood vessels.

Lasers are classified into five classes (1, 2, 3a, 3b and 4) in ascending size of power output. Classes 1 and 2 are relatively low hazard and only emit light in the visible band. Classes 3a, 3b and 4 are more hazardous and the appointment of a laser safety officer is recommended. All lasers should carry information stating their class and any precautions required during use.

The main hazards associated with lasers are eye and skin burns, toxic fumes, electricity and fire. The vast majority of accidents with lasers affect the eyes. Retinal damage is the most common and is irreversible. Cataract development and various forms of conjunctivitis can also result from laser accidents. Skin burning and reddening (erythema) are less common and are reversible.

Infra-red radiation is generated by fires and hot substances and can cause eye and skin damage similar to that produced by ultraviolet radiation. It is a particular problem to fire fighters and those who work in foundries or near furnaces. Eye and skin protection are essential.

Microwaves are used extensively in cookers and mobile telephones and there are ongoing concerns about associated health hazards (and several inquiries are currently underway). The severity of any hazard is proportional to the power of the microwaves. The principal hazard is the heating of body cells, particularly those with little or no blood supply to dissipate the heat. This means that tissues such as the eye lens are most at risk from injury. However, it must be stressed that any risks are higher for items, such as cookers, than for low-powered devices, such as mobile phones.

The measurement of non-ionising radiation normally involves the determination of the power output being received by the worker. Such surveys are best performed by specialists in the field, since the interpretation of the survey results requires considerable technical knowledge.

13.5.4 Radiation protection strategies

Ionising radiation
Protection is obtained by the application of shielding, time and distance either individually or, more commonly, a mixture of all three.

Shielding is the best method because it is an 'engineered' solution. It involves the placing of a physical shield, such as a layer of lead, steel and concrete, between the worker and the radioactive source. The thicker the shield the more effective it is.

Time involves the use of the reduced time exposure principle and thus reduces the accumulated dose.

Distance works on the principle that the effect of radiation reduces as the distance between the worker and the source increases.

Other measures include the following:

➤ effective emergency arrangements
➤ training of employees
➤ the prohibition of eating, drinking and smoking adjacent to exposed areas
➤ a high standard of personal cleanliness and first-aid arrangements
➤ strict adherence to personal protective equipment arrangements, which may include full body protection and respiratory protection equipment
➤ procedures to deal with spillages and other accidents
➤ prominent signs and information regarding the radiation hazards
➤ medical surveillance of employees.

The Ionising Radiations Regulations 1999 specify a range of precautions which must be taken, including the appointment of a Radiation Protection Supervisor and a Radiation Protection Adviser.

The Radiation Protection Supervisor must be appointed by the employer to advise on the necessary measures for compliance with the Regulations and its Approved Code of Practice. The person appointed, who is normally an employee, must be competent to supervise the arrangements in place and have received relevant training.

The Radiation Protection Adviser is appointed by the employer to give advice to the Radiation Protection Supervisor and employer on any aspect of working with ionising radiation including the appointment of the Radiation Protection Supervisor. The Radiation Protection Adviser is often an employee of a national organization with expertise in ionising radiation.

Non-ionising radiation
For ultraviolet and infra-red radiation, eye protection in the form of goggles or a visor is most important, particularly when undertaking arc welding or furnace work. Skin protection is also likely to be necessary for the hands, arms and neck in the form of gloves, sleeves and a collar. For construction and other outdoor workers, protection from sunlight is important, particularly for the head and nose. Barrier creams should also be used.

For laser operations, engineering controls such as fixed shielding and the use of non-reflecting surfaces around the workstation are recommended. For laser in the higher class numbers, special eye protection is recommended. A risk assessment should be undertaken before a laser is used.

Engineering controls are primarily used for protection against microwaves. Typical controls include the enclosure of the whole microwave system in a metal surround and the use of an interlocking device that will not allow the system to operate unless the door is closed.

13.6 The causes and prevention of workplace stress

In 2001, the HSE estimated that stress in the workplace cost approximately 6.7 million days lost each year and society between £3.7 billion and £3.8 billion. This has been accompanied by an increase in civil claims resulting from stress at work.

Stress is not a disease – it is the natural reaction to excessive pressure. Stress can lead to an improved performance, but is not a good thing, since it is likely to lead to both physical and mental ill-health, such as high blood pressure, peptic ulcers, skin disorders and depression.

Most people experience stress at some time during their lives, during an illness or death of a close relative or friend. However, recovery normally occurs after the particular crisis has passed. The position is, however, often different in the workplace because the underlying causes of the stress, known as work-related stressors, are not relieved but continue to build up the stress levels until the employee can no longer cope.

The basic workplace stressors are:

➤ the job itself – boring or repetitive, unrealistic performance targets or insufficient training
➤ individual responsibility – ill-defined roles and too much responsibility with too little power to influence the job outputs
➤ working conditions – unsafe practices, threat of violence, excessive noise or heat, poor lighting, lack of flexibility in working hours to meet domestic requirements
➤ management attitudes – poor communication, consultation or supervision, negative health and safety culture, lack of support in a crisis
➤ relationships – unhappy relationship between workers, bullying, sexual and racial harassment.

Possible solutions to all these stressors have been addressed throughout this book and involve the creation of a positive health and safety culture, effective training and consultation procedures and a set of health and safety arrangements which work on a day-to-day basis.

The following additional measures have also been found to be effective by some employers:

➤ take a positive attitude to stress issues by becoming familiar with its causes and controls;
➤ take employees' concerns seriously and develop a counselling system which will allow a frank, honest and confidential discussion of stress-related problems;
➤ develop an effective system of communication and consultation and ensure that periods of uncertainty are kept to a minimum;
➤ set out a simple policy on work-related stress and include stressors in risk assessments;
➤ ensure that employees are given adequate and relevant training and realistic performance targets;
➤ develop an effective employee appraisal system which includes mutually agreed objectives;

➤ discourage employees from working excessive hours and/or missing break periods (this may involve a detailed job evaluation);

➤ encourage employees to improve their lifestyle (e.g. many local health authorities provide smoking cessation advice sessions);

➤ monitor incidents of bullying, sexual and racial harassment and, where necessary, take disciplinary action;

➤ avoid a blame culture over accidents and incidents of ill-health.

The individual can also take action if he feels that he is becoming over-stressed. Regular exercise, change of job, review of diet and talking to somebody, preferably a trained counsellor, are all possibilities.

Workplace stress has no specific health and safety regulations but is covered by the duties imposed by the Health and Safety at Work Act and the Management of Health and Safety at Work Regulations 1999 to:

➤ ensure so far as is reasonably practicable that workplaces are safe and without risk to health;

➤ carry out a risk assessment relating to the risks to health;

➤ introduce and maintain appropriate control measures.

There have been several successful civil actions for compensatory claims resulting from the effects of workplace stress. However, the Court of Appeal in 2002 redefined the guidelines under which workplace stress compensation claims may be made. Their full guidelines should be consulted and consist of sixteen points. In summary, these guidelines are as follows:

➤ no occupations should be regarded as intrinsically dangerous to mental health

➤ it is reasonable for the employer to assume that the employee can withstand the normal pressures of the job, unless some particular problem or vulnerability has developed

➤ the employer is only in breach of duty if they have failed to take the steps that are reasonable in the circumstances

➤ the size and scope of the employer's operation, its resources and the demands it faces are relevant in deciding what is reasonable; including the interests of other employees and the need to treat them fairly in, for example, the redistribution of duties

➤ if a confidential counselling advice service is offered to the employee, the employer is unlikely to be found in breach of his duty

➤ if the only reasonable and effective action is to dismiss or demote the employee, the employer is not in breach of his duty by allowing a willing worker to continue in the same job

➤ the assessment of damages will take account of any pre-existing disorder or vulnerability and the chance that the claimant would have succumbed to a stress-related disorder in any event.

Stress usually occurs when people feel that they are losing control of a situation. In the workplace, this means that the individual no longer feels that they can cope with the demands made on them. Many such problems can be partly solved by listening to rather than talking at people.

13.7 Causes and prevention of workplace violence

Violence at work, particularly from dissatisfied customers, clients, claimants or patients, causes a lot of stress and in some cases injury. This is not only physical violence as people may face verbal and mental abuse, discrimination, harassment and bullying. Fortunately, physical violence is still rare, but violence of all types has risen significantly in recent years. Violence at work is known to cause pain, suffering, anxiety and stress, leading to financial costs due to absenteeism and higher insurance premiums to cover increased civil claims. It can be very costly to ignore the problem.

In 1999 the Home Office and the Health and Safety Executive published a comprehensive report entitled *Violence at Work: Findings from the British Crime Survey.* This was updated with a joint report *Violence at Work: New Findings of The British Crime Survey 2000,* which was published in July 2001. This report shows the extent of violence at work and how it has changed during the period 1991–1999 (Table 13.4).

The report defines violence at work as:

All assaults or threats which occurred while the victim was working and were perpetuated by members of the public.

Figure 13.11 Security coded access and surveillance CCTV camera. Source HSE. Crown copyright material is reproduced with the permission of the Controller of HMSO and the Queen's Printer for Scotland.

Table 13.4 Trend in physical assaults and threats at work, 1991–1999 (based on working adults of working age)

Number of incidents (000s)	1991	1993	1995	1997	1999
All violence	947	1275	1507	1226	1288
Assaults	451	652	729	523	634
Threats	495	607	779	703	654
Number of victims (000s)					
All violence	472	530	570	649	604
Assaults	227	287	290	275	304
Threats	264	286	352	395	338

Source: British Crime Survey 1999

Physical assaults include the offences of common assault, wounding, robbery and snatch theft. Threats include both verbal threats, made to or against the victim and non-verbal intimidation. These are mainly threats to assault the victim and, in some cases, to damage property.

Excluded from the survey are violent incidents where there was a relationship between the victim and the offender and also where the offender was a work colleague. The latter category was excluded because of the different nature of such incidents.

The British Crime Survey shows that the number of incidents and victims rose rapidly between 1991 and 1995 but then declined to 1997. Unfortunately, since 1997 the decline seems to have reversed as the number of incidents has increased by 5%.

It is interesting that almost half of the assaults and a third of the threats happened after 1800 hrs which suggests that the risks are higher if people work at night or in the late evening. Sixteen per cent of the assaults involved offenders under the age of 16 and were mainly against teachers or other education workers.

Violence at work is defined by the HSE as:

any incident in which an employee is abused, threatened or assaulted in circumstances relating to their work.

In recognition of this, the HSE have produced a useful guide to employers which includes a four stage action plan and some advice on precautionary measures (*Violence at Work a guide for employers*, INDG69 (rev)). The employer is just as responsible, under health and safety legislation, for protecting employees from violence as they are for any other aspects of their safety.

The Health and Safety at Work Act 1974 puts broad general duties on employers and others to protect the health and safety of staff. In particular, section 2 of the HSW Act gives employers a duty to safeguard, so far as is reasonably practicable, the health, safety and welfare at work of their staff.

Employers also have a common law general duty of care towards their staff, which extends to the risk of violence at work. Legal precedents (see *West Bromwich Building Society v Townsend* [1983] IRLR 147 and *Charlton v Forrest Printing Ink Company Limited* [1980] IRLR 331) show that employers have a duty to take reasonable care to see that their staff are not exposed to unnecessary risks at work, including the risk of injury by criminals. In carrying out their duty to provide a safe system of work and a safe working place, employers should, therefore, have regard to, and safeguard their staff against, the risk of injury from violent criminals.

The HSE recommend the following four-point action plan:

1 find out if there is a problem
2 decide on what action to take
3 take the appropriate action
4 check that the action is effective.

13.7.1 Find out if there is a problem

This involves a risk assessment to determine what the real hazards are. It is essential to ask people at the workplace and, in some cases, a short questionnaire may be useful. Record all incidents to get a picture of what is happening over time, making sure that all relevant detail is recorded. The records should include:

➤ a description of what happened
➤ details of who was attacked, the attacker and any witnesses
➤ the outcome, including how people were affected and how much time was lost
➤ information on the location of the event.

Due to the sensitive nature of some aggressive or violent actions, employees may need to be encouraged to report incidents and protected from future aggression.

All incidents should be classified so that an analysis of the trends can be examined. Consider the following:

➤ fatalities
➤ major injury
➤ less severe injury or shock which requires first-aid treatment, outpatient treatment, time off work or expert counselling
➤ threat or feeling of being at risk or in a worried or distressed state.

13.7.2 Decide on what action to take

It is important to evaluate the risks and decide who may be harmed and how this is likely to occur. The threats may be from the public or co-workers at the workplace or it may be as a result of visiting the homes of customers. Consultation with employees or other people at risk will improve their commitment to control measures and will make the precautions much more effective. The level of training and information provided, together with the general working environment and the design of the job, all have a significant influence on the level of risk.

Those people at risk could include those working in:

➤ reception or customer service points
➤ enforcement and inspection
➤ lone working situations and community based activities
➤ front line service delivery
➤ education and welfare
➤ catering and hospitality
➤ retail petrol and late night shopping operations
➤ leisure facilities, especially if alcohol is sold
➤ healthcare and voluntary roles
➤ policing and security
➤ mental health units or in contact with disturbed people
➤ cash handling or control of high value goods.

Consider the following issues:

➤ quality of service provided
➤ design of the operating environment
➤ type of equipment used
➤ designing the job.

Quality of service provided
The type and quality of service provision has a significant effect on the likelihood of violence occurring in the workplace. Frustrated people whose expectations have not

305

been met and who are treated in an unprofessional way may believe they have the justification they seek to cause trouble.

Sometimes circumstances are beyond the control of the staff member and potentially violent situations need to be defused. The use of correct skills can turn a dissatisfied customer into a confirmed supporter simply by careful response to their concerns. The perceived lack of or incorrect information can cause significant frustrations.

Design of the operating environment

Personal safety and service delivery are very closely connected and have been widely researched in recent years. This has resulted in many organizations altering their facilities to reduce customer frustration and enhance sales. It is interesting that most service points experience less violence when they remove barriers or screens, but the transition needs to be carefully planned in consultation with staff and other measures adopted to reduce the risks and improve their protection.

The layout, ambience, colours, lighting, type of background music, furnishings including their comfort, information, things to do while waiting and even smell all have a major impact. Queue jumping causes a lot of anger and frustration and needs effective signs and proper queue management, which can help to reduce the potential for conflict.

Wider desks, raised floors and access for special needs, escape arrangements for staff, carefully arranged furniture and screening for staff areas can all be utilized.

Type of security equipment used

There is a large amount of equipment available and expert advice is necessary to ensure that it is suitable and sufficient for the task. Some measures that could be considered include the following:

➤ **Access control** to protect people and property. There are many variations from staffed and friendly receptions, barriers with swipe-cards and simple coded security locks. The building layout and design may well partly dictate what is chosen. People inside the premises need access passes so they can be identified easily.

➤ **Closed circuit television** is one of the most effective security arrangements to deter crime and violence. Because of the high cost of the equipment, it is essential to ensure that proper independent advice is obtained on the type and the extent of the system required.

➤ **Alarms** – there are three main types:

➤ Intruder alarms fitted in buildings to protect against unlawful entry, particularly after hours

➤ Panic alarms used in areas such as receptions and interview rooms covertly located so that they can be operated by the staff member threatened

➤ Personal alarms carried by an individual to attract attention and to temporarily distract the attacker

➤ **Radios and pagers** can be a great asset to lone workers in particular, but special training is necessary as good radio discipline with a special language and codes are required

➤ **Mobile phones** are an effective means of communicating and keeping colleagues informed of people's movements and problems such as travel delays. Key numbers should be inserted for rapid use in an emergency.

Job design

Many things can be done to improve the way in which the job is carried out to improve security and avoid violence. These include:

➤ using cashless payment methods
➤ keeping money on the premises to a minimum
➤ careful check of customer or client's credentials
➤ careful planning of meetings away from the workplace
➤ team work where suspected aggressors may be involved
➤ regular contact with workers away from their base. There are special services available to provide contact arrangements
➤ avoidance of lone working as far as is reasonably practicable
➤ thinking about how staff who have to work shifts or late hours will get home. Safe transport and/or parking areas may be required
➤ setting up support services to help victims of violence and, if necessary, other staff who could be affected. They may need debriefing, legal assistance, time off work to recover or counselling by experts.

13.7.3 Take the appropriate action

The arrangements for dealing with violence should be included in the safety policy and managed like any other aspect of the health and safety procedures. Action plans should be drawn up and followed through using the consultation arrangements as appropriate. The police should also be consulted to ensure that they are happy with the plan and are prepared to play their part in providing back up and the like.

13.7.4 Check that the action is effective

Ensure that the records are being maintained and any reported incidents are investigated and suitable action taken. The procedures should be regularly audited and changes made if they are not working properly.

Victims should be provided with help and assistance to overcome their distress, through debriefing, counselling, time off to recover, legal advice and support from colleagues.

13.8 Practice NEBOSH questions for Chapter 13

1. (a) Explain the meaning of the term 'daily personal noise exposure' ($L_{EP,d}$).
   (b) Outline the actions required when employees' exposure to noise is found to be in excess of 90 dB(A) $L_{EP,d}$. (March 2001)
2. (a) Define the term 'ergonomics'.
   (b) List six observations made during an inspection of a machine operation which may suggest that the machine has not been ergonomically designed. (March 2001)

3. Outline the factors that could contribute towards the development of work-related upper limb disorders (WRULDs) among employees working at a supermarket checkout. (December 2001)

4. (a) Identify the possible effects on health that may be caused by working in a hot environment such as a foundry.
   (b) Outline measures that may be taken to help prevent the health effects identified in (a). (December 2001)

5. (a) Describe the two main types of personal hearing protection.
   (b) Identify four reasons why personal hearing protection may fail to provide adequate protection against noise. (December 2001)

6. Outline the health, safety and welfare issues that a company might need to consider before introducing a night shift to cope with an increased demand for its products. (June 2001)

7. In relation to the Noise at Work Regulations 1989:
   (i) state, in dB(A), the first and second action levels
   (ii) outline the measures that should be taken when employees are exposed to noise levels in excess of the second action level. (June 2001)

8. (a) For each of the following types of non-ionising radiation, identify a source and state the possible ill-health effects on exposed individuals:
   (i) infrared radiation
   (ii) ultraviolet radiation
   (b) Identify the general methods for protecting people against exposure to non-ionising radiation. (June 2001)

9. (a) Outline the possible health risks associated with working in a seated position for prolonged periods of time.
   (b) Outline the features of a suitable seat for sedentary work. (June 2001)

10. In relation to work-related upper limb disorders (WRULDs):
    (i) identify the typical symptoms that might be experienced by affected individuals
    (ii) Outline the factors that would increase the risk of developing WRULDs. (December 2000)

11. A computer operator has complained of neck and back pain. Outline the features associated with the workstation that might have contributed towards this condition. (June 2000)

12. Inadequate lighting in the workplace may affect the level of stress among employees. Outline eight other factors associated with the physical environment that may increase stress at work. (March 2000)

13. Outline the requirements of the Workplace (Health, Safety and Welfare) Regulations 1992 relating to the provision of welfare facilities. (March 2000)

14. Outline the factors that may lead to unacceptable levels of occupational stress among employees. (June 1999)

15. A pneumatic drill is to be used during extensive repair work to the floor of a busy warehouse.
    (i) Identify, by means of a labelled sketch, three possible transmission paths the noise from the drill could take.
    (ii) Outline appropriate control measures to reduce the noise exposures of the operator and the warehouse staff. (June 1999)

16. (a) Identify two types of non-ionising radiation, giving an occupational source of each.
    (b) Outline the health effects associated with exposure to non-ionising radiation. (June 1999)

17. Explain the meaning of the following terms in relation to noise control:
 (i) silencing
 (ii) absorption
 (iii) damping
 (iv) isolation. (June 1998)

Construction activities – hazards and control

Introduction

The construction industry covers a wide range of activities from large-scale civil engineering projects to very small house extensions. The construction industry has approximately 200 000 firms, of whom only 12 000 employ more than seven people – many of these firms are much smaller. The use of sub-contractors is very common at all levels of the industry.

It is most likely that everybody will be aware or involved with some aspect of the construction industry at their place of work – either in terms of the repair and modification of existing buildings or a major new engineering project. It is, therefore, important that the health and safety practitioner has some basic knowledge of the hazards and health and safety legal requirements associated with construction.

Over many years, the construction industry has had a poor health and safety record. In 1966, there were 292 fatalities in the industry and by 1995 this figure had reduced to 62, but by 2000/2001 the figure had increased to 106. These figures include deaths of members of the public, including children playing on construction sites. Most of these fatalities (over 70%) were caused by falls from height.

At a conference, organized by the Health and Safety Commission in February 2001, to address the problem, it was noted that at least two construction workers are being killed each week. Targets were set to reduce the number of fatalities and major injuries by 40% over a 4-year period.

Due to the fragmented nature of the industry and its accident and ill-health record, the recent construction industry legal framework has concentrated on hazards associated with the industry, welfare issues and the need for management and control at all stages of a construction project. In addition to the Health and Safety at Work Act 1974 and its associated relevant regulations, there are three sets of specific construction regulations which provide this legal framework, as follows:

➤ Construction (Head Protection) Regulations 1989
➤ Construction (Design and Management) Regulations 1994 and Amendment Regulations 2000
➤ Construction (Health, Safety and Welfare) Regulations 1996.

A summary of these regulations is given in Chapter 17.

14.2 The scope of construction

The scope of the construction industry is very wide. The most common activity is general building work which is domestic, commercial or industrial in nature. This work may be new building work, such as a building extension or, more commonly, the refurbishment, maintenance or repair of existing buildings. Larger civil engineering projects involving road and bridge building, water supply and sewage schemes and river and canal work all come within the scope of construction.

The work could involve hazardous operations, such as demolition or roof work, or contact with hazardous materials, such as asbestos or lead. Construction also includes the use of woodworking workshops together with woodworking machines and their associated hazards, painting and decorating and the use of heavy machinery. It will often require work to take place in confined spaces, such as excavations and underground chambers.

Finally, at any given time, there are many young people receiving training on site in the various construction trades. These trainees need supervision and structured training programmes.

14.3 Construction hazards and controls

The Construction (Health, Safety and Welfare) Regulations 1996 deal with the main hazards likely to be found on a building site. In addition to these specific hazards, there will be the more general hazards (e.g. manual handling, electricity, noise, etc.) which have been discussed in more detail in earlier chapters. The hazards and controls identified in the Construction Regulations are as follows.

14.3.1 Safe place of work

Safe access to and egress from the site and the individual places of work on the site are fundamental to a good health and safety environment. This clearly requires that all ladders, scaffolds, gangways, stairways and passenger hoists are safe for use. It further requires that all excavations are fenced, the site is tidy and proper arrangements are in place for the storage of materials and the disposal of waste. The site needs to be adequately lit and secured against intruders, particularly children, when it is unoccupied. Such security will include:

- secure and locked gates with appropriate notices posted
- a secure and undamaged perimeter fence with appropriate notices posted
- all ladders either stored securely or boarded across their rungs
- all excavations covered
- all mobile plant immobilized and fuel removed, where practicable, and services isolated
- secure storage of all inflammable and hazardous substances
- visits to local schools to explain the dangers present on a construction site. This has been shown to reduce the number of child trespassers
- if unauthorized entry persists, then security patrols and closed circuit television may need to be considered.

Figure 14.1 Secure site access gate. Source HSE. Crown copyright material is reproduced with the permission of the Controller of HMSO and the Queen's Printer for Scotland.

14.3.2 Protection against falls

Falls from a height are the most common cause of serious injury or death in the construction industry. The Regulations require that suitable and sufficient steps be taken to prevent falls and specifies that the maximum unprotected gap between the toe board and guardrail of a scaffold is 470 mm. This implies the use of an intermediate guardrail, although other means, such as additional toe boards or screening, may be used. A minimum width of working platforms of 600 mm is also specified, together with requirements for personal suspension equipment and means of arresting falls (such as safety nets).

A hierarchy of control measures is presented in the Regulations for the prevention of injuries due to falls. The hierarchy is:

➤ remove the possibility of falling more than 2 m (for example, by undertaking the work at ground level);
➤ protect against the hazard of falling more than 2 m (for example, by using safety nets);
➤ protect the person from falling more than 2 m (for example, by the provision of safety harnesses);
➤ mitigate the consequences of falling more than 2 m (for example, by the use of air bags).

14.3.3 Fragile roofs

Roof work, particularly work on pitched roofs, is hazardous and requires a specific risk assessment and method statement (see later under the management of construction activities for a definition) prior to the commencement of work. Particular hazards are fragile roofing materials, including those materials which deteriorate and become more brittle with age and exposure to sunlight, exposed edges, unsafe access equipment and

313

Figure 14.2 Proper precautions must always be taken when working on or near fragile roofs.

falls from girders, ridges or purlins. There must be suitable means of access, such as scaffolding, ladders and crawling boards and suitable barriers, guardrails or covers where people work near to fragile materials and roof lights. Suitable warning signs indicating that a roof is fragile should be on display at ground level.

There are other hazards associated with roof work – overhead services and obstructions, the use of equipment, such as gas cylinders and bitumen boilers and manual handling hazards.

It is essential that only trained and competent persons are allowed to work on roofs and that they wear footwear with a good grip. It is good practice to ensure that a person does not work alone on a roof.

14.3.4 Protection against falling objects

Both construction workers and members of the public need to be protected from the hazards associated with falling objects. Both groups should be protected by the use of covered walkways or suitable netting to catch falling debris. Waste material should be brought to ground level by the use of chutes or hoists. Waste should not be thrown and only minimal quantities of building materials should be stored on working platforms. The Construction (Head Protection) Regulations 1989 virtually mandates employers to supply head protection (hard hats) to employees whenever there is a risk of head injury from falling objects. (Sikhs wearing turbans are exempted from this requirement.) The employer is also responsible for ensuring that hard hats are properly maintained and replaced when they are damaged in any way. Self-employed workers must supply and maintain their own head protection. Visitors to construction sites should always be supplied with head protection and mandatory head protection signs displayed around the site.

14.3.5 Demolition

Demolition is one of the most hazardous construction operations and is responsible for more deaths and major injuries than any other activity. The management of demolition work is controlled by the Construction (Design and Management) Regulations 1994 and requires a planning supervisor and a health and safety plan (see the next section of this chapter).

Before any work is started, a full site investigation must be made by a competent person to determine the hazards and associated risks which may affect the demolition workers and members of the public who may pass close to the demolition site. The investigation should cover the following topics:

➤ construction details of the structures or buildings to be demolished and those of neighbouring structures or buildings
➤ the presence of asbestos, lead or other hazardous substances
➤ the location of any underground or overhead services (water, electricity, gas, etc.)
➤ the location of any underground cellars, storage tanks or bunkers, particularly if flammable or explosive substances were previously stored
➤ the location of any public thoroughfares adjacent to the structure or building.

The planning supervisor, who is responsible for notifying the Health and Safety Executive of the proposed demolition work, must ensure that a written risk assessment is made of the design of the structure to be demolished and the influence of that design on the demolition method proposed. This risk assessment will normally be made by the project designer who will also plan the demolition work. A further risk assessment should then be made by the contractor undertaking the demolition – this risk assessment will be used to draw up a method statement for inclusion in the health and safety plan. A written method statement will be required before demolition takes place. The contents of the method statement will include the following:

➤ details of method of demolition to be used, including the means of preventing premature collapse or the collapse of adjacent buildings and the safe removal of debris from upper levels
➤ details of equipment, including access equipment, required and any hazardous substances to be used
➤ arrangements for the protection of the public and the construction workforce, particularly if hazardous substances, such as asbestos or other dust, are likely to be released
➤ details of the isolation methods of any services which may have been supplied to the site and any temporary services required on the site
➤ details of personal protective equipment which must be worn
➤ first aid, emergency and accident arrangements
➤ training and welfare arrangements
➤ arrangements for waste disposal
➤ names of site foremen and those with responsibility for health and safety and the monitoring of the work
➤ COSHH and other risk assessments (personal protective equipment, manual handling, etc.) should be appended to the method statement.

There are two forms of demolition:

➤ piecemeal – where the demolition is done using hand and mechanical tools such as pneumatic drills and demolition balls
➤ deliberate controlled collapse – where explosives are used to demolish the structure. This technique should only be used by trained, specialist competent persons.

A very important element of demolition is the training required by all construction workers involved in the work. Specialist training courses are available for those concerned with the management of the process, from the initial survey to the final demolition. However, induction training, which outlines the hazards and the required control measures, should be given to all workers before the start of the demolition work. The site should be made secure with relevant signs posted to warn members of the public of the dangers.

14.3.6 Excavations

This topic will be covered in more detail later in this chapter. Excavations must be constructed so that they are safe environments for construction work to take place. They must also be fenced and suitable notices posted so that neither people nor vehicles fall into them.

14.3.7 Prevention of drowning

Where construction work takes place over water, steps should be taken to prevent people falling into the water and rescue equipment should be available at all times.

14.3.8 Vehicles and traffic routes

All vehicles used on site should be regularly maintained and records kept. Only trained drivers should be allowed to drive vehicles and the training should be relevant to the particular vehicle (fork lift truck, dumper truck, etc.). Vehicles should be fitted with reversing warning systems. HSE investigations have shown that in over 30% of dumper truck accidents, the drivers had little experience and no training. Common forms of

Figure 14.3 Barriers around excavation by footpath. Source HSE. Crown copyright material is reproduced with the permission of the Controller of HMSO and the Queen's Printer for Scotland.

316

accident include driving into excavations, overturning while driving up steep inclines and runaway vehicles which have been left unattended with the engine running. Many vehicles such as mobile cranes require regular inspection and test certificates.

Traffic routes and loading and storage areas need to be well designed with enforced speed limits, good visibility and the separation of vehicles and pedestrians being considered. The use of one-way systems and separate site access gates for vehicles and pedestrians may be required. Finally, the safety of members of the public must be considered, particularly where vehicles cross public footpaths.

14.3.9 Fire and other emergencies

Emergency procedures relevant to the site should be in place to prevent or reduce injury arising from fire, explosions, flooding or structural collapse. These procedures should include the location of fire points and assembly points, extinguisher provision, site evacuation, contact with the emergency services, accident reporting and investigation and rescue from excavations and confined spaces. There also needs to be training in these procedures at the induction of new workers and ongoing for all workers.

14.3.10 Welfare facilities

The Health and Safety Executive has been concerned for some time at the poor standard of welfare facilities on many construction sites. Sanitary and washing facilities (including showers if necessary) with an adequate supply of drinking water should be provided for everybody working on the site. Accommodation will be required for the changing and storage of clothes and rest facilities for break times. There should be adequate first-aid provision (an accident book) and protective clothing against adverse weather conditions.

14.3.11 Electricity

Electrical hazards have been covered in detail earlier in Chapter 10, and all the control measures mentioned apply on a construction site. However, due to the possibility of wet conditions, it is recommended that only 110V equipment is used on site. Where mains electricity is used (perhaps during the final fitting out of the building), then residual current devices should be used with all equipment. Where workers or tall vehicles are working near or under overhead power lines, either the power should be turned off or 'goal posts' or taped markers used to prevent contact with the lines. Similarly, under-ground supply lines should be located and marked before digging takes place.

14.3.12 Noise

Noisy machinery should be fitted with silencers. When machinery is used in a workshop (such as woodworking machines), a noise survey should be undertaken and, if the noise levels exceed the second action level, the use of ear defenders becomes mandatory.

14.3.13 Health hazards

Health hazards are present on a construction site. These hazards include vibration, dust (including asbestos), cement, solvents and paints and cleaners. A COSHH assessment is essential before work starts with regular updates as new substances are introduced. Copies of the assessment and the related safety data sheets should be kept in the site office for reference after accidents or fires. They will also be required to check that the correct personal protective equipment is available. A manual handling assessment should also be made to ensure that the lifting and handling of heavy objects is kept to a minimum.

14.4 The management of construction activities

The management of construction work, including the selection and control of contractors, is governed by the Construction (Design and Management) Regulations 1994 and Amendment Regulations 2000 – known as the CDM Regulations. The CDM Regulations apply to the whole of a construction project, from the initial feasibility study to the hand-over of the completed structure to the customer.

14.4.1 Definitions of terms used in the CDM Regulations & Approved Code of Practice

Client Clients are those involved in a trade, business or other undertaking (whether for profit or not) and for whom construction work is carried out.

Designer Designers are the organizations or individuals who carry out the design of the project. Designers may include architects, consulting engineers, quantity surveyors, specifiers, principal contractors and specialist sub-contractors.

Planning supervisor The planning supervisor is a company, partnership, organization or an individual who coordinates and manages the health and safety aspects of the design. The planning supervisor also has to ensure that the pre-tender stage of the health and safety plan and the health and safety file are prepared. The Regulations suggest that the planning supervisor will normally be more than one person, except for the smallest of projects.

Principal contractor This is the contractor appointed by the client who has overall responsibility for the management of the site operations. This includes the overall coordination of site health and safety management.

Health and safety plan There are in effect two health and safety plans: the pre-tender health and safety plan prepared before the tendering process brings together the health and safety information obtained from the client and designers and aids selection of the principal contractor and the construction stage health and safety plan details how the construction work will be managed to ensure health and safety.

Health and safety file This is a record of information for the client which focuses on health and safety. It alerts those who are responsible for the structure and equipment in it of the significant health and safety risks that will need to be dealt with during subsequent use, construction, maintenance, repair and cleaning work.

Method statement This is a written document laying out the work procedures and sequences of operations to ensure health and safety. It results from the risk assessment carried out for the task or operation and the control measures identified. If the risk is low, a verbal statement may suffice.

Notifiable work Construction work is notifiable to the Health and Safety Executive if it lasts longer than 30 days or will involve more than 500 person days of work. All demolition work is notifiable. A copy of the notice submitted to the HSE should be posted at the site.

14.4.2 Responsibilities of duty holders

The client
Many clients appoint an agent to undertake their duties and this is allowed under the regulations. However, the client must ensure that the agent is competent to perform the duties and submit a written declaration to the Health and Safety Executive giving the name and address of the agent and the address of the construction site. The duties of the client or his agent are:

➤ select and appoint a planning supervisor and a principal contractor, be assured of their competence and that they will allocate sufficient resources to health and safety issues;
➤ be assured that any designers and any directly appointed contractors are competent and properly funded to cover health and safety issues;
➤ provide adequate and relevant information to the planning supervisor relating to the construction project to aid the compilation of the pre-tender health and safety plan;
➤ ensure that construction work does not begin until the principal contractor has prepared a satisfactory health and safety plan;
➤ ensure that the health and safety file is available for inspection after the project has been completed.

The designer
The designer must:

➤ ensure that no design begins until the client is aware of his duties under the CDM Regulations;
➤ ensure that health and safety issues are considered in the design work (e.g. ease of subsequent maintenance and window cleaning);
➤ when the design process has been completed and there are unavoidable residual risks still present, adequate information on these risks should be provided to those who need it (e.g. the planning supervisor or the principal contractor);
➤ cooperate with the planning supervisor and, where appropriate other designers involved in the project.

The planning supervisor
The planning supervisor is primarily concerned with the planning of the work not the execution of it – this is the duty of the principal contractor. The duties of the planning supervisor are as follows:

➤ ensure that the Health and Safety Executive is notified of the project;
➤ ensure that there is cooperation between designers;
➤ ensure that designers comply with their duties;

➤ ensure that a pre-tender stage health and safety plan is prepared;
➤ advise the client when requested to do so (for example, on the designer's competence and the adequacy of provision of health and safety);
➤ advise the client on the principal contractor's health and safety plan;
➤ ensure that a health and safety file is prepared.

Regulation 8 of CDM requires that the client should only appoint a planning supervisor who has the competence to perform the required functions of a planning supervisor. It is suggested that the following may be included in a competence checklist:

➤ membership of a relevant professional body
➤ knowledge of construction practice, particularly in relation to the nature of the project
➤ familiarity and knowledge of the design function
➤ knowledge of health and safety issues (including fire safety), particularly in preparing a health and safety plan
➤ ability to work with and coordinate the activities of different designers and be a bridge between the design function and construction work on site
➤ the number, experience and qualifications of people to be employed, both internally and from other sources, to perform the various functions in relation to the project
➤ the management system which will be used to monitor the correct allocation of people and other resources in the way agreed at the time when these matters are being finalized
➤ the time to be allowed to carry out the different duties
➤ the technical facilities available to aid the staff in carrying out their duties.

The principal contractor

The principal contractor is responsible for the organization, management and operation of the construction phase of the project. The main duties are to:

➤ develop and implement the health and safety plan from the pre-tender plan and ensure that all site workers comply with its rules;
➤ employ competent sub-contractors who are properly resourced for health and safety, and ensure that there is coordination and cooperation between all contractors;
➤ obtain relevant risk assessments and method statements from all contractors and monitor their health and safety performance;
➤ ensure that only authorized persons are allowed on site and that all workers are informed, consulted and trained on health and safety matters;
➤ display the notification notice to the Health and Safety Executive of the project;
➤ supply relevant information to the planning supervisor for the health and safety file.

14.4.3 The health and safety plan

There are, in effect two plans – the pre-tender health and safety plan and the construction phase health and safety plan. The pre-tender plan is sent to prospective principal contractors so that they are fully aware of the health and safety risks associated with the project and the standard of health and safety performance expected of them. The response of prospective contractors to the plan should provide an indication of health and safety competence of the contractor. The main contents of the pre-tender plan will be:

➤ a general description of the work, location and timescales and relevant drawings
➤ the existing environment including services, confined spaces, hazardous materials, neighbouring buildings or other hazards and soil and ground conditions
➤ significant hazards identified during the design stage, including health hazards associated with essential construction materials
➤ site details, including specific safety rules involving emergency procedures or permits to work or pedestrian access.

The pre-tender health and safety plan is developed into the health and safety plan by the principal contractor and addresses specific construction health and safety issues. Typical contents will include:

➤ safety policies, organization, arrangements, risk assessments and method statements including fire and emergency arrangements
➤ welfare arrangements
➤ any important information for sub-contractors and methods of communication and consultation
➤ health and safety monitoring and inspection arrangements
➤ site health and safety rules
➤ procedures for delivering information to the health and safety file.

14.4.4 The health and safety file

The health and safety file is a record of relevant information for the client or his customer. It should contain details of the location of services and instructions on any equipment or items fitted. It will be developed as the construction progresses with various sub-contractors adding information as it becomes available. Typical contents will include plans and drawings, details of equipment in the building or structure, equipment and maintenance procedures, the location of services and utilities and manuals produced by specialist contractors and suppliers.

14.4.5 Selection and control of contractors

It is important that health and safety factors are considered as well as technical or professional competence when potential contractors are being short listed or employed. The following items will give a guide to health and safety attitudes:

➤ registration with either the HSE or Environmental Health Department of a Local Authority
➤ a current health and safety policy
➤ details of any risk assessments made and control measures introduced
➤ any method statements required to perform the contract
➤ details of competence certification, particularly when working with gas or electricity may be involved
➤ details of insurance arrangements in force at the time of the contract
➤ details of emergency procedures, including fire precautions, for contractor employees
➤ details of any previous accidents or incidents reported under RIDDOR
➤ details of accident reporting procedure

➤ details of previous work undertaken by the contractor
➤ references from previous employers or main contractors
➤ details of any health and safety training undertaken by the contractor and his employees.

On being selected, contractors should be expected to:

➤ familiarize themselves with those parts of the health and safety plan which affect them and their employees and/or sub-contractors;
➤ cooperate with the principal contractor in his health and safety duties to contractors;
➤ comply with their legal health and safety duties.

On arrival at the site, sub-contractors should ensure that:

➤ they report to the site office on arrival on site and report to the site manager;
➤ they abide by any site rules, particularly in respect of personal protective equipment;
➤ the performance of their work does not place others at risk;
➤ they are familiar with the first-aid and accident reporting arrangements of the principal contractor;
➤ they are familiar with all emergency procedures on the site;
➤ any materials brought onto the site are safely handled, stored and disposed of in compliance, where appropriate, with the current Control of Substances Hazardous to Health Regulations;
➤ they adopt adequate fire precaution and prevention measures when using equipment which could cause fires;
➤ they minimize noise and vibration produced by their equipment and activities;
➤ any ladders, scaffolds and other means of access are erected in conformance with good working practice and any statutory requirements;
➤ any welding or burning equipment brought onto the site is in safe operating condition and used safely with a suitable fire extinguisher to hand;
➤ any lifting equipment brought onto the site complies with the current Lifting Operation and Lifting Equipment Regulations;
➤ all electrical equipment complies with the current Electricity at Work Regulations;
➤ connections to the electricity supply is from a point specified by the principal contractor and is by proper cables and connectors. For outside construction work, only 110V equipment should be used;
➤ any restricted access to areas on the site is observed;
➤ welfare facilities provided on site are treated with respect;
➤ any vehicles brought onto the site observe any speed, condition or parking restriction.

The control of sub-contractors can be exercised by monitoring them against the criteria listed above and by regular site inspections. On completion of the contract, the work should be checked to ensure that the agreed standard has been reached and that any waste material has been removed from the site.

14.5 Working above ground level

14.5.1 Hazards and controls associated with working above ground level

Normally 'working above ground level' means at heights above 2 m. The significance of injuries resulting from falls above this height, such as fatalities and other major injuries, have been dealt with earlier in the chapter as has the importance and legal requirements for head protection. Also covered were the many hazards involved in working at height, including fragile roofs and the deterioration of materials, unprotected edges and falling materials. Additional hazards include the weather and unstable or poorly maintained access equipment, such as ladders and various types of scaffold.

The principal means of preventing falls of people or materials includes the use of fencing, guardrails, toe boards, working platforms, access boards, ladder hoops, safety nets and safety harnesses. Safety harnesses arrest the fall by restricting the fall to a given distance due to the fixing of the harness to a point on an adjacent rigid structure. They should only be used when all other possibilities are not practical.

14.5.2 Access equipment

There are many different types of access equipment, but only the following four categories will be considered here:

> ladders
> fixed scaffold
> mobile scaffold towers
> mobile elevated work platforms.

Ladders

The main cause of accidents involving ladders is ladder movement while in use. This occurs when they have not been secured to a fixed point, particularly at the foot. Other causes include over-reaching by the worker, slipping on a rung, ladder defects and, in the case of metal ladders, contact with electricity. The main types of accident are falls from ladders.

There are two common materials used in the construction of ladders – aluminium and timber. Aluminium ladders have the advantage of being light but should not be used in high winds or near live electricity. Timber ladders need regular inspection for damage and should not be painted since this could hide cracks.

The following factors should be considered when using ladders:

> ensure that the use of a ladder is the safest means of access given the work to be done and the height to be climbed;

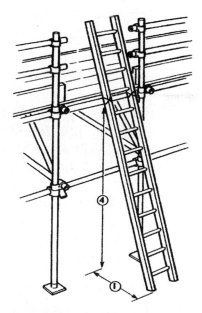

Figure 14.4 Ladders should be correctly angled one out for every four up. Source HSE. Crown copyright material is reproduced with the permission of the Controller of HMSO and the Queen's Printer for Scotland.

➤ the location itself needs to be checked. The supporting wall and supporting ground surface should be dry and slip free. Extra care will be needed if the area is busy with pedestrians or vehicles;

➤ the ladder needs to be stable in use. This means that the inclination should be as near the optimum as possible (1 in 4 ratio of distance from the wall to distance up the wall). The foot of the ladder should be tied to a rigid support. Weather conditions must be suitable (no high winds or heavy rain). The proximity of live electricity should also be checked. (This last point is important when ladders are to be carried beneath power lines);

➤ there should be at least 1 m of ladder above the stepping off point;

➤ the work activity must be considered in some detail. Over-reaching must be eliminated and consideration given to the storage of paints or tools which are to be used from the ladder and any loads to be carried up the ladder. The ladder must be matched to work required;

➤ workers who are to use ladders must be trained in the correct method of use and selection. Such training should include the use of both hands during climbing, clean non-slippery footwear, clean rungs and an undamaged ladder;

➤ ladders should be inspected (particularly for damaged or missing rungs) and maintained on a regular basis and they should only be repaired by competent persons;

➤ the transportation and storage of ladders is important since much damage can occur at these times. They need to be handled carefully and stored in a dry place;

➤ when a ladder is left secured to a structure during non-working hours, a plank should be tied to the rungs to prevent unauthorized access to the structure.

Certain work should not be attempted using ladders. This includes work where:

➤ two hands are required
➤ the work is at an excessive height
➤ where the ladder cannot be secured or made stable
➤ the work is of long duration
➤ the work area is very large
➤ the equipment or materials to be used are heavy or bulky
➤ the weather conditions are adverse
➤ there is no protection from vehicles.

Many of the points outlined above apply to step ladders and trestles where stability and over-reaching are the main hazards. Neither of these should be used as a workplace above 2 m in height unless proper edge protection is provided.

Fixed scaffolds

It is quicker and easier to use a ladder as a means of access, but it is not always the safest. Jobs, such as painting, gutter repair, demolition work or window replacement, are often easier done using a scaffold. Unless the work can be completed comfortably using ladders, then a scaffold should be considered. Scaffolds must be capable of supporting building workers, equipment, materials, tools and any accumulated waste. A common cause of scaffold collapse is the 'borrowing' of boards and tubes from the scaffold, thus weakening it. Falls from scaffolds are often caused by badly constructed working platforms, inadequate guardrails or climbing up the outside of a scaffold. Falls also occur during the assembly or dismantling process.

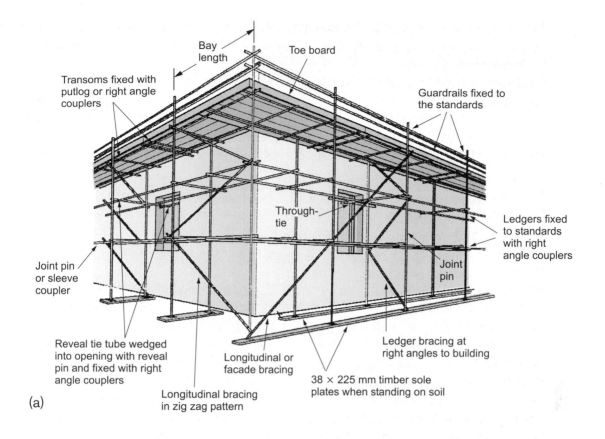

Bay length

Toe board

Transoms fixed with putlog or right angle couplers

Guardrails fixed to the standards

Through-tie

Ledgers fixed to standards with right angle couplers

Joint pin or sleeve coupler

Joint pin

Reveal tie tube wedged into opening with reveal pin and fixed with right angle couplers

Longitudinal or facade bracing

Ledger bracing at right angles to building

38 × 225 mm timber sole plates when standing on soil

Longitudinal bracing in zig zag pattern

(a)

(b)

Figure 14.5 (a) Typical independent tied scaffold. Source HSE. Crown copyright material is reproduced with the permission of the Controller of HMSO and the Queen's Printer for Scotland; (b) fixed scaffold left in place to fit the gutters.

There are two basic types of external scaffold:

> **independent tied** – these are scaffolding structures which are independent of the building but tied to it often using a window or window recess. This is the most common form of scaffolding
> **putlog** – this form of scaffolding is usually used during the construction of a building. A putlog is a scaffold tube which spans horizontally from the scaffold into the building – the end of the tube is flattened and is usually positioned between two brick courses.

The important components of a scaffold have been defined in a guidance note issued by the HSE as follows (Figure 14.5(a)).

Standard an upright tube or pole used as a vertical support in a scaffold.

Ledger a tube spanning horizontally and tying standards longitudinally.

Transom a tube spanning across ledgers to tie a scaffold transversely. It may also support a working platform.

Bracing tubes which span diagonally to strengthen and prevent movement of the scaffold.

Guardrail a horizontal tube fitted to standards along working platforms to prevent persons from falling.

Toe boards these are fitted at the base of working platforms to prevent persons, materials or tools falling from the scaffold.

The following factors must be addressed if a scaffold is being considered for use for construction purposes:

> scaffolding must only be erected by competent people who have attended recognized training courses. Any work carried out on the scaffold must be supervised by a competent person. Any changes to the scaffold must be done by a competent person;
> adequate toe boards, guardrails and intermediate rails must be fitted to prevent people or materials from falling;
> the scaffold must rest on a stable surface, uprights should have base plates and timber sole plates if necessary;
> the scaffold must have safe access and egress;
> work platforms should be fully boarded with no tipping or tripping hazards;
> the scaffold should be sited away from or protected from traffic routes so that it is not damaged by vehicles;
> the scaffold should be properly braced, secured to the building or structure;
> overloading of the scaffold must be avoided;
> the public must be protected at all stages of the work;
> regular inspections of the scaffold must be made and recorded.

Mobile scaffold towers

Mobile scaffold towers are frequently used throughout industry. It is essential that the workers are trained in their use since recent research has revealed that 75% of lightweight mobile tower scaffolding is either erected, used, moved or dismantled in an unsafe manner (Figure 14.6).

The following points must be considered when mobile scaffold towers are to be used:

> the selection, erection and dismantling of mobile scaffold towers must be undertaken by competent and trained persons with maximum height to base ratios not being exceeded

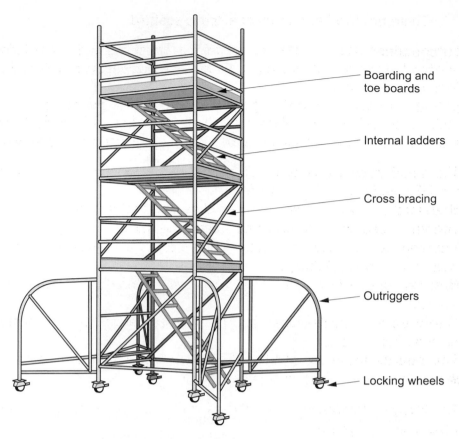

Figure 14.6 Typical tower scaffold.

> diagonal bracing and stabilizers should always be used
> access ladders must be fitted to the narrowest side of the tower or inside the tower and persons should not climb up the frame of the tower
> all wheels must be locked while work is in progress and all persons must vacate the tower before it is moved
> the tower working platform must be boarded, fitted with guardrails and toe boards and not overloaded
> towers must be tied to a rigid structure if exposed to windy weather or to be used for work such as jet blasting
> persons working from a tower must not over-reach or use ladders from the work platform
> safe distances must be maintained between the tower and overhead power lines both during working operations and when the tower is moved
> the tower should be inspected on a regular basis and a report made.

Mobile elevated work platforms

Mobile elevated work platforms are very suitable for high level work such as changing light bulbs in a warehouse. The following factors must be considered when using mobile elevated work platforms:

> the mobile elevated work platform must only be operated by trained and competent persons
> it must never be moved in the elevated position

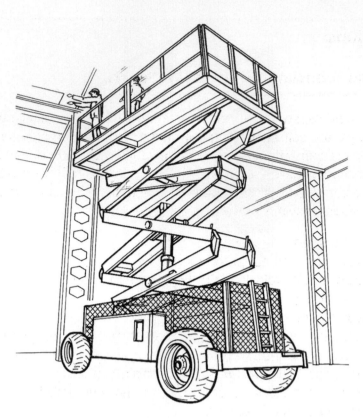

Figure 14.7 Mobile elevating work platform – one type of a wide range. Source HSE. Crown copyright material is reproduced with the permission of the Controller of HMSO and the Queen's Printer for Scotland.

> it must be operated on level and stable ground
> the tyres must be properly inflated and the wheels immobilized
> outriggers should be fully extended and locked in position
> due care must be exercised with overhead power supplies and obstructions
> procedures should be in place in the event of machine failure.

14.5.3 Inspection requirements for scaffolds

Scaffolds must be inspected on a regular basis by a competent person. These inspections should take place before the scaffold is used, after any alteration is made or after adverse weather conditions may have weakened it. In any event an inspection should take place every seven days and any faults rectified.

While there is no longer a specific statutory form to be completed, a record of the inspection should be made which should include the following information:

> any issues which might give rise to a health and safety risk. This is more than a simple fault identification
> details of any action taken to control the identified risks.

The recommended form for inspections is shown in the Appendix to this chapter.

14.6 Excavations

14.6.1 Hazards associated with excavations

There are about seven deaths each year due to work in excavations. Many types of soil, such as clays, are self-supporting but others, such as sands and gravel, are not. Many excavations collapse without any warning resulting in death or serious injury. Many such accidents occur in shallow workings. It is important to note that, although most of these accidents affect workers, members of the public can also be injured. The specific hazards associated with excavations are as follows:

➤ collapse of the sides
➤ materials falling on workers in the excavation
➤ falls of people and/or vehicles into the excavation
➤ workers being struck by plant
➤ specialist equipment such as pneumatic drills
➤ hazardous substances particularly near the site of current or former industrial processes
➤ influx of ground or surface water and entrapment in silt or mud
➤ proximity of stored materials, waste materials or plant
➤ proximity of adjacent buildings or structures and their stability
➤ contact with underground services
➤ access and egress to the excavation
➤ fumes, lack of oxygen and other health hazards (such as Weil's disease).

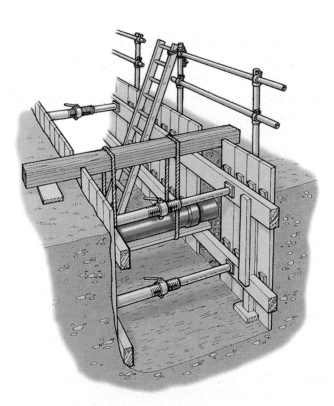

Figure 14.8 Timbered excavation with ladder access and supported services (guard removed on one side for clarity). Source HSE. Crown copyright material is reproduced with the permission of the Controller of HMSO and the Queen's Printer for Scotland.

Clearly, alongside these specific hazards, more general hazards, such as manual hand-ling, electricity, noise and vibrations, will also be present.

14.6.2 Precautions and controls required for excavations

The following precautions and controls should be adopted:

> at all stages of the excavation, a competent person must supervise the work and the workers given clear instructions on working safely in the excavation;
> the sides of the excavation must be prevented from collapsing either by digging them at a safe angle (between 5° and 45° dependent on soil and dryness) or by shoring them up with timber, sheeting or a proprietary support system. Falls of material into the workings can also be prevented by not storing spoil material near the top of the excavation
> the workers should wear hard hats;
> if the excavation is more than 2 m deep, a substantial barrier, consisting of guardrails and toe boards should be provided around the surface of the workings;
> vehicles should be kept away as far as possible using warning signs and barriers. Where a vehicle is tipping materials into the excavation, stop blocks should be placed behind its wheels;
> it is very important that the excavation site is well lit at night;
> all plant and equipment operators must be competent and non-operators should be kept away from moving plant;
> personal protective equipment must be worn by operators of noisy plant;
> nearby structures and buildings may need to be shored up if the excavation may reduce their stability. Scaffolding could also be de-stabilized by adjacent excavation trenches;
> the influx of water can only be controlled by the use of pumps after the water has been channelled into sumps;
> the presence of hazardous substances or health hazards should become apparent during the original survey work and, when possible, removed or suitable control mea-sures adopted. Any such hazards found after work has started, must be reported and noted in the inspection report and remedial measures taken. Exhaust fumes can be dangerous and petrol or diesel plant should not be sited near the top of the excavation;
> the presence of buried services is one of the biggest hazards and the position of such services must be ascertained using all available service location drawings before work commences. Since these will probably not be accurate, service location equipment should be used by specifically trained people. Only hand tools should be used in the vicinity of underground services. Overhead services may also present risks to cranes and other tall equipment;
> safe access by ladders is essential, as are crossing points for pedestrians and vehicles. Whenever possible, the workings should be completely covered outside working hours, particularly if there is a possibility of children entering the site;
> finally, care is needed during the filling in process.

14.6.3 Inspection and reporting requirements

The duty to inspect and prepare a report only applies to excavations which need to be supported to prevent accidental fall of material. Only persons with a recognized and relevant competence should carry out the inspection and write the report. Inspections should take place at the following timing and frequency:

➤ after any event likely to affect the strength or stability of the excavation
➤ before work at the start of every shift
➤ after an accidental fall of any material.

Although an inspection must be made at the start of every shift, only one report is required of such inspections every seven days. However, reports must be completed following all other inspections. The report should be completed before the end of the relevant working period and a copy given to the manager responsible for the excavation within 24 hours. The report must be kept on site until the work is completed and then retained for 3 months at an office of the organization which carried out the work.

A suitable form is shown in the Appendix to this chapter.

14.7 Practice NEBOSH questions for Chapter 14

1. Outline the precautions to be taken when carrying out repairs to the flat roof of a building. (March 2001)
2. (a) Identify the three types of asbestos commonly found in buildings.
   (b) Explain where asbestos is likely to be encountered in a building during renovation work. (March 2001)
3. Outline the precautions that might be taken in order to reduce the risk of injury when using stepladders. (December 2001)
4. (a) Outline the main duties of a planning supervisor under the Construction (Design and Management) Regulations 1994.
   (b) Identify four items of information in the health and safety file for an existing building that might be needed by a contractor carrying out refurbishment work. (December 2001)
5. Outline the main precautions to be taken when carrying out excavation work. (December 2000)
6. Outline eight precautions that should be considered to prevent accidents to children who might be tempted to gain access to a construction site. (December 2000)
7. Identify measures that should be adopted in order to protect against the dangers of people and/or materials falling from a mobile tower scaffold. (December 2000)
8. With reference to the Construction (Design and Management) Regulations 1994:
   (i) identify the circumstances under which a construction project must be notified to an enforcing authority
   (ii) outline the duties of the client under the Regulations. (December 2000)
9. Outline four duties of each of the following persons under the Construction (Design and Management) Regulations 1994:
   (i) the planning supervisor
   (ii) the principle contractor. (June 1999)
10. The exterior paintwork of a row of shops in a busy high street is due to be re-painted. Identify the hazards associated with the work and outline the corresponding precautions to be taken. (June 1999)

Appendix 14.1 – Construction (Health, Safety and Welfare) Regulations 1996

INSPECTION REPORT

Report of results of every inspection made in pursuance of regulation 29(1)

1. Name and address of person for whom inspection was carried out.

2. Site address

3. Date and time of inspection.

4. Location and description of workplace (including any plant, equipment or materials) inspected.

5. Matters which give rise to any health and safety risks.

6. Can work be carried out safely?　　　　Y / N

7. If not, name of person informed.

8. Details of any other action taken as a result of matters identified in 5 above.

9. Details of any further action considered necessary.

10. Name and position of person making the report.

11. Date report handed over.

Construction (Health, Safety and Welfare) Regulations 1996

INSPECTION REPORTS: NOTES

Place of work requiring inspection	Timing and frequency of inspection					
	Before being used for the first time.	After substantial addition, dismantling or alteration.	After any event likely to have affected its strength or stability.	At regular intervals not exceeding 7 days.	Before work at the start of every shift	After accidental fall of rock, earth or any material.
Any working platform or part thereof or any personal suspension equipment.	✔	✔	✔	✔		
Excavations which are supported in pursuit of paragraphs (1), (2) or (3) of regulation 12.			✔		✔	✔
Cofferdams and caissons.			✔		✔	

NOTES

General

1. The inspection report should be completed before the end of the relevant working period.
2. The person who prepares the report should, within 24 hours, provide either the report or a copy to the person on whose behalf the inspection was carried out.
3. The report should be kept on site until work is complete. It should then be retained for three months at an office of the person for whom the inspection was carried out.

Working platforms only

1. An inspection is only required where a person is liable to fall more than 2 metres from a place of work.
2. Any employer or any other person who controls the activities of persons using a scaffold shall ensure that it is stable and of sound construction and that the relevant safeguards are in place before his employees or persons under his control first use the scaffold.
3. No report is required following the inspection of any mobile tower scaffold which remains in the same place for less than 7 days.
4. Where an inspection of a working platform or part thereof or any personal suspension equipment is carried out:
    i. before it is taken into use for the first time; or
    ii. after any substantial addition, dismantling or other alteration; not more than one report is required for any 24 hour period.

Excavations only

1. The duties to inspect and prepare a report apply only to any excavation which needs to be supported to prevent any person being trapped or buried by an accidental collapse, fall or dislodgement of material from its sides, roof or area adjacent to it. Although an excavation must be inspected at the start of every shift, only one report of such inspections is required every 7 days. Reports must be completed for all inspections carried out during this period for other purposes, e.g. after accidental fall of material.

Checklist of typical scaffolding faults

Footings	Standards	Ledgers	Bracing	Putlogs and transoms	Couplings	Bridles	Ties	Boarding	Guardrails and toe-boards	Ladders
Soft and uneven	Not plumb	Not level	Some missing	Wrongly spaced	Wrong fitting	Wrong spacing	Some missing	Bad boards	Wrong height	Damaged
No base plates	Jointed at same height	Joints in same bay	Loose	Loose	Loose	Wrong couplings	Loose	Trap boards	Loose	Insufficient length
No sole plates	Wrong spacing	Loose	Wrong fittings	Wrongly supported	Damaged	No check couplers	Not enough	Incomplete	Some missing	Not tied
Undermined	Damaged	Damaged			No check couplers			Insufficient supports		

Incident investigation, recording and reporting

15.1 Introduction

This chapter is concerned with the recording of incidents and accidents at work; their investigation; the legal reporting requirements; and simple analysis of incidents to help managers benefit from the investigation and recording process.

Incidents and accidents rarely result from a single cause and many turn out to be complex. Most incidents involve multiple, interrelated causal factors. They can occur whenever significant deficiencies, oversights, errors, omissions or unexpected changes occur. Anyone of these can be the precursor for an accident or incident. There is a value on collecting data on all incidents and potential losses as it helps to prevent more serious events. (See Chapter 5 for accident ratios and definitions.)

Incidents and accidents, whether they cause damage to property or more seriously injury and/or ill-health to people, should be properly and thoroughly investigated to allow an organization to take the appropriate action to prevent a recurrence. Good investigation is a key element to making improvements in health and safety performance.

Figure 15.1 A dangerous occurence – fire.

Incident investigation is considered to be part of a **reactive monitoring system** because it is triggered after an event.

The range of events includes:

➤ injuries and ill-health, including sickness absence
➤ damage to property, personal effects, work in progress, etc.
➤ incidents which have the potential to cause injury, ill-health or damage.

Each type of event gives the opportunity to:

➤ check performance;
➤ identify underlying deficiencies in management systems and procedures;
➤ learn from mistakes and add to the corporate memory;
➤ reinforce key health and safety messages;
➤ identify trends and priorities for prevention;
➤ provide valuable information if there is a claim for compensation;
➤ helps to meet legal requirements for reporting certain incidents to the authorities.

15.2 Logic behind incident/accident investigation

15.2.1 Logic

Incident/accident investigation is based on the logic that:

➤ all incidents/accidents have causes . . . eliminate the cause and eliminate future incidents
➤ the direct and indirect causes of an incident/accident can be discovered through investigation
➤ corrective action indicated by the causation can be taken to eliminate future incidents/accidents.

Investigation is not intended to be a mechanism for apportioning blame. There are often strong emotions associated with injury or significant losses. It is all too easy to look for someone to blame without considering the reasons why a person behaved in a particular way. Often short cuts to working procedures that may have contributed to the accident give no personal advantage to the person injured. The short cut may have been taken out of loyalty to the organization or ignorance of a safer method.

15.2.2 What managers need to do

Managers need to:

➤ communicate the type of accident and incident that needs to be reported;
➤ provide a system for reporting and recording;
➤ check that the proper reports are being made;
➤ make appropriate records of accidents and incidents;
➤ investigate all incidents and accidents reported;

> analyse the events routinely to check for trends in performance and the prevalence of types of incident or injury;
> monitor the system to make sure that it is working satisfactorily.

15.3 Which incidents/accidents should be investigated?

15.3.1 Injury accident

Should every accident be investigated or only those that lead to serious injury? In fact the main determinant is the potential of the accident to cause harm rather than the actual harm resulting, for example, a slip can result in just an embarrassing flailing of arms or, just as easily, a broken leg. The frequency of occurrence of the accident type is also important – a stream of minor cuts from paper needs looking into.

As it is not possible to determine the potential for harm simply from the resulting injury, the only really sensible solution is to investigate all accidents. The amount of time and effort spent on the investigation should, however, vary depending on the level of risk (severity of potential harm, frequency of occurrence). The most effort should be focused on significant events involving serious injury, ill-health or losses and events which have the potential for multiple or serious harm to people or substantial losses. These factors should become clear during the accident investigation and be used to guide how much time should be taken.

15.3.2 Incidents (including near miss)

There is a very wide variety of possible incidents, ranging from severe environmental problems to concerns about accidentally knocking into someone else.

All incidents requiring reports to the regulatory authorities should be investigated thoroughly. Investigation of other incidents will need to be judged individually, although if there is likely to be a complaint or an enquiry by a regulatory authority, it is much better to have completed an investigation when the facts were available.

The main difficulty with incident and hazard reporting is persuading people to do it. One of the best ways to encourage this is to treat the report seriously by investigating the incident and reporting back conclusions. Therefore, unless there are good reasons against, all reports of incidents, near misses and hazards should be investigated.

15.4 Investigations and causes of incidents

Who should investigate?

Investigations should be led by line managers or other people with sufficient status and knowledge to make recommendations that will be respected by the organization. The person to lead many investigations will be the department manager or supervisor of the person/area involved because they:

Figure 15.2   Near miss event likely – accident waiting to happen.

➤ know about the situation;
➤ know most about the employees;
➤ have a personal interest in preventing further incidents/accidents affecting 'their' people, equipment, area, materials;
➤ can take immediate action to prevent a similar incident;
➤ can communicate most effectively with the other employees concerned;
➤ can demonstrate practical concern for employees and control over the immediate work situation.

15.4.2 When should the investigation be conducted?

The investigation should be carried out as soon as possible after the incident to allow the maximum amount of information to be obtained. There may be difficulties which should be considered in setting up the investigation quickly – if, for example, the victim is removed from the site of the accident, or if there is a lack of a particular expert. An immediate investigation is advantageous because:

➤ factors are fresh in the minds of witnesses;
➤ witnesses have had less time to talk (there is an almost automatic tendency for people to adjust their story of the events to bring it into line with a consensus view);
➤ physical conditions have had less time to change;
➤ more people are likely to be available, for example, delivery drivers, contractors and visitors who will quickly disperse following an incident making contact very difficult;
➤ there will probably be the opportunity to take immediate action to prevent a recurrence and to demonstrate management commitment to improvement;

> immediate information from the person suffering the accident often proves to be most useful.

Consideration should be given to asking the person to return to site for the accident investigation if they are physically able, rather than wait for them to return to work. A second option, although not as valuable, would be to visit the injured person at home or even in hospital (with their permission) to discuss the accident.

15.4.3 Investigation method

There are four basic elements to a sound investigation:

1 collect facts about what has occurred
2 assemble, and analyse the information obtained
3 compare the information with acceptable industry and company standards and legal requirements to draw conclusions
4 implement the findings and monitor progress.

Information should be gathered from all available sources, for example, witnesses, supervisors, physical conditions, hazard data sheets, written systems of work, training records, etc. The amount of time spent should not, however, be disproportionate to the risk. The aim of the investigation should be to explore the situation for possible underlying factors, in addition to the immediately obvious causes of the accident. For example, in a machinery accident it would not be sufficient to conclude an accident occurred because a machine was inadequately guarded. It is necessary to look into the possible underlying system failure that may have occurred.

Investigations have three facets, which are particularly valuable and can be used to check against each other:

> direct observation of the scene, premises, workplace, relationship of components, materials and substances being used, possible reconstruction of events, and injuries or condition of the person concerned
> documents including written instructions, training records, procedures, safe operating systems, risk assessments, policies, records of inspections or test and examinations carried out
> interviews (including written statements) with persons injured, witnesses, people who have carried out similar functions or examinations and tests on the equipment involved and people with specialist knowledge.

Immediate causes

A detailed investigation should look at the following factors as they can provide useful information about **immediate causes** that have been manifested in the incident/accident.

> Personal factors:
>> behaviour of the people involved
>> suitability of people doing the work
>> training and competence
> Task factors:
>> workplace conditions and precautions or controls

➤ actual method of work adopted at the time
➤ ergonomic factors
➤ normal working practice either written or customary.

Underlying causes

A thorough investigation should also look at the following factors as they can provide useful information about **underlying causes** that have been manifested in the incident/accident:

➤ Management and organizational factors:
  ➤ previous similar incidents
  ➤ supervision
  ➤ the control and coordination of work
  ➤ quality of the health and safety policy and procedures
  ➤ quality of consultation and cooperation of employees
  ➤ the adequacy and quality of communications and information
  ➤ deficiencies in risk assessments, plans and control systems
  ➤ deficiencies in monitoring and measurement of work activities
  ➤ quality and frequency of reviews and audits.

15.4.4 Investigation interview techniques

It must be made clear at the outset and during the course of the interview that the aim is not to apportion blame but to discover the facts and use them to prevent similar accidents in the future.

A witness should be given the opportunity to explain what happened in their own way without too much interruption and suggestion. Questions should then be asked to elicit more information. These should be of the open type, which do not suggest the answer. Questions starting with the words in Figure 15.3 are useful.

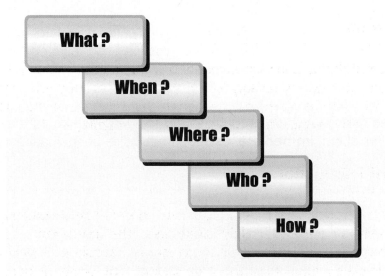

Figure 15.3 Questions to be asked in an investigation.

'Why' should not be used at this stage. The facts should be gathered first, with notes being taken at the end of the explanation. The investigator should then read them or give a summary back to the witness, indicating clearly that they are prepared to alter the notes, if the witness is not content with them.

If possible, indication should be given to the witness about immediate actions that will be taken to prevent a similar occurrence and that there could be further improvements depending on the outcome of the investigation.

Accidents can often be very upsetting for witnesses, which should be borne in mind. This does not mean they will not be prepared to talk about what has happened. They may in fact wish to help, but questions should be sensitive; upsetting the witness further should be avoided.

15.4.5 Comparison with relevant standards

There are usually suitable and relevant standards which may come from the HSE, industry or the organization itself. These should be carefully considered to see if:

> suitable standards are available to cover legal standards and the controls required by the risk assessments
> the standards are sufficient and available to the organization
> the standards were implemented in practice
> the standards were implemented, why was there a failure?
> changes should be made to the standards.

15.4.6 Recommendations

The investigation should have highlighted both immediate causes and underlying causes. Recommendations, both for immediate action and for longer-term improvements, should come out of this, but it may be necessary to ensure that the report goes further up the management chain if the improvements recommended require authorization, which cannot be given by the investigating team.

15.4.7 Follow-up

It is essential that a follow-up is made to check on the implementation of the recommendations. It is also necessary to review the effect of the recommendations to check whether they have achieved the desired result and whether they have had unforeseen 'knock-on' effects, creating additional risks and problems.

15.4.8 Use of information

The accident investigation should be used to generate recommendations but should also be used to generate safety awareness. The investigation report or a summary should therefore be circulated locally to relevant people and, when appropriate, summaries circulated throughout the organization. The accident does not need to have resulted in a three-day lost time injury for this system to be used.

15.4.9 Training

A number of people will potentially be involved in accident investigation. For most of these people it will only be necessary on very few occasions. Training guidance and help will therefore be required. Training can be provided in accident investigation in courses run on site and also in numerous off-site venues. Computer-based training courses are also available. These are intended to provide refresher training on an individual basis or complete training at office sites, for example, where it may not be feasible to provide practical training.

15.4.10 Investigation form

Headings which could be used to compile an accident/incident investigation form are given below:

- date and location of accident
- circumstances of accident
- immediate cause of accident
- underlying cause of accident
- immediate action taken
- recommendation for further improvement
- report circulation list
- date of investigation
- signature of investigation team leader
- names of investigating team.

Follow-up
- were the recommendations implemented?
- were the recommendations effective?

An example of an investigation form using 16 different causes of accidents for analysis purposes is shown in Appendix 15.1.

15.5 Legal recording and reporting requirements

15.5.1 Accident book

Under the Social Security (Claims and Payments) Regulations 1979, regulation 25, employers must keep a record of accidents at premises where more than ten people are employed. Anyone injured at work is required to inform the employer and record information on the accident in an accident book, including a statement on how the accident happened.

The employer is required to investigate the cause and enter this in the accident book if they discover anything that differs from the entry made by the employee. The purpose of this record is to ensure that information is available if a claim is made for compensation.

The HSE has produced a new Accident Book BI 510 in May 2003 with notes on these Regulations, and the Reporting of Injuries Diseases and Dangerous Occurrences Regulations 1995 (RIDDOR), (see Figure 15.4) and now complies with the Data Protection Act 1998

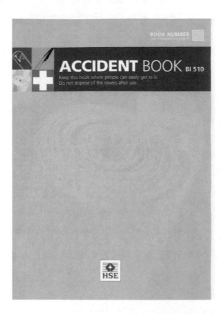

Figure 15.4 The Accident Book BI 510 ISBN 0-7176-2603-2

15.5.2 RIDDOR

RIDDOR requires employers, the self-employed and those in control of premises, to report certain more serious accidents and incidents to the HSE or other enforcing authority and to keep a record. There are no exemptions for small organizations. A full summary of the Regulations is given in Chapter 17. The reporting and recording requirements are as follows:

Death or major injury
If an accident occurs at work and:

➤ an employee, or self-employed person working on the premise is killed or suffers a major injury (including the effects of physical violence) (see Chapter 17 under RIDDOR for definition of major injury)
➤ a member of the public is killed or taken to hospital.

The responsible person must notify the enforcing authority without delay by the quickest practicable means, for example, telephone. They will need to give brief details about the organization, the injured person(s) and the circumstances of the accident and, within ten days, the responsible person must also send a completed accident report form, F2508.

Over three-day lost time injury
If there is an accident connected with work (including physical violence) and an employee, or self-employed person working on the premises, suffers an injury and is away from work or not doing their normal duties for more than three days (including

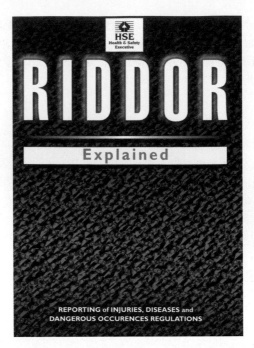

Figure 15.5   RIDDOR leaflet from HSE Books.

weekends, rest days or holidays) but not counting the day of the accident, the responsible person must send a completed accident report form, F2508, to the enforcing authority within ten days.

Disease

If a doctor notifies the responsible person that an employee suffers from a reportable work-related disease a completed disease report form, F2508A, must be sent to the enforcing authority. A summary is included in Chapter 17 and a full list is included with a pad of report forms. The HSE InfoLine or the Incident Contact Centre can be contacted to check whether a particular disease is reportable.

Dangerous occurrence

If an incident happens which does not result in a reportable injury, but obviously could have done, it could be a Dangerous Occurrence as defined by a list in the regulations (see Chapter 17 for a summary of Dangerous Occurrences). All Dangerous Occurrences must be reported immediately by, for example, telephone, to the enforcing authorities. The HSE InfoLine or the Incident Contact Centre can be contacted to check whether a dangerous occurrence is reportable.

A completed accident report form, F2508, must be sent to the enforcing authorities within ten days.

Whom to report to

Until 2001 all reports had to be made to the local HSE office or local authority. This can still be done, but there is a centralized national system, called the Incident Contact Centre (ICC), which is a joint venture between the HSE and local authorities. All reports sent locally are now passed on to the ICC.

This means that employers no longer need be concerned about which authority to report to, as all incidents can be reported to the ICC directly. Reports can be sent by telephone, fax, the Internet, or by post. If reporting by the Internet or telephone a copy of the report is sent to the responsible person for correction, if necessary, and for their records.

➤ Postal reports should be sent to;
> Incident Contact Centre
> Caerphilly Business Park
> Caerphilly
> CF83 3GG

➤ For Internet reports go to:
> **www.riddor.gov.uk**
> or link through
> **www.hse.gov.uk**

➤ By telephone (charged at local rates)
> 0845 300 9923

➤ By fax (charged at local rates)
> 0845 300 9924

➤ By email: **riddor@natbrit.com**

15.6 Internal systems for collecting and analysing incident data

Introduction

Managers need effective internal systems to know whether the organization is getting better or worse; to know what is happening and why; and to assess whether objectives are being achieved. Chapter 16 deals with monitoring generally, but here, the basic requirements of a collection and analysis system for incidents are discussed.

The incident report form (discussed earlier) is the basic starting point for any internal system. Each organization needs to lay down what the system involves and who is responsible to do each part of the procedure. This will involve:

➤ what type of incidents should be reported
➤ who completes the incident report form – normally the manager responsible for the investigation
➤ how copies should be circulated in the organization
➤ who is responsible to provide management measurement data
➤ how the incident data should be analysed and at what intervals
➤ the arrangements to ensure that action is taken on the data provided.

The data should seek to answer the following questions:

➤ are failure incidents occurring, including injuries, ill-health and other loss incidents?
➤ where are they occurring?
➤ what is the nature of the failures?
➤ how serious are they?
➤ what were the potential consequences?
➤ what are the reasons for the failures?

- how much has it cost?
- what improvements in controls and the management system are required?
- how do these issues vary with time?
- is the organization getting better or worse?

Type of incident

Most organizations will want to collect:

- all injury accidents
- cases of ill-health
- sickness absence
- damage to property, personal effects and work in progress
- incidents with the potential to cause serious injury, ill-health or damage.

Not all of these are required by law, but this should not deter the organization that wishes to control hazards effectively.

Analysis

All the information, whether in accident books or report forms, will need to be analysed so that useful management data can be prepared. Many organizations look at analysis each month and annually. However, where there are very few incidents, quarterly may be sufficient. The health and safety information should be used alongside other business measures and should receive equal status.

There are several ways in which data can be analysed and presented. The most common ways are:

- by causation using the classification used on the RIDDOR form F2508. This has been used on the example accident report form (see Appendix 15.1)
- by nature of the injury, such as cuts, abrasions, asphyxiation, amputations
- by the part of the body affected, such as hands, arms, feet, lower leg, upper leg, head, eyes, back and so on. Sub-divisions of these categories could be useful if there were sufficient incidents
- by age and experience at the job
- by time of day
- by occupation or location of the job
- by type of equipment used.

There are a number of up-to-date computer recording programs which can be used to manipulate the data if significant numbers are involved. The trends can be shown against monthly, quarterly and annual past performance of, preferably, the same organization. If indices are calculated, such as Incident Rate, comparisons can be made nationally with HSE figures and with other similar organizations or businesses in the same industrial group. This is really only of major value to larger organizations with significant numbers of events. (See Chapter 4 for more information on definitions and calculation of Incident Rates.)

The HSE produces annual bulletins of national performance plus a detailed statistical report, which can be used for comparisons. There are difficulties in comparisons across Europe and say, with the USA, where the definitions of accidents or time lost vary.

Reports should be prepared with simple tables and graphs showing trends and comparisons. Line graphs, bar charts and pie charts are all used quite extensively with good effect. All analysis reports should be made available to employees as well as managers. This can often be done through the health and safety committee and safety representatives, where they exist, or directly to all employees in small organizations. Other routine meetings, team briefings and notice boards can all be used to communicate the message.

It is particularly important to make sure that any actions recommended or highlighted by the reports are taken quickly and employees kept informed.

15.7 Compensation and insurance issues

Accidents arising out of the organization's activities resulting in injuries to people and incidents resulting in damage to property can lead to compensation claims. The second objective of an investigation should be to collect and record relevant information for the purposes of dealing with any claim. It must be remembered that, in the longer term, prevention is the best way to reduce claims and must be the first objective in the investigation. An overzealous approach to gathering information concentrating on the compensation aspect can, in fact, prompt a claim from the injured party where there was no particular intention to take this route before the investigation. Nevertheless, relevant information should be collected. Sticking to the collection of facts is usually the best approach.

Appendix 15.2 provides a checklist of headings, which may assist in the collection of information. It is not expected that all accidents and incidents will be investigated in depth and a dossier with full information prepared. Judgement has to be applied as to which incidents might give rise to a claim and when a full record of information is required. All accident report forms should include the names of all witnesses as a minimum. Where the injury is likely to give rise to lost time, a photograph(s) of the situation should be taken.

15.8 Practice NEBOSH questions for Chapter 15

1. With reference to the Reporting of Injuries, Diseases and Dangerous Occurrences Regulations 1995:
    (i) state the legal requirements for reporting a fatality resulting from an accident at work to an enforcing authority.
    (ii) Outline three further categories of work-related injury (other than fatal injuries) that are reportable. (March 2001)
2. (a) State the requirements for reporting an 'over three-day' injury under the Reporting of Injuries, Diseases and Dangerous Occurrences Regulations 1995.
    (b) Giving reasons in each case, identify three categories of persons who may be considered a useful member of an internal accident investigation team. (June 2001)
3. With reference to the Reporting of Injuries, Diseases and Dangerous Occurrences Regulations 1995:
    (i) list four types of major injury
    (ii) outline the procedures for reporting a major injury to an enforcing authority. (December 2000)

4. (a) Give four reasons why an organization should have a system for the internal reporting of accidents.
   (b) Outline factors that may discourage employees from reporting accidents at work. (March 2000)
5. Outline the key points that should be covered in a training session for employees on the reporting of accidents/incidents. (June 1998)

Appendix 15.1 – Injury report form

INCIDENT / ACCIDENT REPORT

INJURED PERSON:.. Date of Accident: /   /20 Time........am/pm
POSITION: .... Place of Incident: ....
DEPARTMENT:.. Details of Injury: ....
Investigation carried out by: ....
Position:.. Estimated Absence: ....

Brief details of Accident (A detailed report together with diagrams, photographs and any witness statements should be attached where necessary. Please complete all details requested overleaf.)

Immediate Causes	**Underlying Cause**

Conclusions (How can we prevent this kind of incident/accident occurring again?)

Action to be taken: **Completion Date: /   /20**

IMPORTANT

Please ensure that an accident investigation and report is completed and forwarded to Human Resources within 48 hours of the accident occurring.

Remember that accidents involving major injuries or dangerous occurrences have to be notified immediately by telephone to the HSE/ Local authority.

Signature of manager making report: ....   Copies: Personnel Manager
                                                                                Health & Safety Manager
                          Date: /   /20   Payroll Controller

INJURED PERSON: Surname .... Forenames ....

                                                                                Male/Female

Home address .... Age ....

Employee ☐   *Agency Temp* ☐   *Contractor* ☐   *Visitor* ☐   *Youth Trainee* ☐   (Tick one box)

Kind of Accident Indicate what kind of accident led to the injury or condition (tick one box)

Contact with moving machinery or material being machined **1**	Injury while handling lifting or carrying **5**	Drowning or asphyxiation **9**	Contact with electricity or an electrical discharge **13**
Struck by moving, flying, or falling object **2**	Slip, trip or fall on same level **6**	Exposure to or contact with harmful substance **10**	Injured by an animal **14**
Struck by moving vehicle **3**	Fall from height indicate approx distance of fall..........m **7**	Exposure to fire **11**	Violence, physically assaulted by a person **15**
Struck against something fixed or stationary **4**	Trapped by something collapsing **8**	Exposure to an explosion **12**	Other kind of accident **16**

Detail any machinery, chemicals, tools etc involved

Accident first reported to: Name ....

Position & ....   Dept ....

First Aid/medical attention by: First Aider Name ....   Dept ....

                               Doctor Name ....

                               Medical centre ....   Hospital ....

WITNESSES

Name	Position & Dept	Statement obtained(yes/no) Attach all statements taken
....	....	yes/no
....	....	yes/no
....	....	yes/no
....	....	yes/no

For Office use only

If relevant: Date reported to HSE/LA   a) by telephone   ..../..../20

                                      b) on form F2508   ..../..../20

           Date reported to organization's Insurers   ..../..../20

Were the recommendations effective? Yes/No
If no say what action should be taken.

Appendix 15.2 – Information for insurance/compensation

Information for insurance/compensation purposes following accident or incident

Factual information needs to be collected where there is the likelihood of some form of claim either against the organization or by the organization (e.g. damage to equipment). This aspect should be considered as a second objective in accident investigation, the first being to learn from the accident to reduce the possibility of accidents occurring in the future.

Information to be collected

➤ Details of machinery, plant or other equipment involved in the incident. Description should include details, e.g. size, age, serial number, maker, ownership.

➤ Description of the state of any safety related equipment, e.g. guard, trip systems.

➤ Description of any maintenance schedule where appropriate.

➤ Description of conditions, including weather conditions or internal climate, state of the floor, general layout, etc.

➤ Descriptions and details of any hazardous substances used or involved in the accident.

➤ Relevant measurements, sketches can be very useful. Photographs where possible and, with manual handling, the weight of any articles involved.

➤ Circumstances of the incident. This should be a factual account derived from witness reports.

➤ Witness statements can be useful, these should be brief and factual. Try to avoid including hearsay, speculation and opinion.

➤ Description of the training received by relevant personnel and in relevant subjects. Remember training does not always have to be formal. If training has been received, for example, by simply being shown what to do working alongside others, this is often relevant.

➤ Take details of relevant qualifications from personnel involved.

➤ Record the age and experience of relevant personnel.

➤ Record the supervision being provided, including details of the job position, qualifications and experience of the supervisor.

➤ Refer to any written risk assessments, for example, safety systems of work, method statements, where these are relevant.

16

Monitoring, review and audit

16.1 Introduction

This chapter concerns the monitoring of health and safety performance, including both positive measures like inspections and negative measures like injury statistics. It is about reviewing progress to see if something better can be done and auditing to ensure that what has been planned is being implemented.

Measurement is a key step in any management process and forms the basis of continuous improvement. If measurement is not carried out correctly, the effectiveness of the health and safety management system is undermined and there is no reliable information to show managers how well the health and safety risks are controlled.

Managers should ask key questions to ensure that arrangements for health and safety risk control are in place, comply with the law as a minimum, and operate effectively.

Proactive monitoring, by taking the initiative before things go wrong, involves routine inspections and checks to make sure that standards and policies are being implemented and that controls are working. **Reactive monitoring**, after things go wrong, involves looking at historical events to learn from mistakes and see what can be put right to prevent a recurrence.

The HSE's experience is that organizations find health and safety performance measurement a difficult subject. They struggle to develop health and safety performance measures which are not based solely on injury and ill-health statistics.

16.2 The traditional approach to measuring health and safety performance

Senior managers often measure company performance by using, for example, percentage profit, return on investment or market share. A common feature of the measures would be that they are generally positive in nature – which demonstrates achievement – rather than negative, which demonstrates failure.

Yet, if senior managers are asked how they measure their companies' health and safety performance, it is likely that the only measure would be accident or injury statistics. While the general business performance of an organization is subject to a range of positive measures, for health and safety it too often comes down to one negative measure, injury and ill-health statistics – measures of failures.

Health and safety differs from many areas measured by managers because improvement in performance means fewer outcomes from the measure (injuries or ill-health) rather than more. A low injury or ill-health rate trend over years is still no guarantee that risks are being controlled and that incidents will not happen in the future. This is particularly true in organizations where major hazards are present but there is a low probability of accidents.

There is no single reliable measure of health and safety performance. What is required is a 'basket' of measures, providing information on a range of health and safety issues.

There are some significant problems with the use of injury/ill-health statistics in isolation:

➤ there may be under-reporting – focusing on injury and ill-health rates as a measure, especially if a reward system is involved, can lead to non-reporting to keep up performance;

➤ it is often a matter of chance whether a particular incident causes an injury, and they may not show whether or not a hazard is under control. Luck or a reduction in the number of people exposed, may produce a low injury/accident rate rather than good health and safety management;

➤ an injury is the particular consequence of an incident and often does not reflect the potential severity. For example, an unguarded machine could result in a cut finger or an amputation;

➤ people can be absent from work for reasons which are not related to the severity of the incident;

➤ there is evidence to show that there is little relationship between 'occupational' injury statistics (e.g. slips, trip and falls) and the reasons for the lack of control of major accident hazards (e.g. loss of containment of flammable or toxic material);

➤ a small number of accidents may lead to complacency;

➤ injury statistics demonstrate outcomes not causes.

Because of the potential shortcomings related to the use of accident/injury and ill-health data as a single measure of performance, more proactive or 'up stream' measures are required. These require a systematic approach to deriving positive measures and how they link to the overall risk control process, rather than a quick-fix based on things that can be easily counted, such as the numbers of training courses or numbers of inspections, which has limited value. The resultant data provide no information on how the figure was arrived at, whether it is 'acceptable' (i.e. good/bad) or the quality and effectiveness of the activity. A more disciplined approach to health and safety performance measurement is required. This needs to develop as the health and safety management system develops.

16.3 Why measure performance?

16.3.1 Introduction

You can't manage what you can't measure – Drucker

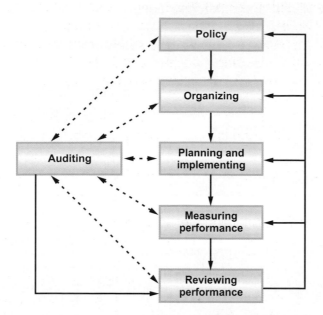

Figure 16.1 The health and safety management system. Source HSE. Crown copyright material is reproduced with the permission of the Controller of HMSO and the Queen's Printer for Scotland.

Measurement is an accepted part of the 'plan-do-check-act' management process. Measuring performance is as much part of a health and safety management system as financial, production or service delivery management. The HSG 65 framework for managing health and safety, discussed at the end of Chapter 1 and illustrated in Figure 16.1, shows where measuring performance fits within the overall health and safety management system.

The main purpose of measuring health and safety performance is to provide information on the progress and current status of the strategies, processes and activities employed to control health and safety risks. Effective measurement not only provides information on what the levels are but also why they are at this level, so that corrective action can be taken.

16.3.2 Answering questions

Health and safety monitoring or performance measurement should seek to answer such questions as:

> where is the position relative to the overall health and safety aims and objectives?
> where is the position relative to the control of hazards and risks?
> how does the organization compare with others?
> what is the reason for the current position?
> is the organization getting better or worse over time?
> is the management of health and safety doing the right things?
> is the management of health and safety doing things right consistently?
> is the management of health and safety proportionate to the hazards and risks?
> is the management of health and safety efficient?

> is an effective health and safety management system in place across all parts of the organization?

> is the culture supportive of health and safety, particularly in the face of competing demands?

These questions should be asked at all management levels throughout the organization. The aim of monitoring should be to provide a complete picture of an organization's health and safety performance.

16.3.3 Decision making

The measurement information helps in deciding:

> where the organization is in relation to where it wants to be
> what progress is necessary and reasonable in the circumstances
> how that progress might be achieved against particular restraints (e.g. resources or time)
> priorities – what should be done first and what is most important
> effective use of resources.

16.3.4 Addressing different information needs

Information from the performance measurement is needed by a variety of people. These will include directors, senior managers, line managers, supervisors, health and safety professionals and employees/safety representatives. They each need information appropriate to their position and responsibilities within the health and safety management system.

For example, what the chief executive officer of a large organization needs to know from the performance measurement system will differ in detail and nature from the information needs of the manager of a particular location.

A coordinated approach is required so that individual measuring activities fit within the general performance measurement framework.

Although the primary focus for performance measurement is to meet the internal needs of an organization, there is an increasing need to demonstrate to external stakeholders (regulators, insurance companies, shareholders, suppliers, contractors, members of the public, etc.) that arrangements to control health and safety risks are in place, operating correctly and effectively.

16.4 What to measure

16.4.1 Introduction

In order to achieve an outcome of no injuries or work-related ill-health, and to satisfy stakeholders, health and safety risks need to be controlled. Effective risk control is founded on an effective health and safety management system. This is illustrated in Figure 16.2.

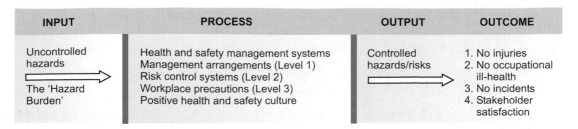

INPUT	PROCESS	OUTPUT	OUTCOME
Uncontrolled hazards

The 'Hazard Burden' | Health and safety management systems
Management arrangements (Level 1)
Risk control systems (Level 2)
Workplace precautions (Level 3)
Positive health and safety culture | Controlled hazards/risks | 1. No injuries
2. No occupational ill-health
3. No incidents
4. Stakeholder satisfaction |

Figure 16.2 Health and safety management system. Source HSE. Crown copyright material is reproduced with the permission of the Controller of HMSO and the Queen's Printer for Scotland.

16.4.2 Effective risk control

The health and safety management system comprises three levels of control (see Figure 16.2):

> level 3 – effective workplace precautions provided and maintained to prevent harm to people who are exposed to the risks

> level 2 – risk control systems (RCSs): the basis for ensuring that adequate workplace precautions are provided and maintained

> level 1 – the key elements of the health and safety management system: the management arrangements (including plans and objectives) necessary to organize, plan, control and monitor the design and implementation of RCSs.

The health and safety culture must be positive to support each level.

Performance measurement should cover all elements of Figure 16.2 and be based on a balanced approach which combines:

> **input: monitoring** the scale, nature and distribution of hazards created by the organization's activities – measures of the hazard burden

> **process: active monitoring** of the adequacy, development, implementation and deployment of the health and safety management system and the activities to promote a positive health and safety culture – measures of success

> **outcomes: reactive monitoring** of adverse outcomes resulting in injuries, ill-health, loss and accidents with the potential to cause injuries, ill-health or loss – measures of failure.

16.5 Measuring failure – reactive monitoring

So far, this chapter has concentrated on measuring activities designed to prevent the occurrence of injuries and work-related ill-health (active monitoring). Failures in risk control also need to be measured (reactive monitoring), to provide opportunities to check performance, learn from failures and improve the health and safety management system.

Reactive monitoring arrangements include systems to identify and report:

> injuries and work-related ill-health (details of the incident rate calculation is given in Chapter 4)

> other losses such as damage to property

➤ incidents, including those with the potential to cause injury, ill-health or loss (near misses)

➤ hazards and faults

➤ weaknesses or omissions in performance standards and systems, including complaints from employees and enforcement action by the authorities.

Guidance on investigating and analysing these incidents is given in Chapter 15.

16.6 Proactive monitoring – how to measure performance

16.6.1 Introduction

The measurement process can gather information through:

➤ direct observation of conditions and of people's behaviour (sometimes referred to as unsafe acts and unsafe conditions monitoring);

➤ talking to people to elicit facts and their experiences as well as gauging their views and opinions;

➤ examining written reports, documents and records.

These information sources can be used independently or in combination. Direct observation includes inspection activities and the monitoring of the work environment (e.g. temperature, dust levels, solvent levels, noise levels) and people's behaviour. Each risk control system should have a built-in monitoring element that will define the frequency of monitoring; these can be combined to form a common inspection system.

16.6.2 Inspections

General

This may be achieved by developing a checklist or inspection form that covers the key issues to be monitored in a particular department or area of the organization within a particular period. It might be useful to structure this checklist using the 'four Ps' (note that the examples are not a definitive list):

➤ **premises**, including:
  ➤ access/escape
  ➤ housekeeping
  ➤ services like gas and electricity
  ➤ working environment
  ➤ fire precautions
➤ **plant and substances**, including:
  ➤ machinery guarding
  ➤ tools and equipment
  ➤ local exhaust ventilation
  ➤ use/storage/separation of materials/chemicals
➤ **procedures**, including:
  ➤ safe systems of work

(a)

(b)

Figure 16.3   (a) Poor conditions – inspection needed; (b) inspection in progress.

➤ permits to work
➤ use of personal protective equipment
➤ procedures followed
➤ **people**, including:
   ➤ health surveillance
   ➤ people's behaviour
   ➤ training and supervision
   ➤ appropriate authorized person.

It is essential that people carrying out an inspection do not in any way put themselves or anyone else at risk. Particular care must be taken with regard to safe access. In carrying out these safety inspections, the safety of people's actions should be considered, in addition to the safety of the conditions they are working in – a ladder might be in perfect condition but it has to be used properly too.

Key points in becoming a good observer

To improve health and safety performance, managers and supervisors must eliminate unsafe acts by observing them, taking immediate corrective action, and following up to prevent recurrence. To become a good observer, they must improve their observation skills and must learn how to observe effectively. Effective observation includes the following key points:

➤ be selective
➤ know what to look for
➤ practice
➤ keep an open mind
➤ guard against habit and familiarity
➤ do not be satisfied with general impressions
➤ record observations systematically.

Observation techniques

In addition, to become a good observer, a person must:

➤ stop for 10 to 30 seconds before entering a new area to ascertain where employees are working;
➤ be alert for unsafe practices that are corrected as soon as you enter an area;
➤ observe activity – do not avoid the action;
➤ remember ABBI – look Above, Below, Behind, Inside;
➤ develop a questioning attitude to determine what injuries might occur if the unexpected happened and how the job might be accomplished more safely. Ask 'why?' and 'what could happen if . . . ?';
➤ use all senses: sight, hearing, smell, touch;
➤ maintain a balanced approach. Observe all phases of the job;
➤ be inquisitive;
➤ observe for ideas – not just to determine problems;
➤ recognize good performance.

Daily/weekly/monthly safety inspections

These will be aimed at checking conditions in a specified area against a fixed checklist drawn up by local management. It will cover specific items, such as the guards at particular machines, whether access/agreed routes are clear, whether fire extinguishers are in place, etc. The checks should be carried out by staff of the department who should sign off the checklist. It should not last more than half an hour, perhaps less. This is not a specific hazard spotting operation, but there should be a space on the checklist for the inspectors to note down any particular problems encountered.

Reports from inspections

Some of the items arising from safety inspections will have been dealt with immediately, other items will require action by specified people. Where there is some doubt about the problem, and what exactly is required, advice should be sought from the site safety advisor or external expert. A brief report of the inspection and any resulting action list should be submitted to the safety committee. While the committee may not have the time available to consider all reports in detail, it will want to be satisfied that appropriate action is taken to resolve all matters; it will be necessary for the committee to follow up the reports until all matters are resolved.

In order to get maximum value from inspection checklists, they should be designed so that they require objective rather than subjective judgements of conditions. For example, asking the people undertaking a general inspection of the workplace to rate housekeeping as good or bad, begs questions as to what does good and bad mean, and what criteria should be used to judge this. If good housekeeping means there is no rubbish left on the floor; all waste bins are regularly emptied and not overflowing; floors are swept each day and cleaned once a week; decorations should be in good condition with no peeling paint, then this should be stated. Adequate expected standards should be provided in separate notes so that those inspecting know the standards that are required.

The checklist or inspection form should facilitate:

➤ the planning and initiation of remedial action, by requiring those doing the inspection to rank deficiencies in priority order (those actions which are most important rather than those which can be easily done quickly);

➤ laying down those responsible for taking remedial actions, with sensible timescales to track progress on implementation;

➤ periodic monitoring to identify common themes which might reveal underlying problems in the system;

➤ information to help managers decide on the frequency or nature of the monitoring arrangements.

Appendix 16.1 gives examples of poor workplaces which can be used for practice exercises. Appendix 16.2 shows a basic list of inspection issues, which can be adapted to any particular organization's needs. A policy checklist is given in Chapter 2.

16.6.3 Safety sampling

Safety sampling is a helpful technique that helps organizations to concentrate on particular areas or subjects at a time. A specific area is chosen which can be inspected in about thirty minutes. A checklist is drawn up to facilitate the inspection looking at specific issues. These may be different types of hazard or they may be say unsafe acts or conditions noted, they may be positive, good behaviour noted.

The inspection team or person then carries out the sampling at the same time each day or week in the specified period. The results are recorded and analysed to see if the changes are good or bad over time. Of course, defects noted must be brought to the notice of the appropriate person for action on each occasion.

16.7 Who should monitor performance?

Performance should be measured at each management level from directors downwards. It is not sufficient to monitor by exception, where unless problems are raised, it is assumed to be satisfactory. Senior managers must satisfy themselves that the correct arrangements are in place and working properly. Responsibilities for both active and reactive monitoring must be laid down and managers need to be personally involved in making sure that plans and objectives are met and compliance with standards is achieved. Although systems may be set up with the guidance of safety professionals, managers should be personally involved and given sufficient training to be competent to make informed judgements about monitoring performance.

Other people like safety representatives will also have the right to inspect the workplace. Each employee should be encouraged to inspect their own workplace frequently to check for obvious problems and rectify them if possible or report hazards to their supervisors.

Specific statutory (or thorough) examinations of, for example, lifting equipment or pressure vessels, have to be carried out at intervals laid down in written schemes by competent persons – usually specially trained and experienced inspection/insurance company personnel.

16.8 Frequency of monitoring and inspections

This will depend on the level of risk and any statutory inspection requirement. Directors may be expected to examine the premises formally at an annual audit, while departmental supervisors may be expected to carry out inspection each week. Senior managers should regularly monitor the health and safety plan to ensure that objectives are being met and to make any changes to the plan as necessary.

Data from reactive monitoring should be considered by senior managers at least once a month. In most organizations serious events would be closely monitored as they happen.

16.9 Report writing

General

Report writing is about communication. The aim of a report writer is threefold:

➤ to get a message through to the reader
➤ to make the message and the arguments clearly understandable
➤ to make the arguments and conclusions persuasive.

Presentation is a vital part of this; so while a handwritten report is better than nothing if the situation is urgent, a well-organized, typed report is clearly much more effective. To the reader of the report, who may well be very busy with a great deal of written information to wade through, a clear, well presented report will produce a positive attitude from the outset, conferring instant benefit on the report writer. These five factors help to make reports more effective:

➤ the way the report is structured
➤ the way that arguments are presented
➤ the way the report is written
➤ the way that the data are presented – use of graphics
➤ the way that the report itself is presented.

16.9.2 Structure

The structure of a report is the key to its professionalism. Not only will good structure help the reader to understand the information and follow the arguments contained in the report, it also increases the writer's credibility and helps in the organization of material to the best advantage. Sometimes the report will be a simple short document on a single accident or incident, or it may be a substantial review or audit on a large organization or site.

The following structure will help to produce a clear and useful piece of report writing:

1 Title page
2 Summary
3 Contents list
4 Introduction
5 Main body of the report
6 Conclusions
7 Recommendations
8 Appendices.

NB: although this is a useful and frequently used format, it is important to check with the organization requesting the report in case their in-house format is different.

Title page
This will contain:

➤ a title and often a subtitle
➤ the name of the person or organization to whom the report is addressed
➤ the name of the writer(s) and their organization
➤ the date on which the report was submitted.

Remembering that report writing is about communication, it is often more effective to produce a title that is eye-catching and memorable as well as being informative.

Summary
This should be limited to between 150 and 500 words. It should not include any evidence or data, which should be kept for the main report. A summary should include the writers' main conclusions and principal recommendations, and should be situated near the front of the report.

Contents list
This should also be placed early in the report. It is not necessary for short reports, but if there are several headings it does help the reader to grasp the overall content of the report in a short time.

Introduction
The introduction should contain the following:

➤ information about who commissioned the report and when
➤ the purpose of the report
➤ objectives of the report
➤ terms of reference

➤ the preparation of the report (type of data, research used/undertaken, relevant legislation, witnesses, other people interviewed, etc.)
➤ methodology used in analysis, if appropriate
➤ any problems and the methods used to tackle them
➤ details about consultation with clients, employees, etc.

There may be other items which are specific to the subject of the report.

The main body of the report

The function of this part of the report is to describe, in detail, the facts, analysis and evaluation of the subject of the report. (In other words, what was discovered, the significance of these discoveries and their importance.) In this part of the report, it can often be useful to illuminate the findings with the use of sketches, graphs, tables and charts. These graphics should have the function of summarizing information rather than giving large amounts of detail. The more detailed graphics should be made into appendices.

To make it more digestible, this part of the report should be divided into sections, using numbered headings and subheadings. Very long and complicated reports will need to be broken down into chapters.

Conclusions

The concluding part of the report should be a reasonably detailed 'summing up'. It should give the conclusions arrived at by the writer and explain why the writer has reached these conclusions.

Recommendations

The use of this section depends on the requirements of the person commissioning the report. If recommendations are required, it is better not to make too many as the focus will be lost. Report writers are often asked only to provide the facts.

Appendices

This part of the report should contain sections which may be useful to a reader who requires more detail. Examples would be the detailed charts, graphs and tables mentioned in the main body of the report, any questionnaires used in constructing the evidence, lists, forms, case studies and so on. The appendices are the background material of the report.

16.9.3 Making a report more emphatic and persuasive

Because the reader is likely to be a person with some degree of expertise in the subject, a report must be reliable, credible, relevant and thorough. It is therefore important to avoid emotional language, opinions presented as facts and arguments that have no supporting evidence. To make a report more persuasive, it is essential to:

➤ present the information clearly
➤ provide reliable evidence
➤ present arguments logically
➤ avoid falsifying, tampering with or concealing facts.

Expertise in an area of knowledge means that distortions, errors and omissions are quickly spotted by the reader and the presence of any of these will cast doubt on the credibility of the whole.

Normally, reports are used as part of a decision-making process. In this case the main requirement will be the facts. Interpretations and opinions are not usually sought. Exceptions to this would be where the report is a proposal document, or where a recommendation is specifically requested. Unless this is the case, it is better not to make recommendations.

A report should play a key role in organizing information for the use of decision makers. It should review a complex and/or extensive body of information and make a summary of all the important issues.

Reports should be straightforward to produce if the writer keeps to a clear format. This format is designed to tell the reader, as clearly as possible:

➤ what happened
➤ what the legal liability is
➤ what it cost
➤ what the result was etc.

There may be a request for a special report and this is likely to be longer and more difficult to produce. Often it will relate to a 'critical incident' and the decision makers will be looking for information to help them:

➤ decide whether this is a problem or an opportunity
➤ decide whether to take action
➤ decide what action, if any, to take.

Finally, report writing should be kept simple. Nothing is gained, in fact much is lost, in the use of long, complicated sentences, jargon and official-sounding language. When the report is finished, it is helpful to run through with the express intention of simplifying the language and making sure that it says what was intended in a clear manner. A good way to achieve this is to keep a notice pinned up in a prominent place saying:

> **KEEP IT SHORT AND SIMPLE**

16.10 Review and audit

16.10.1 Audits – purpose

The final steps in the health and safety management control cycle are auditing and performance review. Organizations need to be able to reinforce, maintain and develop the ability to reduce risks. The 'feedback loop' produced by this final stage in the process enables them to do this and to ensure continuing effectiveness of the health and safety management system.

Audit is a business discipline which is frequently used, for example, in finance, environmental matters and quality. It can equally well be applied to health and safety.

The term is often used to mean inspection or other monitoring activity. Here, the following definition is used, which follows HSG 65:

The structured process of collecting independent information on the efficiency, effectiveness and reliability of the total health and safety management system and drawing up plans for corrective action.

Over time, it is inevitable that control systems will decay and may even become obsolete as things change. Auditing is a way of supporting monitoring by providing managers with information. It will show how effectively plans and the components of health and safety management systems are being implemented. In addition, it will provide a check on the adequacy and effectiveness of the management arrangements and risk control systems (RCSs).

Auditing is critical to a health and safety management system, but it is not a substitute for other essential parts of the system. Companies need systems in place to manage cash flow and pay the bills – this cannot be managed through an annual audit. In the same way, health and safety needs to be managed on a day-to-day basis and for this organizations need to have systems in place. A periodic audit will not achieve this.

The aims of auditing should be to establish that the three major components of a safety management system are in place and operating effectively. It should show that:

➤ appropriate management arrangements are in place
➤ adequate risk control systems exist, are implemented, and consistent with the hazard profile of the organization
➤ appropriate workplace precautions are in place.

Where the organization is spread over a number of sites, the management arrangements linking the centre with the business units and sites should be covered by the audit.

There are a number of ways in which this can be achieved and some parts of the system do not need auditing as often as others. For example, an audit to verify the implementation of risk control systems would be made more frequently than a more overall audit; for example one of overall capability of the organization or of the management arrangements for health and safety. Critical risk control systems, which control the principle hazards of the business, would need to be audited more frequently. Where there are complex workplace precautions, it may be necessary to undertake technical audits. An example would be chemical process plant integrity and control systems.

A well-structured auditing programme will give a comprehensive picture of the effectiveness of the health and safety management system in controlling risks. Such a programme will indicate when and how each component part will be audited. Managers, safety representatives and employees, working as a team, will effectively widen involvement and cooperation needed to put together the programme and implement it.

The process of auditing involves:

➤ gathering information from all levels of an organization about the health and safety management system;
➤ making informed judgements about its adequacy and performance.

16.10.2 Gathering information

Decisions will need to be made about the level and detail of the audit before starting to gather information about the health and safety management of an organization. Auditing involves sampling, so initially it is necessary to decide how much sampling is needed for the assessment to be reliable. The type of audit and its complexity will relate to its objectives and scope, to the size and complexity of the organization and to the length of time that the existing health and safety management system has been in operation.

Information sources of interviewing people, looking at documents and checking physical conditions are usually approached in the following order:

Preparatory work
> meet with relevant managers and employee representatives to discuss and agree the objectives and scope of the audit;
> gather and consider documentation;
> prepare and agree the audit procedure with managers.

On-site
> interviewing;
> review and assessment of additional documents;
> observation of physical conditions and work activities.

Conclusion
> assemble the evidence;
> evaluate the evidence;
> write an audit report.

16.10.3 Making judgements

It is essential to start with a relevant standard or benchmark against which the adequacy of a health and safety management system can be judged. If standards are not clear, assessment cannot be reliable. Audit judgements should be informed by legal standards, HSE guidance and applicable industry standards. HSG 65 sets out benchmarks for management arrangements and for the design of risk control systems. This book follows the same concepts.

Auditing should not be seen as a fault-finding activity. It should make a valuable contribution to the health and safety management system and to learning. It should recognize achievement as well as highlight areas where more needs to be done.

Scoring systems can be used in auditing along with judgements and recommendations. This can be seen as a useful way to compare sites or monitor progress over time. However, there is no evidence that quantified results produce a more effective response than the use of qualitative evidence. Indeed, the introduction of a scoring system can, the HSE believes, have a negative effect as it may encourage managers to place more emphasis on high scoring questions which may not be as relevant to the development of an effective health and safety management system.

To achieve the best results, auditors should be competent people who are independent of the area and of the activities being audited. External consultants can be used or staff from other areas of the organization. An organization can use its own auditing system or one of the proprietary systems on the market or, since it is unlikely that any ready made system will provide a perfect fit, a combination of both. With any scheme, cost and benefits have to be taken into account. Common problems include:

➤ systems which are too general in their approach. These may need considerable work to make them fit the needs and risks of the organization
➤ systems which are too cumbersome for the size and culture of the organization
➤ scoring systems may conceal problems in underlying detail
➤ organizations may design their management system to gain maximum points rather than using one which suits the needs and hazard profile of the business.

HSE encourages organizations to assess their health and safety management systems using in-house or proprietary schemes but without endorsing any particular one.

16.10.4 Performance review

When performance is reviewed, judgements are made about its adequacy and decisions are taken about how and when to rectify problems. The feedback loop is needed by organizations so that they can see whether the health and safety management system is working as intended. The information for review of performance comes from audits of risk control systems and workplace precautions, and from the measurement of activities. There may be other influences, both internal and external, such as re-organization, new legislation or changes in current good practice. These may result in the necessity to redesign or change parts of the health and safety management system or to alter its direction or objectives.

Performance standards need to be established which will identify the systems requiring change, responsibilities, and completion dates. It is essential to feed back the information about success and failure so that employees are motivated to maintain and improve performance.

In a review, the following areas will need to be examined:

➤ the operation and maintenance of the existing system
➤ how the safety management system is designed, developed and installed to accommodate changing circumstances.

Reviewing is a continuous process. It should be undertaken at different levels within the organization. Responses will be needed as follows:

➤ by first-line supervisors or other managers to remedy failures to implement workplace precautions which they observe in the course of routine activities
➤ to remedy sub-standard performance identified by active and reactive monitoring
➤ to the assessment of plans at individual, departmental, site, group or organizational level
➤ to the results of audits.

The frequency of review at each level should be decided upon by the organization and reviewing activities should be devised which will suit the measuring and auditing activities. The review will identify specific remedial actions which establish who is responsible for implementation and set deadlines for completion.

16.11 Practice NEBOSH questions for Chapter 16

1. Identify eight measures that could be used by an organization in order to monitor its health and safety performance. (March 2001)
2. Identify the main topic areas that should be included in a planned health and safety inspection of a workplace. (September 2000)
3. Outline four proactive monitoring methods that can be used in assessing the health and safety performance of an organization. (March 2000)
4. Outline two reactive measures and two proactive measures that can be used in monitoring an organization's health and safety performance. (June 1998)
5. Explain how the following may be used to improve safety performance within an organization:
   (i) accident data
   (ii) safety inspections. (March 1998)
6. Outline ways in which an organization can monitor its health and safety performance. (June 1997)

Appendix 16.1 Workplace inspection exercises

The following four examples are given of workplaces with numerous inadequately controlled hazards. They can be used to practise workplace inspections and risk assessments.

Office

Figure 16.4

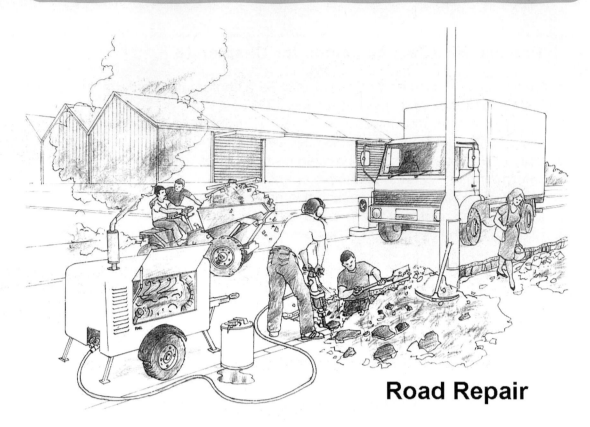

Road Repair

Figure 16.5

Workshop

Figure 16.6

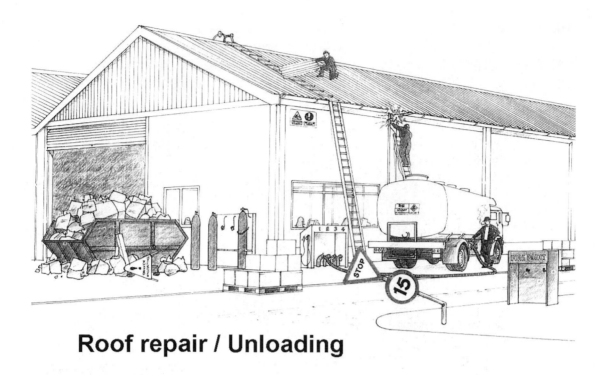

Roof repair / Unloading

Figure 16.7

Appendix 16.2 – Basic checklist of items to be covered in an area inspection

1.	Access equipment	Right equipment for the job?
		Properly erected? Maintained?
2.	Access ways	Unobstructed ?
3.	Chemicals	COSHH assessments OK?
		Exposures adequately controlled?
		Data sheet information available?
		Spillage procedure available?
		Properly stored? Properly disposed of?
4.	Cleaning	Slip risk controlled?
5.	Contractors	Are there control rules and procedures?
		Are they followed?
6.	Eating facilities	Clean and adequate/ Means of heating food?
7.	Electrical equipment	Portable equipment tested?
		Leads tidy not damaged?
8.	Ergonomics	Tasks require uncomfortable postures or actions?
		Frequent repetitive actions accompanying muscular strain?
9.	Eye-wash bottles	Full? In date?
10.	Employers' liability insurance	Notice displayed? In date?
11.	Fire exits	Unobstructed? Easily opened?
12.	Fire extinguishers	In place? Full? Correct type?
13.	Fire instructions	Posted up? Not defaced or damaged?

14. First Aid — Suitably placed and provisioned? Appointed person? Trained first aider?

15. Flammable liquids — Stored properly? Used properly?

16. Flooring — Even and in good condition?

17. Hand tools — Right tool for the job? In good condition?

18. Health and Safety law poster — Displayed/completed?

19. Housekeeping — Tidy, clean, well organized?

20. Ladders/step ladders — Right equipment for the job? Secure base? Correct angle? Equipment in good condition?

21. Machinery — Guards in place? Effective? Service schedules available?

22. Manual handling — Moving excessive weight? Assessments carried out? Using correct technique? Could it be eliminated or reduced?

23. New or nursing mothers — Facilities for resting?

24. Noise — Normal conversation possible? Noise assessment needed/not needed? Noise areas designated?

25. PPE — Correct type? Worn correctly? Good condition?

26. Risk assessments — Carried out? General and fire? Suitable and sufficient?

27. Safe systems of work — Available? Satisfactory? Followed?

28. Sharps — Safety knives used? Knives/needles/glass properly used/disposed of?

29. Vehicles — Speeding? Following correct route? Driver looks where he is going? Properly maintained? Passengers only where specifically intended with suitable seat?

30. Visual display units — Workstation assessments needed/not needed Chairs adjustable/comfortable/maintained properly? Cables properly controlled? Lighting OK? No glare?

31. Working environment — Crowded? Too hot/cold? Ventilation? Humidity? Dusty? Lighting?

32. Welfare — Washing and toilet facilities satisfactory? Kept clean with soap and towels/ Adequate changing facilities

33. Young persons — Employed? Special risk assessments?

17

Summary of the main legal requirements

17.1 Introduction

The achievement of an understanding of the basic legal requirements can be a daunting task. However, the health and safety student should not despair because at NEBOSH National Certificate level they do not need to know the full history of UK health and safety legislation, the complete range of regulations made under the Health and Safety at Work Act 1974 or the details of the appropriate European Directives. These summaries, which are correct up to the 1st May 2003, cover those Acts and Regulations which are essential for the Certificate student and will provide the essential foundation of knowledge. They will also cover many of the requirements for similar awards at Certificate level and below. Students should check the latest edition of the NEBOSH Certificate guide to ensure that the regulations summarized here are the latest for the course being undertaken. Remember that NEBOSH does not examine on legislation until six months after it has been published.

Many managers will find these summaries a useful quick reference to the legal requirements. However, anyone involved in achieving compliance with legal requirements should ensure they read the regulations themselves, any approved codes of practice and HSE guidance.

17.2 The legal framework

17.2.1 General

The *Health and Safety at Work etc. Act 1974* is the foundation of British Health and Safety law. It describes the general duties that employers have towards their employees and to members of the public, and also the duties that employees have to themselves and to each other.

The term '*so far as is reasonably practicable*' qualifies the duties in the HSW Act. In other words, the degree of risk in a particular job or workplace needs to be balanced against the time, trouble, cost and physical difficulty of taking measures to avoid or reduce the risk. The law simply expects employers to behave in a way that demonstrates good management and common sense. They are required to look at what the hazards are and take sensible measures to tackle them.

The Management of Health and Safety at Work Regulations 1999 (the Management Regulations) clarifies what employers are required to do to manage health and safety under the Health and Safety at Work Act. Like the Act, they apply to every work activity. The law requires every employer to carry out a *risk assessment*. If there are five or more employees in the workplace, the significant findings of the risk assessment need to be recorded.

In a place like an office, risk assessment should be straightforward; but where there are serious hazards, such as those in a chemical plant, laboratory or an oil rig, it is likely to be more complicated.

The Factories Act 1961 and The Office Shops and Railway Premises Act of 1963, while still partly remaining on the Statute Book for most practical health and safety work can be ignored. However, strangely, the notification to the authorities of premises which come within the definition of these two acts are carried out on Form 9 for factories, Form 10 for construction and OSR 1 for office shop and railway premises. In addition, the Highly Flammable Liquids regulations 1972 are still operated under the Factories Act. They are due to be replaced in 2003 by the Dangerous Substances and Explosive Atmospheres Regulations under the HSW Act.

17.2.2 The relationship between the regulator and industry

The HSC consults widely with those affected by its proposals.

The HSC/E work through:

> the HSC's Industry and Subject Advisory Committees, which have members drawn from the areas of work they cover, and focus on health and safety issues in particular industries (such as the textile industry, construction and education or areas such as toxic substances and genetic modification)
> intermediaries, such as small firms' organizations
> providing information and advice to employers and others with responsibilities under the Health and Safety at Work Act
> guidance to enforcers, both HSE inspectors and those of local authorities
> the day-to-day contact which inspectors have with people at work.

The HSC consults with small firms through Small Firms Forums. It also seeks views in detail from representatives of small firms about the impact on them of proposed legislation.

17.3 List of Acts and Regulations summarized

The following Acts and Regulations are covered in this summary:

> The Health and Safety at Work etc. Act 1974 (HSW Act)
> The Fire Precautions Act 1971 (FPA)
> The Environmental Protection Act 1990.

The most important of these for health and safety is the Health and Safety at Work Act. Most of the relevant regulations covering health and safety at work have been made under this Act since 1974. These are included here and are all relevant to the Certificate student. The first list is alphabetical and some titles have been modified to allow an easier search. They are:

17.3.1 Alphabetical list of Regulations summarized

Confined Spaces Regulations
Construction (Design and Management) Regulations
Construction (Head Protection) Regulations
Construction (Health, Safety and Welfare) Regulations
Consultation with Employees Regulations
Control of Substances Hazardous to Health Regulations
Display Screen Equipment Regulations
Electricity at Work Regulations
Employers Liability (Compulsory Insurance) Regulations
Fire Precautions (Workplace) Regulations
First Aid Regulations
Information for Employees Regulations
Ionising Radiations Regulations
Lifting Operations and Lifting Equipment Regulations
Management of Health and Safety at Work Regulations
Manual Handling Operations Regulations
Noise at Work Regulations
Personal Protective Equipment at Work Regulations
Provision and Use of Work Equipment Regulations (except Part IV – Power Presses)
Reporting of Injuries Diseases and Dangerous Occurrences Regulations
Safety Representatives and Safety Committees Regulations
Safety Signs and Signals Regulations
Supply of Machinery (Safety) Regulations
Workplace (Health, Safety and Welfare) Regulations

The following Acts and Regulations have also been included even though they are not in the NEBOSH General Certificate syllabus. Very brief summaries only are given:

Asbestos (Licensing) Regulations
Asbestos at Work Regulations
Chemicals (Hazard Information and Packaging for Supply) Regulations
Dangerous Substances and Explosive Atmospheres Regulations
Electrical Equipment (Safety) Regulations
Gas Appliances (Safety) Regulations
Gas Safety (Installation and Use) Regulations
Lead at Work Regulations
Occupiers Liability Acts
Pesticides Regulations
Pressure Systems Safety Regulations
Working Time Regulations.

17.3.2 Chronological list of Regulations summarized

The list below gives the correct titles, year produced and Statutory Instrument number.

Year	SI no	Title
1977	0500	Safety Representatives and Safety Committees Regulations
1981	0917	Health and Safety (First Aid) Regulations
1989	0635	Electricity at Work Regulations
1989	0682	Health and Safety (Information for Employees) Regulations
1989	1790	Noise at Work Regulations
1989	2209	Construction (Head Protection) Regulations
1992	2792	Health and Safety (Display Screen Equipment) Regulations
1992	2793	Manual Handling Operations Regulations
1992	2966	Personal Protective Equipment at Work Regulations
1992	3004	Workplace (Health, Safety and Welfare) Regulations
1992	3073	Supply of Machinery (Safety) Regulations
1994	3140	Construction (Design and Management) Regulations
1995	3246	Reporting of Injuries Diseases and Dangerous Occurrences Regulations
1996	0341	Health and Safety (Safety Signs and Signals) Regulations
1996	1513	Health and Safety (Consultation with Employees) Regulations
1996	1592	Construction (Health, Safety and Welfare) Regulations
1997	1713	Confined Spaces Regulations
1997	1840	Fire Precautions (Workplace) Regulations
1998	2306	Provision and Use of Work Equipment Regulations (except Part IV – Power Presses)
1998	2307	Lifting Operations and Lifting Equipment Regulations
1998	2573	Employers Liability (Compulsory Insurance) Regulations
1999	3232	Ionising Radiations Regulations
1999	3242	Management of Health and Safety at Work Regulations
2002	2677	Control of Substances Hazardous to Health Regulations

Very brief summaries only for:

1957	Ch 31	Occupiers Liability Acts
1983	1649	Asbestos (Licensing) Regulations
1986	1510	Control of Pesticides Regulations
1992	0711	Gas Appliances (Safety) Regulations
1998	2451	Gas Safety (Installation and Use) Regulations
1994	3260	Electrical Equipment (Safety) Regulations
1998	1833	Working Time Regulations
2000	128	Pressure Systems Safety Regulations
2002	1689	Chemicals (Hazard Information and Packaging for Supply) Regulations
2002	2675	Control of Asbestos at Work Regulations
2002	2676	Control of Lead at Work Regulations
2002	2776	Dangerous Substances and Explosive Atmospheres Regulation

17.4 Health and Safety at Work etc. Act 1974

The HSW Act was introduced to provide a comprehensive and integrated piece of legislation dealing with the health and safety of people at work and the protection of the public from work activities. A small amount of pre-1974 legislation is still in place despite strenuous efforts to repeal and replace it with updated legislation under the HSW Act.

The Act imposes a duty of care on everyone at work related to their roles. This includes employers, employees, owners, occupiers, designers, suppliers and manufacturers of articles and substances for use at work. It also includes self-employed people. The detailed requirements are spelt out in Regulations.

The Act basically consists of four parts:

Part 1 covers:

➤ the health and safety of people at work
➤ protection of other people affected by work activities
➤ the control of risks to health and safety from articles and substances used at work
➤ the control of some atmospheric emissions.

Part 2: sets up the Employment Medical Advisory Service.

Part 3: makes amendments to the safety aspects of building regulations.

Part 4: general and miscellaneous provisions.

17.4.1 Duties of employers – Section 2

The employers' main general duties are to ensure, so far as is reasonably practicable, the health, safety and welfare at work of all their employees, in particular:

➤ the provision of safe plant and systems of work
➤ the safe use, handling, storage and transport of articles and substances
➤ the provision of any required information, instruction, training and supervision
➤ a safe place of work including safe access and egress
➤ A safe working environment with adequate welfare facilities.

When more than four people are employed the employer must:

➤ prepare a written general health and safety policy;
➤ set down the organization and arrangements for putting that policy into effect;
➤ revise and update the policy as necessary;
➤ bring the policy and arrangements to the notice of all employees;
➤ consult safety representatives appointed by recognized trade unions*;
➤ establish a safety committee if requested to do so by recognized safety representatives*.

*Note that this has been enacted by the Safety Representatives and Safety Committees Regulations 1977 and enhanced by the Health and Safety (Consultation with Employees) Regulations 1996 (see later summaries).

17.4.2 Duties of owners/occupiers – Sections 3–5

Every employer and self-employed person is under a duty to conduct their undertaking in such a way as to ensure, so far as is reasonably practicable, that persons not in their employment (and themselves for self-employed) who may be affected, are not exposed to risks to their health and safety.

Those in control of non-domestic premises have a duty to ensure, so far as is reasonably practicable, that the premises, the means of access and exit, and any plant or substances are safe and without risks to health. The common parts of residential premises are non-domestic.

Section 5 in relation to the harmful emissions into atmosphere is policed by the Environment Agency and the Scottish Environment Protection Agency. Section 5 was repealed by the Environmental Protection Act 1990 subject to Commencement Orders for the 1990 Act.

17.4.3 Duties of manufacturers/suppliers – Section 6

Persons who design, manufacture, import or supply any article or substance for use at work must ensure, so far as is reasonably practicable, that:

➤ they are safe and without risks to health when properly used (i.e. according to manufacturers' instructions);
➤ carry out such tests or examinations as are necessary for the performance of their duties;
➤ provide adequate information (including revisions) to perform their duties;
➤ carry out any necessary research to discover, eliminate or minimize any risks to health or safety;
➤ the installer or erector must ensure that nothing about the way in which the article is installed or erected makes it unsafe or a risk to health.

17.4.4 Duties of employees – Section 7

Two main duties are placed on employees:

➤ to take reasonable care for the health and safety of themselves and others who may be affected by their acts or omissions at work
➤ to cooperate with their employer and others to enable them to fulfil their legal obligations.

17.4.5 Other duties – Sections 8–9

No person may misuse or interfere with anything provided in the interest of health, safety or welfare in pursuance of any of the relevant statutory provisions.

Employees cannot be charged for anything done, or provided, to comply with the relevant statutory provisions. For example personal protective equipment required by a health and safety regulation.

17.4.6 Powers of inspectors – Sections 20–25

Inspectors appointed under this Act have the authorization to enter premises at any reasonable time (or anytime in a dangerous situation), and to:

➤ take a constable with them if necessary;
➤ take with them another authorized person and necessary equipment;
➤ examine and investigate;
➤ require premises or anything in them to remain undisturbed for purposes of examination or investigation;

- take measurements, photographs and recordings;
- cause an article or substance to be dismantled or subjected to any test;
- take possession of or retain anything for examination or legal proceedings;
- take samples as long as a comparable sample is left behind;
- require any person who can give information to answer questions and sign a statement. Evidence given under this Act cannot be used against that person or their spouse;
- require information, facilities, records or assistance;
- do anything else necessary to enable them to carry out their duties;
- issue an **Improvement Notice(s),** which is a notice identifying a contravention of the law and specifying a date by which the situation is remedied. There is an appeal procedure which must be triggered within 21 days. The notice is suspended pending the outcome of the appeal;
- issue a **Prohibition Notice(s),** which is a notice identifying and halting a situation which involves or will involve a risk of serious personal injury to which the relevant statutory provisions apply. A contravention need not have been committed. The notice can have immediate effect or be deferred, for example to allow a process to be shut down safely. Again there is provision to appeal against the notice but the order still stands until altered or rescinded by an Industrial tribunal;
- initiate prosecutions;
- seize, destroy or render harmless any article or substance which is a source of imminent danger.

17.4.7 Offences – Section 33

Offences prosecuted under the HSW Act or Regulations attract a maximum fine of £5000 if tried on summary conviction in a Magistrates Court. Exceptionally this is a maximum of £20 000 for breaches of Sections 2–6 (General duties) and £20 000 and/ or 6 months imprisonment for contravening a prohibition notice, an improvement notice or order of the court.

A number of cases can also be tried on indictment, in the Crown Court where the fines are unlimited and the maximum prison sentence can be up to 2 years. However, prison sentences are limited to certain cases only. Table 17.1 summarizes the position. Note that offences relating to breaches of general health and safety requirements (shown in bold) do *not* attract a prison sentence.

If a Regulation has been contravened, failure to comply with an approved code of practice is admissible in evidence as failure to comply. Where an offence is committed by a corporate body with the knowledge, connivance or neglect of a responsible person, both that person and the body corporate are liable to prosecution.

In proceedings the onus of proving the limits of what is reasonably practicable rests with the accused.

17.4.8 The Health and Safety Commission/Executive – Section 10

Section 10 of the Act establishes these two bodies whose chief functions are:

- The Health and Safety Commission (HSC) has the prime responsibility for administering the law and practice on occupational health and safety. Its nine members are appointed by the Government from industry, trade unions, local authorities and others. The HSC

Table 17.1 Offences and penalties under HSW Act

Offence	Summary conviction	Conviction on indictment
HSW Act s.33(1A) • breach of HSW Act ss. 2, 3, 4 or 6	**Max. £20 000 fine**	**Fine (unlimited)**
HSW Act s.33(2) • obstructing an inspector, impersonating an inspector, contravening requirement of inspector under s.20, or similar	Max. £5000 fine	
HSW Act s.33(2A) • failure to comply with enforcement notice or court order	Max. £20 000 and/or up to six months' imprisonment	Fine (unlimited) and/or up to two years' imprisonment

HSW Act s.33(3)
• all other offences (except those specified in s.33(4))
   • breach of HSW Act ss. 7, 8 or 9
   • breach of Regulations made under the Act
   • making false statements or entries
   • using, or possessing a document with the intent to deceive
	Max. £5000 fine	**Fine (unlimited)**

HSW Act s.33(3) − offences specified in s.33(4) • operating without, or contravening the terms or conditions of, a licence when one is required (e.g. Under Asbestos (Licensing) Regulations 1983) • acquiring or attempting to acquire, possessing or using an explosive article or substance in contravention of the relevant statutory provisions • using or disclosing information in contravention of sections 27(4) and 28 (relates to information provided to HSC/E etc. by notice or under relevant statutory provisions)	Max. £5000 fine	Fine (unlimited) and/or up to two years' imprisonment

has wide powers to do anything (except borrow money) which is calculated to facilitate, or is conducive or incidental to, the performance of its functions.

➤ The Health and Safety Executive is the executive and enforcement arm of the HSC. The HSC may direct its work except where it involves enforcement of the law in individual cases. Enforcement of the law in certain premises like offices and shops is the responsibility of local authority environmental health departments, whose officers have identical powers to HSE inspectors.

17.4.9 References

Health and Safety at Work etc. Act 1974, Chapter 37, London, The Stationery Office
Stating Your Business: Guidance on preparing a Health and Safety Policy Document for small firms, INDG324, HSE Books
The health and safety system in Great Britain, HSC, 2002 HSE Books
ISBN 0-7176-2243-6

17.5 Fire Precautions Act 1971

17.5.1 Introduction

The purpose of the Fire Precautions Act 1971 is to protect people from the hazard of fire. This is achieved through a Fire Certificate procedure issued by local fire authorities for premises which have a designated use. There are two designating orders in force, the first for hotels and boarding houses, and the second for factories, offices, shops and railway premises.

17.5.2 When is a fire certificate required?

Under the first designating order (the Fire Precautions (Hotels and Boarding Houses) Order 1972) a fire certificate is required for premises used as a hotel or boarding house which will provide sleeping accommodation for more than six people (including employees), or if they provide sleeping accommodation elsewhere than on the ground or first floor of the premises.

Under the Fire Precautions (Factories, Offices, Shops and Railway Premises) Order 1989, the second designating order, a certificate must be applied for when:

➤ More than 20 people are at work or;
➤ More than 10 are at work other than on the ground floor.

Also for factory premises a fire certificates is required when:

➤ Explosive or highly flammable materials are stored or used, unless the fire authority determines otherwise.

It is an offence to use premises for which a fire certificate is required unless an application for a certificate has been made and a certificate has not been refused. The Fire Authority can grant an exemption for premises if they consider them to be of low risk.

It should be noted that the removal of regulation 3(5)(a) (which exempted premise for which a fire certificate was in force or an application pending) of the 1997 Fire Precautions (Workplace) Regulations does not in any way override the requirement to apply for a fire certificate if the premises concerned meet the numerical and other thresholds set out in the designation Orders made in pursuance of section 1 of the 1971 Act (the Fire Precautions (Hotels and Boarding Houses) Order 1972 and the Fire Precautions (Factories, Offices, Shops and Railway Premises) Order 1989). The Fire Regulations apply *in full* to such workplaces, whether an application for a fire certificate is pending or a fire certificate is in force under section 1 of the 1971 Act, or where the premises are exempt from the requirement for a fire certificate by virtue of the grant of exemption by a fire authority under section 5A of the 1971 Act (powers for fire authority to grant exemption in particular cases). The Regulations also continue to apply to those workplace premises which were issued with a means of escape certificate under the Factories Act 1961 or the Offices, Shops and Railway Premises Act 1963.

17.5.3 Application for a fire certificate

The owner/occupier of premises which require a fire certificate must make an application on form FP1(Rev) obtainable from the fire authority.

Note that where people are at work, the owner/occupier must comply with the 1997 Fire Regulations regardless of the application for a fire certificate. The fire risk assessment may well be required by the fire officer who visits to consider the certificate.

17.5.4 Powers of inspectors

Fire Authority inspectors have similar powers to those inspectors under the HSW Act. These include entry to premise and to prosecute if there is a breach of the Act. Under section 10 of the Act, inspectors can issue a Prohibition Notice to close down immediately an unsafe building or part of a building used as a place of work. Under subsequent amendments, through the Fire Safety and Safety of Sports Grounds Act 1987, the powers of inspectors have been widened to include Improvement Notices.

17.6 Environmental Protection Act 1990

17.6.2 Introduction

The Environmental Protection Act 1990 is the centrepiece of current UK legislation on environmental protection. It is divided into nine parts, corresponding to the wide range of subjects dealt with by the Act.

Integrated pollution control (IPC) is a system established by Part 1 of the Act. Part 1 introduces *Part A Processes,* which are the most potentially polluting or technologically complex processes. In England and Wales this is enforced by the Environment Agency. In Scotland there is a parallel system enforced by the Scottish Environment Protection Agency. Less polluting industry may be classified as Part B, with only emissions released to air being subject to regulatory control. For such processes local authorities are the enforcing body and the system is known as Local Air Pollution Control (LAPC).

Both IPC and LAPC will eventually be replaced by a new Pollution Prevention and Control (PPC) regime that will implement the requirements of the EC Directive 96/61 on integrated pollution prevention and control. This will be phased in before 2007.

17.6.2 Definitions

The Act defines the following:

➤ Pollutants as
  ➤ solid wastes discharging onto land
  ➤ liquid wastes discharging onto land or into water
  ➤ discharges into the atmosphere
  ➤ noise in the community.

➤ Controlled waste as:
  ➤ waste from households
  ➤ waste from industry
  ➤ commercial waste.
➤ Special waste as:
  ➤ controlled waste that is so hazardous that it can only be disposed of using special procedures. Examples would be where there is a high risk to life or liquids with a flash point of 21°C or less.

17.6.3 Main features of integrated pollution control

➤ It covers around 2000 processes from the major industry sectors.
➤ Operators of the potentially most polluting processes ('prescribed processes' which are specified in the amended Environmental Protection (Prescribed Processes and Substances) Regulations 1991/472) have to apply for prior authorization from the Environment Agency to operate the process. IPC requires operators to consider the total impact of all releases to air, water and land when making an application.
➤ The operator has to advertise that an application has been made. Details of this are held in a public register (obtainable from local authority and Environment Agency offices) which is available for the public and statutory consultees to inspect. The Environment Protection Act only allows for exclusion of the application from the public register on grounds of commercial confidentiality or National Security. Commercial Confidentiality is allowed only when the operator can prove that release of the information would reduce a commercial advantage or produce an unreasonable commercial disadvantage (the information is placed on a public register after four years unless it is determined that it is still commercially confidential).
➤ After careful consideration, the Environment Agency may either grant an authorization or reject the application. In granting an authorization the Environment Agency must include conditions to ensure that:
  ➤ Best Available Techniques Not Entailing Excessive Cost are used (see Box 17.1)

Box 17.1   BATNEEC – Best Available Techniques Not Entailing Excessive Cost

These techniques are used to prevent, and where that is not practicable, to minimize release of certain substances (prescribed in regulations) into any environmental medium (air, water, land); and to render harmless both any such substances which are released and any other substances which may cause harm if released into any environmental medium. BATNEEC is usually expressed as emission limits for the prescribed substances released by the process.

> if a process involves releases to more than one environmental medium, the operator must use the Best Practicable Environmental Option (see Box 17.2) to achieve the best environmental solution overall

Box 17.2 BPEO – Best Practicable Environmental Option

This option is used when emissions are released to more than one environmental medium and ensures that there is least impact to the environment as a whole.

> there is compliance with any direction given by the Secretary of State for the Environment to implement European Community or international obligations, or any statutory environmental quality standards or objectives, or other statutory limits, plans or other requirements

> Operators have to monitor their emissions and report them to the Environment Agency on a yearly basis. Information on levels of pollution released from individual IPC processes is available from the Environment Agency's Pollution Inventory.

> If the Environment Agency believes that the operator is breaching the conditions of the authorization, the Agency has enforcement options available and can serve the following notices:

> revocation: when the regulator has reason to believe that withdrawing an authorization is necessary

> enforcement: if an operator is not adhering to the conditions set out by the authorization, the regulator specifies the steps to take and time limit for the problem to be remedied

> prohibition: if the regulator believes that operation of a process under authorization involves imminent risk of serious pollution, the authorization will be suspended until the risk is removed.

> The Secretary of State has the authority to direct the Environment Agency on authorization and conditions to be included in authorization.

> Operators can appeal: if the Environment Agency refuses to issue or vary an authorization; against the conditions of the authorization; against all the different types of notice the Environment Agency may serve against an authorization.

17.6.4 Duty of care

Waste and the duty of care

The duty of care is covered in Part II of the Environmental Protection Act 1990. The duty of care applies to anyone who produces or imports, keeps or stores, transports, treats or disposes of waste. It also applies if they act as a broker and arrange these things.

The duty holder is required to take all reasonable steps to keep waste safe. If they give waste to someone else, the duty holder must be sure they are authorized to take it and can transport, recycle or dispose of it safely.

Penalty for breach of this law is an unlimited fine.

Waste can be anything owned, or a business produces, which a duty holder wants to get rid of. Controlled waste is defined in Box 17.3.

Box 17.3

Controlled waste means household, commercial or industrial waste. It includes any waste from a house, school, university, hospital, residential or nursing home, shop, office, factory or any other trade or business premises. It is controlled waste whether it is solid or liquid and even if it is not hazardous or toxic.

If the waste comes from a person's own home, the duty of care **does not** apply to them. But if the waste is not from the house they live in, for example, if it is waste from their workplace or waste from someone else's house, the duty of care **does** apply.

Animal waste collected and transported under the Animal By-Products Order 1992 is not subject to the duty of care.

Duty holders must take all reasonable steps to fulfil the duty and complete some paperwork. What is reasonable depends on what is done with the waste.

Steps to take if the duty of care applies
When a duty holder has waste they must:

➤ stop it escaping from their control and store it safely and securely. They must prevent it causing pollution or harming anyone;

➤ keep it in a suitable container. Loose waste in a skip or on a lorry must be covered;

➤ if the duty holder gives waste to someone else, they must check that they have authority to take it. The law says the person to whom they give the waste must be authorized to take it. Box 17.4 explains who is allowed to take waste and how the duty holder can check;

➤ describe the waste in writing. The duty holder must fill in and sign a transfer note for it and keep a copy. To save on paperwork, the description of the waste can be written on the transfer note (see Box 17.5).

When a person takes waste from someone else they must
➤ be sure the law allows them to take it. Box 17.4 explains who is allowed to take waste.

➤ make sure the person giving them the waste describes it in writing. The waste receiver must fill in and sign a transfer note and keep a copy (see Box 17.5).

Box 17.4 Who has authority to take waste?

Council waste collectors

The duty holder does not have to do any checking, but if they are not a householder, there is some paperwork to complete. This is explained in Box 17.5.

Registered waste carriers

Most carriers of waste have to be registered with the Environment Agency or the Scottish Environment Protection Agency. Look at the carrier's certificate of registration or check with the Agencies.

Exempt waste carriers

The main people who are exempt are charities and voluntary organizations. Most exempt carriers need to register their exemption with the Environment Agency or the Scottish Environment Protection Agency. If someone says they are exempt, ask them why. Check with the Agencies that their exemption is registered.

Holders of waste management licences

Some licences are valid only for certain kinds of waste or certain activities. Ask to see the licence. Check that it covers the kind of waste being consigned.

Businesses exempt from waste management licences

There are exemptions from licensing for certain activities and kinds of waste. For example, the recycling of scrap metal or the dismantling of scrap cars. Most exempt businesses need to register their exemption with the Environment Agency or the Scottish Environment Protection Agency. Check with the Agencies that their exemption is registered.

Authorized transport purposes

Waste can also be transferred to someone for 'Authorized transport purposes'. This means:

➤ the transfer of controlled waste between different places within the same premises;

➤ the transport of controlled waste into Great Britain from outside Great Britain; and

➤ the transport by air or sea of controlled waste from a place in Great Britain to a place outside Great Britain.

Registered waste brokers

Anyone who arranges the recycling or disposal of waste, on behalf of someone else, must be registered as a waste broker. Check with the Environment Agency or the Scottish Environment Protection Agency that the broker is registered.

Exempt waste brokers

Most exempt waste brokers need to register with the Environment Agency or the Scottish Environment Protection Agency. Those who are exempt are mainly charities and voluntary organizations. If someone tells you they are exempt, ask them why. You can check with the Environment Agencies that their exemption is registered.

Box 17.5 Filling in paperwork

When waste is passed from one person to another the person taking the waste must have a written description of it. A transfer note must also be filled in and signed by both persons involved in the transfer.

The duty holder can write the description of the waste on the transfer note. Who provides the transfer note is not important as long as it contains the right information. The Government has published a model transfer note with the Code of Practice which can be used if desired.

Repeated transfers of the same kind of waste between the same parties can be covered by one transfer note for up to a year. For example, weekly collections from shops.

The transfer note to be completed and signed by both persons involved in the transfer, must include:

➤ what the waste is and how much there is

➤ what sort of containers it is in

➤ the time and date the waste was transferred

➤ where the transfer took place

➤ the names and addresses of both persons involved in the transfer

➤ whether the person transferring the waste is the importer or the producer of the waste

➤ details of which category of authorized person each one is. If the waste is passed to someone for authorized transport purposes, you must say which of those purposes applies

➤ if either or both persons is a registered waste carrier, the certificate number and the name of the Environment Agency which issued it

➤ if either or both persons has a waste management licence, the licence number and the name of the Environment Agency which issued it

➤ the reasons for any exemption from the requirement to register or have a licence

➤ where appropriate, the name and address of any broker involved in the transfer of waste.

The written description The written description must provide as much information as someone else might need to handle the waste safely.

Keeping the papers Both persons involved in the transfer must keep copies of the transfer note and the description of the waste for two years. They may have to prove in Court where the waste came from and what they did with it. A copy of the transfer note must also be made available to the Environment Agency or the Scottish Environment Protection Agency if they ask to see it.

17.6.5 Applying for a waste management licence

Introduction

A waste management licence is a legal document issued under the Environmental Protection Act 1990. There are two types of waste management licence:

➤ a site licence – authorizing the deposit, recovery or disposal of controlled waste in or on land
➤ a mobile plant licence – authorizing the recovery or disposal of controlled waste using certain types of mobile plant.

A licence has conditions to make sure that the authorized activities do not cause pollution of the environment, harm to human health or serious detriment to local amenities.

Anyone who deposits, recovers or disposes of **controlled waste** must do so either:

➤ within the conditions of a waste management licence, or
➤ within the conditions of an exemption from licensing

and must not cause pollution of the environment, harm to human health or serious detriment to local amenities. Otherwise they could be fined and sent to prison.

The conditions of the exemptions from licensing are in the Waste Management Licensing Regulations 1994. If an activity may be exempt this should be discussed with the Environment Agency.

Before applying for a licence contact the Environment Agency for advice. If a licence is needed the Environment Agency will:

➤ meet to discuss your proposals;
➤ explain the application process and provide an application package;
➤ provide guidance on preparing a working plan;
➤ set out in writing any additional information required specifically for the proposals;
➤ help work out the application fee, and;
➤ say who will deal with the application.

There are financial and legal implications to holding a licence and the licensee may want to get independent advice about these.

Box 17.6   What is a working plan?

As part of the licence application a working plan will need to be prepared. This is a document describing how the licensee intends to prepare, develop, operate and restore (where relevant) the site or plant. If they write a comprehensive working plan it will help to avoid delays in processing the application. It may also mean that some of the licence conditions could be less restrictive, giving more opportunity for flexibility in the operations.

Applying for a licence

To apply for a waste management licence the licensee must send:

➤ a completed application form
➤ the fee for processing the application (this cannot be refunded)
➤ any additional information which the Agency has asked for in writing and
➤ the supporting information in the checklist on the application form to the Environment Agency at the address shown on the application form.

When all the information has been received the Environment Agency will write to confirm that the application is complete and that the 4-month period allowed in law for deciding the application has started.

The application form will be put on the public register. If the licensee feels that any information on the form should not be made public because it is commercially confidential, they can apply to have that information withheld. The application form tells them how to do this.

Processing the application

Licence applications are processed by the Environment Agency using the waste management licensing process handbook which helps to ensure the licensing service is efficient and consistent.

The Environment Agency will consider whether it can write licence conditions which will make sure that the proposed activities do not cause pollution of the environment, harm to human health or serious detriment to local amenities.

When processing an application the Agency will also:

➤ check that the activities to be licensed have planning permission or equivalent (where it is needed). It is the licensee's responsibility to make sure that the activities have the necessary planning permission;
➤ consult with other regulatory bodies and;
➤ decide whether the licensee is a fit and proper person.

Box 17.7   What is a fit and proper person?

When considering whether the licensee is a fit and proper person the Agency will look at:
➤ whether they have made adequate financial provision to cover the licence obligations. If the Agency plans to issue a licence a draft of the licence conditions will be sent to the licensee. They should use this draft to help calculate the amount of financial provision needed
➤ whether a technically competent person will be managing the licensed activities
➤ whether the licensee, or another relevant person, have been convicted of a relevant offence.

If the Environment Agency is not satisfied on any aspect, it will write and say what it needs. If the licensee is still unable to satisfy the Agency the application will be rejected.

After a licence is issued

Once the site is licensed, the licensee must comply with the licence conditions at all times. The Environment Agency will make visits to check that the licence conditions are being met. There will be an annual fee to cover the costs of these visits. The fees for waste management licensing are in the Waste Management Licensing (Charges) Scheme.

The licence and working plan will be put on the public register. If the licensee thinks any of the information should not be made public because it is commercially confidential, they can apply to have that information withheld.

The licensee will be responsible for the obligations arising from that licence until:

➤ the Agency accepts an application to transfer the licence to another person
➤ for a site licence – the Agency accepts the surrender of the licence
➤ for a mobile plant licence – the licensee surrenders the licence.

Changes to licence conditions can only be made by the Environment Agency. If the licensee wants to change their operations, this should be discussed with the Agency.

17.6.6 Further information

The law
The Environmental Protection Act 1990
    ISBN 0-10-544390-5
The Waste Management Licensing Regulations 1994 Statutory Instrument 1994 No. 1056 (as amended)
    ISBN 0-11-044056-0

Government guidance
Waste Management Paper 4 *Licensing of Waste Management Facilities*
    ISBN 0-11-752727-0
Waste Management Paper 4A *Licensing of Metal Recycling Facilities*
    ISBN 0-11-753064-6
DOE Circular 11/94 Environmental Protection Act 1990: Part II Waste Management Licensing The Framework Directive on Waste (Welsh Office Circular 26/94, Scottish Office Environment Department Circular 10/94)
    ISBN 0-11-752975-3
DOE Circular 6/95 Environmental Protection Act 1990: Part II Waste Management Licensing The Framework Directive on Waste (Welsh Office Circular 25/95, Scottish Office Environment Department Circular 8/95)
    ISBN 0-11-753098-0
Practical guidance for everyone who is under the duty is in *Waste Management, The Duty of Care, A Code of Practice*, ISBN 0-11-753210-X, published by HMSO in March 1996 and available from HMSO bookshops or by telephoning 0870 600 5522.

17.7 Confined Spaces Regulations 1997

17.7.1 Introduction

These Regulations concern any work which is carried on in a place which is substantially (but not always entirely) enclosed, where there is a reasonably foreseeable risk of serious injury from conditions and/or hazardous substances in the space or nearby. Every year about 15 people are killed and a number seriously injured across a wide range of industries ranging from simple open top pits to complex chemical plants. Rescuers without proper training and equipment, often become the victims.

17.7.2 Definitions

Confined space – means any place, including any chamber, tank, vat, silo, pit, trench, pipe, sewer, flue, well or similar space in which, by virtue of its enclosed nature, there arises a reasonably foreseeable specified risk.

    Specified risk – means a risk to any person at work of:

➤ serious injury arising from a fire or explosion
➤ loss of consciousness arising from an increase in body temperature
➤ loss of consciousness or asphyxiation arising from gas, fume, vapour or the lack of oxygen
➤ drowning arising from the increase in the level of liquid
➤ asphyxiation arising from a free flowing solid or because of entrapment by it.

17.7.3 Employers duties – Regulation 3

Duties are placed on employers to:

➤ comply regarding any work carried out by employees; and
➤ ensure, so far as is reasonably practicable, that other persons (e.g. use competent contractors) comply regarding work in the employer's control.

The self-employed also have duties to comply.

17.7.4 Work in confined space – Regulation 4

1 No person at work shall enter a confined space for any purpose unless it is not reasonably practicable to achieve that purpose without such entry
2 Other than in an emergency, no person shall enter, carry out work or leave a confined space otherwise than in accordance with a safe system of work, relevant to the specified risks.

17.7.5 Risk assessment

Risk assessment is an essential part of complying with these Regulations and must be done (under the Management of Health and Safety at Work Regulations 1999) to determine a safe system of work. The risk assessment needs to follow a hierarchy of controls to comply. This should start with the measures, both in design and procedures, that can be adopted to enable any work to be carried out outside the confined space.

The assessment must be done by a competent person and will form the basis of a safe system of work. This will normally be formalized into a specific permit to work, applicable to a particular task. The assessment will involve:

(a) The general conditions to assess what may or may not be present. Consider:
➤ what has been the previous contents of the space
➤ residues that have been left in the space, for example, sludge, rust, scale and what may be given off if they are disturbed
➤ contamination which may arise from adjacent plant, processes, services, pipes or surrounding land, soil or strata
➤ oxygen deficiency and enrichment. There are very high risks if the oxygen content differs significantly from the normal level of 20.8%. If it is above this level increased flammability exists, if it is below then impaired mental ability occurs, with loss of consciousness under 16%
➤ physical dimensions and layout of the space can affect air quality.
(b) Hazards arising directly from the work being undertaken, consider:
➤ the use of cleaning chemicals and their direct effect or interaction with other substances
➤ sources of ignition for flammable dusts, gases, vapours, plastics and the like.
(c) the need to isolate the confined space from outside services or from substances held inside such as liquids, gases, steam, water, fire extinguishing media, exhaust gases, raw materials and energy sources.
(d) the requirement for emergency rescue arrangements including trained people and equipment.

17.7.6 Safe system of work

The detailed precautions required will depend on the nature of the confined space and the actual work being carried out. The main elements of a safe system of work which may form the basis of a 'permit-to-work' are:

➤ the type and extent of supervision
➤ competence and training of people doing the work
➤ communications between people inside, from inside to outside and to summon help
➤ testing and monitoring the atmosphere for hazardous gas, fume, vapour, dust, etc. and for concentration of oxygen
➤ gas purging of toxic or flammable substances with air or inert gas, such as nitrogen
➤ good ventilation, sometimes by mechanical means
➤ careful removal of residues using equipment which does not cause additional hazards
➤ effective isolation from gases, liquids and other flowing materials by removal of pieces of pipe, blanked off pipes, locked off valves

➤ effective isolation from electrical and mechanical equipment to ensure complete isolation with lock off and a tag system with key security. Need to secure against stored energy or gravity fall of heavy presses etc.

➤ if it is not possible to make the confined space safe, the provision of personal and respiratory protective equipment

➤ supply of gas via pipes and hoses should be carefully controlled

➤ access and egress to give quick unobstructed and ready access and escape

➤ fire prevention

➤ lighting, including emergency lighting

➤ prohibition of smoking

➤ emergencies and rescue

➤ limiting of working periods and the suitability of individuals.

17.7.7 Emergency arrangements – regulation 5

Before people enter a confined space suitable and sufficient rescue arrangements must be set up. These must:

➤ reduce the risks to rescuers so far as is reasonably practicable;

➤ include the provision and maintenance of suitable resuscitation equipment designed to meet the specified risks.

To be suitable and sufficient arrangements will need to cover:

➤ rescue and resuscitation equipment

➤ raising the alarm, alerting rescue and watch keeping

➤ safeguarding the rescuers

➤ fire safety precautions and procedures

➤ control of adjacent plant

➤ first-aid arrangements

➤ notification and consultation with emergency services

➤ training of rescuers and simulations of emergencies

➤ size of access openings to permit rescue with full breathing apparatus, harnesses, fall arrest gear and lifelines, which is the normal suitable respiratory protection and rescue equipment for confined spaces.

17.7.8 Training

Specific, detailed and frequent training is necessary for all people concerned with confined spaces, whether they are acting as rescuers, watchers or those carrying out the actual work inside the confined space. The training will need to cover all procedures and the use of equipment under realistic simulated conditions.

17.7.9 References

Safe work in confined spaces. Approved Code of Practice, Regulations and Guidance. L101 1997 HSE Books
ISBN 0 7176 1405 0

The Selection Use and Maintenance of Respiratory Protective Equipment: A Practical Guide. HSG53 1998 second edition (reprinted 2001 with minor amendments, due to be updated) HSE Books
ISBN 0 7176 61537 5

Respiratory Protective Equipment, Legislative requirements and lists of HSE approved standards and type approved equipment. Fourth edition 1995 HSE Books
ISBN 0 7176 1036 5

17.8 Construction (Design and Management) (CDM) Regulations 1994 and Amendment Regulations 2000

17.8.1 Application of the Regulations

The Regulations apply to most common building, civil engineering and construction work. They do not apply to construction work where the local authority is the enforcing authority for health and safety purposes. This means where work is not notifiable and is either:

➤ carried out inside offices, shops and similar premises where the construction work is done without interrupting the normal activities in the premises and without separating the construction activities from the other activities; or
➤ the maintenance or removal of insulation on pipes, boilers or other parts of heating or water systems

that it is not subject to the CDM Regulations.

Apart from the above exceptions, the CDM Regulations apply to all design work carried out for construction purposes (including demolition and dismantling).

The Regulations impose responsibilities on clients who are acting in connection with a trade or business. They do not cover house occupiers carrying out work on domestic premises or instructing an architect or other professional.

As far as contractors are concerned, the CDM Regulations apply:

➤ to all demolition and dismantling work
➤ to other construction work unless
   ➤ the work will last 30 days or less and
   ➤ it involves less than five people on site at any one time
➤ the work is being done for a non-domestic client. (Duties to notify HSE and those placed on designers, however, do apply.)

17.8.2 Clients' responsibilities

Clients should, so far as is reasonably practicable:

➤ select and appoint a planning supervisor and principal contractor;
➤ be satisfied that the planning supervisor and principal contractor are competent and will allocate adequate resources to health and safety;

➤ be satisfied that designers and any contractors they appoint directly are competent and properly resourced;

➤ provide the planning supervisor with information relevant to health and safety on the project. If a health and safety file already exists relevant sections should be provided;

➤ ensure that construction work does not start until the principal contractor has prepared a satisfactory health and safety plan;

➤ ensure that the health and safety file is available for inspection, after the project is complete.

If an agent is appointed by the client, reasonable enquiries should be made to check that the agent is competent.

17.8.3 The planning supervisor's responsibilities

The planning supervisor is appointed by the client and may be taken on by an individual or by a company. It can be an independent person or discharged from within the client's, designer's or contractor's teams. The role involves:

➤ coordination of health and safety issues during the design and planning phases of the project to ensure that risks are properly assessed and steps taken to avoid the risks or reduce them so far as is reasonably practicable;

➤ making sure so far as is reasonable that designers cooperate with each other for the purposes of health and safety;

➤ ensuring that the pre-tender stage health and safety plan is produced in time for the contractors bidding for the work to use in their preparation of a tender;

➤ giving advice about health and safety competences and resources required during the project;

➤ making sure the project is notified to the HSE;

➤ arranging to collect information for inclusion in the health and safety file which they must ensure is prepared and delivered to the client at the end of the project.

Once construction work has started the planning supervisor is not required to visit the site or assess performance of the principal contractor.

17.8.4 The designer's responsibilities

Designers are required to play a key role under CDM to ensure that the design allows the construction work to be carried out in a safe manner. Often, by careful design, risks can be eliminated or significantly reduced. The key responsibilities are to:

➤ alert clients to their duties under CDM Regulations;

➤ consider the hazards and risks which may arise to those constructing or maintaining the premises or structure;

➤ design in such a way that risks are avoided so far as is reasonably practicable;

➤ reduce risks at source if avoidance is not practicable;

➤ consider measures to protect all workers if avoidance or reduction is not possible;

➤ ensure that the design provides adequate information on health and safety;

➤ pass health and safety information to the planning supervisor for inclusion in the safety file;

➤ cooperate with the planning supervisor and other designers, if necessary.

17.8.5 The principal contractor

The principal contractor is appointed by the client to plan, manage and control health and safety during the construction phase of the project.

The key responsibilities of the principal contractor are to:

➤ develop and implement the health and safety plan from the pre-tender plan;
➤ ensure that competent and properly resourced sub-contractors are used where necessary;
➤ ensure the coordination and cooperation of all contractors;
➤ ensure so far as is reasonably practicable that all contractors comply with the rules laid down in the Health and Safety Plan;
➤ obtain risk assessments and method statements from contractors;
➤ ensure that workers have received adequate training;
➤ monitor health and safety performance;
➤ make sure that only authorized people are allowed on site and that all workers are properly informed and consulted;
➤ display the notification of the project to the HSE;
➤ give information to the planning supervisor for the Health and Safety File.

17.8.6 Pre-tender stage Health and Safety Plan

The planning supervisor is responsible to ensure that the pre-tender stage health and safety plan is prepared so that prospective principal contractors can be fully aware of the project's health, safety and welfare requirements. It also helps the client to assess the resources being allocated to health and safety and the competence of the prospective principal contractor.

The actual contents will depend on the project, but the following should be considered:

➤ a general description of the work including name of the client, location, nature of construction work and timescales
➤ the existing environment including services, surrounding land use and restrictions, planning restrictions, traffic problems and access, existing structures and materials of construction, and ground conditions
➤ existing drawings including drawings of buildings being demolished, buildings being incorporated in the new structure, services and where available the health and safety file
➤ the design including significant hazards which cannot be designed out or avoided, principles of the structural design and precautions, specific problems where proposals will be required to manage the problem
➤ construction materials including health hazards which cannot be avoided or designed out
➤ site-wide elements including position of site access and egress, location of temporary site accommodation, location of unloading and storage areas, traffic and pedestrian routes

➤ overlap with client's undertaking where the work is to be undertaken in an occupied or partially occupied premises

➤ site rules which the client or planning supervisor may wish to lay down including permits to work, emergency procedures and other safety rules

➤ continuing liaison including procedures to deal with the health and safety implications of the contractors' packages and for dealing with unforeseen issues during the construction phase.

17.8.7 Construction stage – Health and Safety Plan

The principal contractor must develop the health and safety plan so that it addresses the specific issues of the project during construction. These will include:

➤ how health and safety will be managed during construction to protect the health and safety of all persons at work or those who may be affected by the operation

➤ how information will be passed to contractors and their activities coordinated

➤ contractors' risk assessments and health and safety method statements for high hazard activities

➤ information about welfare arrangements in enough detail to allow contractors on site to understand what they need to do to comply with their welfare requirements

➤ common arrangements including, for example, site hoardings, welfare and emergency requirements

➤ how consultation with workers will be coordinated

➤ necessary health and safety training required for all those working on the project

➤ arrangements for delivery of plant and materials and how common suppliers will be selected

➤ arrangements for monitoring compliance with health and safety law

➤ site health and safety rules and standards expected, particularly if clients are requiring standards above the legal minimum

➤ procedures for delivering information for the health and safety file.

17.8.8 Health and Safety File

The health and safety file is a record of information for the client or end user. The planning supervisor ensures that the file is produced at the end of the contract and handed to the client. It should include:

➤ record or 'as built' drawings and plans used and produced throughout the construction stage together with the design criteria

➤ general details of the construction methods and materials used

➤ details of the structure's equipment and maintenance facilities

➤ maintenance procedures and requirements for the structure

➤ manuals produced by specialist contractors and suppliers which outline operating and maintenance procedures and schedules for plant and equipment installed as part of the structure

➤ details on the location and nature of utilities and services, including emergency and fire-fighting systems.

The file can be produced and stored in an electronic version as a paper copy is not required by law.

17.8.9 Notification to the HSE

The following particulars must be notified to the HSE in compliance with Regulation 7. Form 10 (revised) can be used. The particulars are:

➤ date of forwarding
➤ exact address of the construction site
➤ name and address of the client or clients
➤ type of project
➤ name and address of the planning supervisor
➤ a declaration signed on behalf of the planning supervisor that he has been appointed as such
➤ name and address of the principal contractor
➤ a declaration signed by or on behalf of the principal contractor that he has been appointed as such
➤ date planned for the start of the construction phase
➤ planned duration of the construction phase
➤ estimated maximum number of people at work on the construction site
➤ planned number of contractors on the construction site
➤ name and address of any contractor or contractors already chosen.

17.8.10 References

Managing Health and Safety in Construction, Construction (Design and Management) Regulations 1994, Approved Code of Practice and Guidance, HSG 224, 2001 HSE Books
     ISBN 0-7176-2139-1
Health and Safety in Construction HSG150(rev1) 2001 HSE Books
     ISBN 0-7176-2106-5
A Guide to Managing Health and Safety in Construction, HSC Construction Industry Advisory Committee 1998 HSE Books
     ISBN 0-7176-0755-0

17.9 Construction (Head Protection) Regulations 1989

17.9.1 Application

These Regulations apply to 'building operations' and 'works of engineering construction' as defined originally by the Factories Act 1961.

17.9.2 Provision and maintenance

Every employer must provide suitable head protection for each employee and shall maintain it and replace it whenever necessary. A similar duty is placed on self-employed people. The employer must make an assessment and review it as necessary, to determine whether the head protection is suitable. Accommodation must be provided for head protection when it is not in use.

17.9.3 Ensuring that head protection is worn – Regulation 4

Every employer shall ensure, so far as is reasonably practicable, that each employee (and any other person over whom they have control) at work wears suitable head protection, unless there is no foreseeable risk of injury to their head other than through falling.

17.9.4 Rules and directions – Regulation 5

The person in control of a site **may** make rules regulating the wearing of suitable head protection. These must be in writing. An employer may give directions requiring their employees to wear head protection.

17.9.5 Wearing of suitable head protection, reporting loss – Regulations 6,7,9

Every employee who has been provided with suitable head protection shall wear it when required to do so by rules made or directions given under Regulation 5 of these Regulations. They must make full and proper use of it and return it to the accommodation.

Employees must take reasonable care of head protection, and report any loss or obvious defect.

Exemption certificates, issued by the HSE, are allowed.

17.9.6 References

Head protection for Sikhs wearing turbans INDG262
Construction (Head Protection) Regulations 1989 Guidance on regulations L 102 (rev) HSE Books.

17.10 Construction (Health, Safety and Welfare) (CHSW) Regulations 1996

17.10.1 Introduction

The CHSW Regulations complete the implementation of the EC Directive on construction which was started by the CDM Regulations 1994. They explain the detailed requirements for working in construction activities, such as using access equipment or work in excavations, although they do not include lifting operations which are covered by LOLER 1998.

CHSW is aimed at protecting the people who work in construction and those who may be affected by the work. The schedules referred to are attached to the Regulations.

17.10.2 Who has duties? – Regulation 4

The main duty holders are employers, self-employed and persons who control the way in which construction work is carried out.

Employees have duties to cooperate on health and safety issues, work safely and report any defects to persons in control.

17.10.3 Safe place of work – Regulation 5

This regulation requires so far as is reasonably practicable:

➤ safe access and egress to places of work
➤ maintenance of access and egress
➤ provision of safe places of work with adequate space and suitable for workers
➤ that people are prevented from gaining access to unsafe access or workplaces.

Risks to health have to be minimized so far as is reasonably practicable.

17.10.4 Falls – Regulation 6

Suitable and sufficient steps shall be taken to prevent persons falling from any height but specific steps must be taken for work with a possibility of falling over 2 m. These include:

➤ guardrails and toe boards, barriers or similar which comply with schedule 1
➤ working platforms which comply with schedule 2
➤ where it is not practicable or the work duration is very short, personal suspension equipment which complies with schedule 3
➤ where it is not practicable to comply with above fall arrest equipment which complies with schedule 4
➤ a ladder shall not be used unless it is reasonable to do so and where used must comply with schedule 5
➤ equipment must be properly maintained

➤ installations etc. of scaffolds and personal suspension equipment must be done under the supervision of a competent person.

17.10.5 Fragile material – Regulation 7

Suitable and sufficient steps must be taken to prevent falls through fragile materials. This is a general requirement, but specific steps must be taken for work with a possibility of falling over 2 m. This involves:

➤ platforms, guardrails, coverings or other similar means of support being provided and used to prevent persons passing, working or nearby from falling through fragile materials and
➤ prominent warning signs being posted at the approach to fragile materials.

17.10.6 Falling objects – Regulation 8

Where necessary to prevent danger suitable and sufficient precautions must be taken to:

➤ prevent the fall of materials.

Or where it is not reasonably practicable:

➤ prevent persons being struck, for example, by covered walk ways;
➤ stop material being thrown from a height where it is liable to cause injury;
➤ store materials and equipment safely.

17.10.7 Stability and demolition of structures – Regulations 9 and 10

Suitable and sufficient steps must be taken to ensure that:

➤ any new or existing structure which may be or become weak does not collapse accidentally;
➤ a structure is not overloaded;
➤ temporary supporting structures are erected and dismantled under competent supervision;
➤ demolition or dismantling work is planned and carried out safely under competent supervision.

17.10.8 Explosives – Regulation 11

Explosives may only be used when suitable and sufficient steps have been taken to prevent people getting hurt.

17.10.9 Excavations – Regulation 12

All practicable steps shall be taken to:

➤ ensure that excavations do not collapse accidentally;

➤ prevent, so far as is reasonably practicable, a person from being buried or trapped by a fall of material;

➤ support, so far as is reasonably practicable, the sides, roof or adjacent area to prevent a collapse;

➤ provide suitable and sufficient equipment to support the excavation;

➤ work on supports for an excavation only under competent supervision;

➤ prevent persons, vehicles, plant, equipment or any accumulation of material falling into the excavation;

➤ stop any person, material vehicle or equipment being near to an excavation where it is likely to cause collapse;

➤ prevent any risk, so far as is reasonably practicable, of injury from underground cables or other services;

➤ design, construct and maintain cofferdams or caissons properly and of suitable and sound materials;

➤ work on cofferdams or caissons under competent supervision.

17.10.10 Prevention of drowning – Regulation 14

Where persons could fall into water or other liquid steps must be taken to:

➤ prevent, so far as is reasonably practicable, people from falling;

➤ minimize the risk of drowning if people fall;

➤ provide, use and maintain suitable rescue equipment.

Transport to work by water must be:

➤ suitable

➤ maintained

➤ under competent supervision

➤ not overloaded.

17.10.11 Traffic routes – Regulation 15

So far as is reasonably practicable traffic routes shall satisfy the following:

➤ organized so that vehicles and people can move safely

➤ sufficient in number, size and position

➤ designed so that any pedestrian using a door or gate leading onto a traffic route can see approaching vehicles

➤ organized so there is separation between people and vehicles, or where this is not reasonably practicable, people are protected and warning of approaching vehicles is given

➤ loading bays to have at least one exit for pedestrians only

➤ separate access gates for pedestrians are required where it is unsafe to use vehicle gate

➤ vehicles can only be used where routes are free of obstructions, there is sufficient room or warning has been given to the driver

➤ suitable signs should be erected.

17.10.12 Doors and gates – Regulation 16

Doors and gates should have safety devices to prevent injury including:

➤ derailing sliding doors
➤ falling doors etc.
➤ powered doors, must also have means of opening if power fails.

17.10.13 Vehicles – Regulation 17

Steps should be taken to:

➤ stop unintended movement of vehicles;
➤ give warnings to people who might be endangered, for example, reversing vehicles;
➤ ensure safe operation of vehicles, including the prohibition of riding or remaining in unsafe positions.

17.10.14 Fire and other emergencies – Regulations 18, 19, 20 and 21

Steps shall be taken, so far as is reasonably practicable, to:

➤ prevent injury arising from fires, explosions, flooding, or substances liable to cause asphyxiation;
➤ provide unobstructed, properly signed emergency routes and exits to a place of safety;
➤ set up appropriate emergency procedures with designated people;
➤ make people on site familiar with the procedures and test the procedures at suitable intervals;
➤ provide suitable and sufficient fire-fighting equipment, fire detectors and alarm systems;
➤ maintain the equipment and train people who may need to use it;
➤ provide suitable signs to indicate the fire-fighting equipment.

17.10.15 Welfare facilities – Regulation 22

Any person who is in control of a site, in addition to employers and the self-employed, shall ensure, so far as is reasonably practicable, that they provide or ensure are provided:

➤ sanitary and washing (including showers if necessary) facilities with an adequate supply of drinking water
➤ accommodation for clothing and changing facilities
➤ rest facilities.

Schedule 6 lays down the detailed requirements for welfare.

17.10.16 Site-wide issues – Regulations 23, 24, 25 and 26

Suitable and sufficient steps shall be taken to ensure that:

➤ sufficient fresh or purified air is available at every workplace and that associated plant gives a warning in case of failure

➤ a reasonable temperature is maintained at indoor workplaces

➤ protective clothing against adverse weather is provided where necessary for persons working outdoors

➤ every workplace, traffic route and access is properly lit with lighting which does not adversely affect coloured signals etc.

➤ suitable emergency lighting is provided

➤ the site is kept in good order and in a reasonable state of cleanliness with no projecting nails

➤ the perimeter where necessary is identified by signs.

17.10.17 Training, inspection and reports – Regulations 28, 29 and 30

➤ Construction work where training or technical knowledge is required to protect people shall only be carried out by competent people or under competent supervision.

➤ Where the place of work is part of a scaffold, excavation, cofferdam or caisson make sure it is inspected before work and at subsequent specified intervals, by a competent person, who must be satisfied that the work can be carried out safely;

➤ Reports of inspections must be produced by the competent person before the end of the working period and provided to the person responsible for the equipment within 24 hours. Schedule 8 has the particulars to be included in the report.

17.10.18 References

Health and safety in construction, HSG150(rev1), 2001 HSE Books
    ISBN 0-7176-2106-5
Electrical safety on construction sites, HSG141, 1995 HSE Books
    ISBN 0-7176-1000-4
Guide to the Construction (Health, Safety and Welfare) Regulations, 1996 HSE Books
IND(G)220L
Fire safety in construction work, HSG 168, 1997 HSE Books
    ISBN 0-7176-1332-1
Health and safety in excavations, HSG 185, 1999 HSE Books
    ISBN 0-7176-1563-4
Health and safety in roof work, revised, HSG33, 1999 HSE Books
    ISBN 0-7176-1425-5
Safe use of vehicles on construction sites, HSG 144, 1998 HSE Books
    ISBN 0-7176-1610-X
Backs for the future, HSG 149, HSE Books
    ISBN 0-7176-1122-1

17.11 Health and Safety (Consultation with Employees) Regulations 1996

17.11.1 Application

These Regulations apply to all employers and employees in Great Britain except:

➤ where employees are covered by safety representatives appointed by recognized trade unions under the Safety Representatives and Safety Committees Regulations 1977
➤ domestic staff employed in private households
➤ crew of a ship under the direction of the master.

17.11.2 Employers' duty to consult – Regulation 3

The employer must consult relevant employees in good time with regard to:

➤ the introduction of any measure which may substantially affect their health and safety
➤ the employer's arrangements for appointing or nominating competent persons under the Management of Health and Safety at Work Regulations 1999
➤ any information required to be provided by legislation
➤ the planning and organization of any health and safety training required by legislation
➤ the health and safety consequences to employees of the introduction of new technologies into the workplace.

The guidance emphasizes the difference between informing and consulting. Consultation involves listening to employees' views and taking account of what they say before any decision is taken.

17.11.3 Persons to be consulted – Regulation 4

Employers must consult with either:

➤ the employees directly or
➤ one or more persons from a group of employees, who were elected by employees in that group to represent them under these regulations. They are known as 'Representatives of Employee Safety' (ROES).

Where ROES are consulted, all employees represented must be informed of:

➤ the names of ROES
➤ the group of employees represented.

Employers shall not consult a ROES if:

➤ the ROES does not wish to represent the group
➤ the ROES has ceased to be employed in that group
➤ the election period has expired
➤ the ROES has become incapacitated from carrying out their functions.

If an employer decides to consult directly with employees he must inform them and ROES of that fact. Where no ROES is elected employers will have to consult directly.

17.11.4 Duty to provide information – Regulation 5

Employers must provide enough information to enable ROES to participate fully and carry out their functions. This will include:

➤ what the likely risks and hazards arising from their work may be
➤ reported accidents and diseases etc. under RIDDOR 95
➤ the measures in place, or which will be introduced, to eliminate or reduce the risks
➤ what employees ought to do when encountering risks and hazards.

An employer need not disclose information which:

➤ could endanger national security;
➤ violates a legal prohibition;
➤ relates specifically to an individual without their consent;
➤ could substantially hurt the employers undertaking or infringe commercial security;
➤ was obtained in connection with legal proceedings.

17.11.5 Functions of ROES – Regulation 6

ROES have the following functions (but no legal duties):

➤ to make representations to the employer on potential hazards and dangerous occurrences related to the group of employees represented;
➤ to make representations to the employer on matters affecting the general health and safety of relevant employees;
➤ to represent the group of employees for which they are the ROES in consultation with inspectors.

17.11.6 Training, time off and facilities – Regulation 7

Where an employer consults ROES, he shall:

➤ ensure that each ROES receives reasonable training at the employers' expense;
➤ allow time off with pay during the ROES working hours to perform the duties of a ROES and while a candidate for election;
➤ such other facilities and assistance that ROES may reasonably require.

17.11.7 Civil liability and complaints – Regulation 9 and Schedule 2

Breach of these Regulations does not confer any right of action in civil proceedings subject to Regulation 7(3) and schedule 2 relating to complaints to industrial tribunals. A ROES can complain to an industrial tribunal that:

➤ their employer has failed to permit time off for training or to be a candidate for election
➤ their employer has failed to pay them as set out in Schedule 1 to the Regulations.

17.11.8 Elections

The guidance lays down some ideas for the elections, although there are no strict rules. ROES do not need to be confined to consultation related to these Regulations. Some employers have ROES sitting on safety committees and taking part in accident investigation similar to union Safety Representatives.

17.11.9 References

A guide to the Health and Safety (consultation with Employees) Regulations 1996. L95, HSE Books
      ISBN 0-7176-1234-1

17.12 Control of Substances Hazardous to Health Regulations (COSHH) 2002

17.12.1 Introduction

The 2002 COSHH Regulations have updated the 1999 Regulations with a few changes which include: additional definitions like 'inhalable dust' and 'health surveillance'; clarify and extend the steps required under risk assessment; introduces a duty to deal with accidents and emergencies.

COSHH covers most substances hazardous to health found in workplaces of all types. The substances covered by COSHH include:

➤ substances used directly in work activities (e.g. solvents, paints, adhesives, cleaners)
➤ substances generated during processes or work activities (e.g. dust from sanding, fumes from welding)
➤ naturally occurring substances (e.g. grain dust).

But COSHH does **not** include:

➤ asbestos and lead, which have specific regulations
➤ substances which are hazardous only because they are:
  ➤ radioactive
  ➤ simple asphyxiants
  ➤ at high pressure
  ➤ at extreme temperatures
  ➤ have explosive or flammable properties (separate regulations cover these)
  ➤ biological agents if they are not directly connected with work and are not in the employer's control, such as catching flu from a workmate.

17.12.2 Definition of substance hazardous to health – Regulation 2

The range of substances regarded as hazardous under COSHH are:

➤ substances or mixtures of substances classified as dangerous to health under the Chemicals (Hazard, Information and Packaging for Supply) Regulations 1994 (as amended) (CHIP). These have COSHH warning labels and manufacturers must supply data sheets. They cover substances that are: very toxic, toxic, harmful, corrosive or irritant under CHIP

➤ substances with occupational exposure limits as listed in EH40 published by the HSE

➤ biological agents (bacteria and other microorganisms) if they are directly connected with the work

➤ any kind of dust in a concentration specified in the regulations, i.e.

  ➤ 10 mg/m^3, as a time-weighted average over an 8-hour period, of total inhalable dust

  ➤ 4 mg/m^3, as a time-weighted average over an 8-hour period, of respirable dust

➤ any other substance which has comparable hazards to people's health but which, for technical reasons, is not covered by CHIP.

17.12.3 Duties under COSHH – Regulation 3

The duties placed on employers under these Regulations are extended to (except for health surveillance, monitoring and information and training) persons who may be on the premises but not employed whether they are at work or not, for example, visitors and contractors.

17.12.4 General requirements

There are seven basic steps to comply with COSHH, they are:

1 assess the risks to health
2 decide what precautions are needed
3 prevent or adequately control exposure
4 ensure that control measures are used and maintained
5 monitor the exposure of employees to hazardous substances
6 carry out appropriate health surveillance where necessary
7 ensure employees are properly informed, trained and supervised.

17.12.5 Steps 1 and 2 – assessment of health risk – Regulation 6

No work may be carried where employees are liable to be exposed to substances hazardous to health unless:

➤ a suitable and sufficient risk assessment, including the steps needed to meet COSHH, has been made.

The assessment must be reviewed and changes made regularly and immediately if:

➤ it is suspected that it is no longer valid
➤ there has been a significant change in the work to which it relates.

The assessment involves identifying the substance present in the workplace; assessing the risk it presents in the way it is used; and deciding what precautions (and health surveillance) are needed. Where more than four employees are employed the significant findings must be recorded.

17.12.6 Step 3 – prevention or control of exposure – Regulation 7

Every employer must ensure that exposure to substances hazardous to health is either:

➤ prevented or, where this is not reasonably practicable
➤ adequately controlled.

Preference must be given to substituting with a safer substance. Where it is not reasonably practicable to prevent exposure, protection measures must be adopted in the following order of priority:

➤ the design and use of appropriate work processes, systems and engineering controls and the provision and use of suitable work equipment;
➤ the control of exposure at source, including adequate ventilation systems and appropriate organizational measures; and
➤ where adequate control of exposure cannot be achieved by other means, the additional provision of suitable personal protective equipment.

PPE must conform to the Personal Protective Equipment (EC Directive) Regulations 2002.

Additionally respiratory protective equipment must, if there is no requirement imposed by these regulations, be of a type or shall conform to a standard, approved by the HSE. Where exposure to a carcinogen is involved, additional precautions are laid down in Regulation 7.

17.12.7 Step 4 – use, maintenance, examination and test of control measures – Regulations 8 and 9

➤ Every employer must take all reasonable steps to ensure that control measures, PPE or anything else provided under COSHH, are properly used or applied.
➤ Employees must make full and proper use of control measures, PPE or anything else provided. Employees must as far as is reasonable return items to accommodation and report defects immediately.

Employers shall also:

➤ properly maintain controls and, in the case of PPE, keep them clean and stored properly;
➤ carry out thorough examination and tests on engineering controls:

> in the case of Local Exhaust Ventilation (LEV) at least once every 14 months (except those in Schedule 4 which covers blasting of castings – monthly; dry grinding, polishing or abrading of metal for more than 12 hours per week – 6 months; and jute manufacture – 6 months);

> in any other case, at suitable intervals;

> carry out thorough examination and tests where appropriate (except on disposable items) on respiratory protective equipment at suitable intervals;

> keep a record of examination and tests for at least 5 years.

17.12.8 Step 5 – monitoring exposure – Regulation 10

Employers must monitor exposure:

> where this is necessary to ensure maintenance of control measures or the protection of health

> specifically for vinyl chloride monomer or chromium plating as required by schedule 5 and

> keep a record of identifiable personal exposures for 40 years and any other exposures for 5 years

> allow an employee access to their own records.

17.12.9 Step 6 – health surveillance – Regulation 11

Employers must ensure employees, where appropriate, who are exposed or liable to be exposed are under health surveillance. It is considered appropriate when:

> employee is exposed (if significant) to substances or processes in schedule 6

> an identifiable disease or adverse health effect may be related to the exposure, and there is a reasonable likelihood that disease may occur and there are valid disease indication or effect detection methods.

Records of health surveillance containing approved particulars must be kept for 40 years or offered to HSE if trading ceases. If a person is exposed to a substance and/ or process in schedule 6, the health surveillance shall include medical surveillance by an employment medical adviser or appointed doctor at intervals not exceeding 12 months.

If a medical adviser certifies that a person should not be engaged in particular work they must not be permitted to carry out that work except under specified conditions. Health records must be available for the individual employee to see after reasonable notice.

Employees must present themselves, during working hours at the employer's expense, for appropriate health surveillance.

An employment medical adviser or appointed doctor has the power to inspect the workplace or look at records for the purpose of carrying out functions under COSHH

17.12.10 Step 7 – information, instruction, training and emergencies – Regulation 12 and 13

Where employees are likely to be exposed to substances hazardous to health employers must provide:

> information, instruction and training on the risks to health and the precautions which should be taken (this duty is extended to anyone who may be affected)
> information on any monitoring of exposure (particularly if there is a maximum exposure limit where the employee or their representative must be informed immediately)
> information on collective results of health surveillance (designed so that individuals cannot be identified)
> procedures for accidents and emergencies.

17.12.11 Defence – Regulation 16

It is a defence under these regulations for a person to show that they have taken all reasonable precautions and exercised all due diligence to avoid the commission of an offence.

17.12.12 References

The Control of Substances Hazardous to Health Regulations 2002, SI 2002 2677, Stationery Office

General COSH ACOP (Control of substances hazardous to health) and Carcinogens ACOP (Control of carcinogenic substances) and Biological agents ACOP (Control of biological agents). L5, 1999 HSE Books
ISBN 0-7176-1670-3

COSHH Essentials: easy steps to control chemicals. HSG193, 1999 HSE Books
ISBN 0-7176-2421-8

Health surveillance under COSHH: guidance for employers. 1990 HSE Books
ISBN 0-7176-0491-8

Occupational Exposure Limits EH40/2001. 2001 HSE Books
ISBN 0-7176-1730-0 (revised annually)

The complete idiot's guide to CHIP. Chemicals (Hazard Information and Packaging for Supply) Regulations. 2002 INDG350, 2002 HSE Books

8 steps to successful substitution of hazardous substances. HSG110, 1994 HSE Books
ISBN 0-7176-1485-9

A step by step guide to COSHH assessment, revised, HSG97, 1997 HSE Books
ISBN 0-7176-1446-8

17.13 Health and Safety (Display Screen Equipment) Regulations 1992

17.13.1 General

These Regulations cover the minimum health and safety requirements for the use of display screen equipment and are accompanied by a guidance note. They typically apply to computer equipment with either a cathode ray tube or liquid crystal monitors. But any type of display is covered with some exceptions, for example, on board a means of transport, or where the main purpose is for screening a film or for a television. Multi-media equipment would generally be covered.

Equipment not specifically covered by these Regulations or where it is not being used by a defined 'user' are, nevertheless, covered by other requirements under the Management of Health and Safety at Work Regulations and the Provision and Use of Work Equipment Regulations.

17.13.2 Definitions – Regulation 1

(a) *Display Screen Equipment* (DSE) refers to any alphanumeric or graphic display screen, regardless of the display process involved

(b) A *user* is an employee and an *operator* is a self-employed person, both of whom habitually use display screen equipment as a significant part of their normal work. Both would be people to whom most or all of the following apply. A person:

➤ who depends on the display screen equipment to do their job
➤ who has no discretion as to use or non-use
➤ who needs particular training and/or skills in the use of display screen equipment to do their job
➤ who uses display screen equipment for continuous spell of an hour or more at a time
➤ who does so on a more or less daily basis
➤ for whom fast transfer of data is important for the job
➤ of whom a high level of attention and concentration is required, in particular to prevent critical errors.

(c) A *workstation* is an assembly comprising:

➤ display screen equipment with or without a keyboard, software or input device
➤ optional accessories
➤ disk drive, telephone, modem, printer, document holder, chair, desk, work surface, etc.
➤ the immediate working environment.

17.13.3 Exemptions – Regulation 1(4)

Exemptions include display screen equipment used in connection with:

➤ drivers' cabs or control cabs for vehicles or machinery
➤ on board a means of transport
➤ mainly intended for public operation
➤ portable systems not in prolonged use

➤ calculators, cash registers and small displays related to the direct use of this type of equipment
➤ window typewriters.

17.13.4 Assessment of risk – Regulation 2

Possible hazards associated with display screen equipment are physical (musculoskeletal) problems, visual fatigue and mental stress. They are not unique to display screen work nor an inevitable consequence of it, and indeed research shows that the risk to the individual from typical display screen work is low. However, as in other types of work, ill-health can result from poor work organization, working environment, job design and posture, and from inappropriate working methods.

Employers must carry out a suitable and sufficient analysis of users' and operators' workstations in order to assess the risks to health and safety. The guidance gives detailed information on workstation minimum standards and possible effects on health.

The assessment should be reviewed when major changes are made to software, hardware, furniture, environment or work requirements.

17.13.5 Workstations – Regulation 3

All workstations must meet the requirements laid down in the schedule to the regulations. This schedule lays down minimum requirements for display screen workstations, covering the equipment, the working environment, and the interface between computer and user/operator.

17.13.6 Daily work routine of users – Regulation 4

The activities of users should be organized so that their daily work on display screen equipment is periodically interrupted by breaks or changes of activity that reduce their workload at the equipment.

In most tasks, natural breaks or pauses occur from time to time during the day. If such breaks do not occur, deliberate breaks or pauses must be introduced. The guidance requires that breaks should be taken before the onset of fatigue and must be included in the working time. Short breaks are better than occasional long ones, for example, a 5–10 minute break after 50–60 minutes continuous screen and/or keyboard work is likely to be better than a 15-minute break every two hours. If possible, breaks should be taken **away** from the screen. Informal breaks, with time on other tasks, appear to be more effective in relieving visual fatigue than formal rest breaks.

17.13.7 Eyes and eyesight – Regulation 5

Initially on request, employees have the right to a free eye and eyesight test conducted by a competent person where they:

➤ are already users
➤ before becoming a user.

The employer must provide a further eye and eyesight test at regular intervals thereafter or when a user is experiencing visual difficulties which could be caused by working with display screen equipment.

There is no reliable evidence that work with display screen equipment causes any permanent damage to eyes or eyesight, but it may make users with pre-existing vision defects more aware of them.

An *eye and eyesight test* means a *sight test* as defined in the Opticians Act 1989. These should be carried out by a registered ophthalmic optician or medical practitioner (normally only those with an ophthalmic qualification do so).

Employers shall provide special corrective appliances to users where:

➤ normal corrective appliances cannot be used
➤ the result of the eye and eyesight test shows that such provision is necessary.

The guidance indicates that the liability of the employer extends only to the provision of corrective appliances, which are of a style and quality adequate for its function. If an employee chooses a more expensive design or multi-function correction appliances, the employer need only pay a proportion of the cost.

Employers are free to specify that users' eye and eyesight tests and correction appliances are provided by a nominated company or optician.

The confidential clinical information from the tests can only be supplied to the employer with the employee's consent.

Vision screen tests can be used to identify people with defects but they are not a substitute for the full eyesight test and employees have the right to opt for the full test from the outset.

17.13.8 Training – Regulation 6

Employers shall ensure that adequate health and safety training is provided to users and potential users in the use of any workstation and refresh the training following any re-organization. The guidance suggests a range of topics to be covered in the training. In summary these involve:

➤ the recognition of hazards and risks, including the absence of desirable features and the presence of undesirable ones
➤ causes of risk and how harm may occur
➤ what the user can do to correct them
➤ how problems can be communicated to management
➤ information on the regulations
➤ the user's contribution to assessments.

17.13.9 Information – Regulation 7

Operators and users shall be provided with adequate information on all aspects of health and safety relating to their workstation and what steps the employer has taken to comply with the Regulations (insofar as the action taken relates to that operator or user and their work). Under Regulation 7 specific information should be provided as outlined in Table 17.2.

Table 17.2 Provision of information under Regulation 7

Does employer have to provide information to display screen workers who are:	Information on:					
	Risks from DSE and workstation	Risk assessment and measures to reduce the risks Reg 2 & 3	Breaks and activity changes Reg 4	Eye and eyesight tests Reg 5	Initial training Reg 6(1)	Training when workstation modified Reg 6(2)
Users employed by the undertaking	Yes	Yes	Yes	Yes	Yes	Yes
Users employed by other employer	Yes	Yes	Yes	No	No	Yes
Operators in the undertaking	Yes	Yes	No	No	No	No

Source HSE

17.13.10 References

Display screen equipment work, Guidance on Regulations, Health and Safety (Display Screen Equipment) Regulations 1992 L26, HSE Books
ISBN 0-7176-0410-1
VDU's an Easy Guide to the Regulations, HS(G)90, 1994 HSE Books
ISBN 0-7176-0735-6
Working with VDU's IND(G)36(rev1) 1998 HSE Books
ISBN 0-7176-1504-9

17.14 Electricity at Work Regulations 1989

The purpose of these Regulations is to require precautions to be taken against the risk of death or personal injury from electricity in work activities. The regulations impose duties on persons ('duty holders') in respect of systems, electrical equipment and conductors and in respect of work activities on or near electrical equipment. They apply to almost all places of work and electrical systems at all voltages.

Guidance on the Regulations is contained in the *Memorandum of guidance on the Electricity at Work Regulations 1989*, published by the HSE.

17.14.1 Definitions

(a) Electrical equipment – includes anything used to generate, provide, transmit, rectify, convert, conduct, distribute, control, store, measure or use electrical energy.

(b) Conductor – means a conductor of electrical energy. It means any material (solid liquid or gas), capable of conducting electricity.

(c) System – means an electrical system in which all the electrical equipment is, or may be, electrically connected to a common source of electrical energy, and includes the source and equipment. It includes portable generators and systems on vehicles.

(d) Circuit conductor – term used in Regulations 8 and 9 only, means a conductor in a system which is intended to carry electric current in normal conditions. It would include a combined neutral and earth conductor, but does not include a conductor provided solely to perform a protective connection to earth or other reference point and energized only during abnormal conditions.

(e) Danger – in the context of these Regulations means a risk of injury from any electrical hazard.

Every year about 30 people die from electric shock or electric burns at work. Each year several hundred serious burns are caused by arcing when the heat generated can be very intense. In addition, intense ultraviolet radiation from an electric arc can cause damage to the eyes – known as arc-eye. These hazards are all included in the definition of **Danger**.

17.14.2 Duties – Regulation 3

Duties are imposed on employers, self-employed and employees. The particular duties on employees are intended to emphasize the level of responsibility which many employees in the electrical trades and professions are expected to take on as part of their job. They are:

➤ to cooperate with their employer so far as is necessary to enable any duty placed on the employer to be complied with (this reiterates section 7(b) of the HSW Act);

➤ to comply with the provisions of these Regulations in so far as they relate to matters which are within their control. (This is equivalent to duties placed on employers and self-employed where these matters are within their control.)

17.14.3 Systems, work activities and protective equipment – Regulation 4

Systems must, at all times, be of such construction as to prevent danger. Construction covers the physical condition, arrangement of components and design of the system and equipment.

All systems must be maintained so as to prevent danger.

Every work activity, including operation, use and maintenance or work near a system shall be carried out in a way, which prevents danger.

Protective equipment shall be suitable, suitably maintained and used properly.

17.14.4 Strength and capability of equipment – Regulation 5

No electrical equipment may be put into use where its strength and capability may be exceeded in such a way as may give rise to danger, in normal transient or fault conditions.

17.14.5 Adverse or hazardous environments – Regulation 6

Electrical equipment which may be exposed to:

➤ mechanical damage
➤ the effects of weather, natural hazards, temperature or pressure
➤ the effects of wet, dirty, dusty or corrosive conditions
➤ any flammable or explosive substances including dusts, vapours or gases

shall be so constructed and protected that it prevents danger.

17.14.6 Insulation, protection and placing of conductors – Regulation 7

All conductors in a system which may give rise to danger shall either be suitably covered with insulating material and further protected as necessary, for example, against mechanical damage, using trunking or sheathing; or have precautions taken that will prevent danger, for example, being suitably placed like overhead electric power cables, or by having strictly controlled working practices.

17.14.7 Earthing, integrity and other suitable precautions – Regulation 8

Precautions shall be taken, either by earthing or by other suitable means, for example, double insulation, use of safe voltages and earth-free non-conducting environments, where a conductor, other than a circuit conductor, could become charged as a result of either the use of, or a fault in, a system.

An earth conductor shall be of sufficient strength and capability to discharge electrical energy to earth. The conductive part of equipment, which is not normally live but energized in a fault condition, could be a conductor.

If a circuit conductor is connected to earth or to any other reference point nothing which could break electrical continuity or introduce high impedance, for example, fuse, thyristor or transistor, is allowed in the conductor unless suitable precautions are taken. Permitted devices would include a joint or bolted link, but not a removable link or manually operated knife switch without bonding of all exposed metal work and multiple earthing.

17.14.8 Connections – Regulation 10

Every joint and connection in a system shall be mechanically and electrically suitable for its use. This includes terminals, plugs and sockets.

17.14.9 Excess current protection – Regulation 11

Every part of a system shall be protected from excess current, for example, short circuit or overload, by a suitably located efficient means such as a fuse or circuit breaker.

17.14.10 Cutting off supply and isolation – Regulation 12

There should be suitably located and identified means of cutting off (switch) the supply of electricity to any electrical equipment and also isolating any electrical equipment. Although these are separate requirements, they could be effected by a single means. The isolator should be capable of being locked off to allow maintenance to be done safely.

Sources of electrical energy (accumulators, capacitors and generators) are exempt from this requirement, but precautions must be taken to prevent danger.

17.14.11 Work on equipment made dead – Regulation 13

Adequate precautions shall be taken to prevent electrical equipment that has been made dead from becoming live while work is carried out on or near the equipment. This will include means of locking off isolators, tagging equipment, permits to work and removing fuses.

17.14.12 Work on or near live conductors – Regulation 14

No person shall work near a live conductor, except if it is insulated, unless:

> it is unreasonable in all the circumstances for it to be dead;
> it is reasonable in **all** the circumstances for them to be at work on or near it while it is live;
> suitable precautions (including where necessary the provision of suitable protective equipment) are taken to prevent injury.

17.14.13 Working space access and lighting – Regulation 15

Adequate working space means of access and lighting shall be provided for all electrical equipment at which or near which work is being done in circumstances which may give rise to danger. This covers work of any kind. However, when the work is on live conductors the access space must be sufficient for a person to fall back out of danger and if needed for persons to pass one another with ease and without hazard.

17.14.14 Competence – Regulation 16

Where technical knowledge or experience is necessary to prevent danger, all persons must possess such knowledge or experience or be under appropriate supervision.

17.14.15 References

HS (R) 25 Memorandum of guidance on the Electricity at Work Regulations, Guidance on Regulations, 1989 HSE Books
     ISBN 0-7176-1602-9
HS(G)85 Electricity at Work: safe working practices, 1993 HSE Books
     ISBN 0-7176-0442-X
HS(G)107 Maintaining portable and transportable electrical equipment, 1994 HSE Books
     ISBN 0-7176-0715-1
Electrical safety and you, INDG231, 1996 HSE Books
     ISBN 0-7176-1207-4
Maintaining portable electrical equipment in offices and other low-risk environments, INDG236, 1996 HSE Books
     ISBN 0-7176-1272-4
Maintaining portable electrical equipment in hotels and tourist accommodation, INDG237, 1997 HSE Books

17.15 Employers' Liability (Compulsory Insurance) Act 1969 and Regulations 1998

17.15.1 Introduction

Employers are responsible for the health and safety of employees while they are at work. Employees may be injured at work, or they or former employees may become ill as a result of their work while employed. They may try to claim compensation from the employer if they believe them to be responsible. The Employers' Liability Compulsory Insurance Act 1969 ensures that an employer has at least a minimum level of insurance cover against any such claims.

Employers' liability insurance will enable employers to meet the cost of compensation for employees' injuries or illnesses whether they are caused on or off site. However, any injuries or illnesses relating to motor accidents that occur while employees are working for them, may be covered separately by motor insurance.

Public liability insurance is different. It covers for claims made against a person/company by members of the public or other businesses, but not for claims made by employees. While public liability insurance is generally voluntary, employers' liability insurance is compulsory. Employers can be fined if they do not hold a current employers' liability insurance policy which complies with the law.

17.15.2 Application

An employer needs employers' liability insurance unless they are exempt from the Employers' Liability Compulsory Insurance Act. The following employers are exempt:

➤ most public organizations including government departments and agencies, local authorities, police authorities and nationalized industries
➤ health service bodies, including National Health Service trusts, health authorities, Family Health Services Authorities and Scottish Health Boards and State Hospital Management Committees
➤ some other organizations which are financed through public funds, such as passenger transport executives and magistrates' courts committees
➤ family businesses, i.e. if employees are closely related to the employer (as husband, wife, father, mother, grandfather, grandmother, stepfather, stepmother, son, daughter, grandson, granddaughter, stepson, stepdaughter, brother, sister, half-brother or half-sister). However, this exemption does not apply to family businesses that are incorporated as limited companies.

A full list of employers who are exempt from the need to have employers' liability insurance is shown at Schedule 2 of the Employers' Liability (Compulsory Insurance) Regulations 1998.

17.15.3 Coverage

Employers are only required by law to have employers' liability insurance for people whom they employ. However, people who are normally thought of as self-employed may be considered to be employees for the purposes of employers' liability insurance.

Whether or not an employer needs employers' liability insurance for someone who works for them depends on the terms of the contract. This contract can be spoken, written or implied. It does not matter whether someone is usually called an employee or self-employed or what their tax status is. Whether the contract is called a contract of employment or a contract for services is largely irrelevant. What matters is the real nature of the employer/employee relationship and the degree of control the employer has over the work they do.

There are no hard and fast rules about who counts as employee for the purposes of Employers' liability insurance. The following paragraphs may help to give some indication.

In general, employers' liability insurance may be needed for a worker if:

➤ national insurance and income tax is deducted from the money paid to them;
➤ the employer has the right to control where and when they work and how they do it;
➤ most materials and equipment are supplied by the employer;
➤ the employer has a right to any profit workers make even though the employer may choose to share this with them through commission, performance pay or shares in the company. Similarly, the employer will be responsible for any losses;
➤ that person is required to deliver the service personally and they cannot employ a substitute if they are unable to do the work;
➤ they are treated in the same way as other employees, for example, if they do the same work under the same conditions as some other employee.

In most cases employers' liability insurance is needed for volunteers. Although, in general, the law may not require an employer to have insurance for:

➤ students who work unpaid
➤ people who are not employed but are taking part in youth or adult training programmes
➤ schoolchildren on work experience programmes.

However, in certain cases, these groups might be classed as employees. In practice, many insurance companies will provide cover for people in these situations.

One difficult area is domestic help. In general, an employer will probably not need employers' liability insurance for people such as cleaners or gardeners if they work for more than one person. However, if they only work for one employer, that employer may be required to take out insurance to protect them.

17.15.4 Display of certificate

Under the regulations, employers must display in a suitable convenient location, a current copy of the certificate of insurance at each place of business where they employ relevant people.

17.15.5 Retention of certificates

An employer must retain for at least 40 years copies of certificates of insurance which have expired. This is because claims for diseases can be made many years after the disease is caused. Copies can be kept electronically if this is more convenient than paper. An employer must make these available to health and safety inspectors on request.

These requirements do not apply to policies which expired before 1 January 1999. However, it is still very important to keep full records of previous insurance policies for the employer's own protection.

17.15.6 Penalties

The Health and Safety Executive enforces the law on employers' liability insurance and HSE inspectors will check that employers have employers' liability insurance with an approved insurer for at least £5m. They will ask to see the certificate of insurance and other insurance details.

Employers can be fined up to £2500 for any day they are without suitable insurance. If they do not display the certificate of insurance or refuse to make it available to HSE inspectors when they ask, employers can be fined up to £1000.

17.16 Fire Precautions (Workplace) Regulations 1997

17.16.1 Introduction

These Fire Regulations and the Fire Precautions Act 1971 are the responsibility of the Home Office and are enforced by the fire authorities. In Crown occupied and Crown owned buildings enforcement is carried out by the Fire Services Inspectorates of the Home Departments.

Fire precaution regulations deal with general fire precautions. These include:

➤ means of detection and giving warning in case of fire
➤ the provision of means of escape
➤ means of fighting fire
➤ the training of staff in fire safety.

The requirement to undertake an assessment of fire risks is now contained in the Management of Health and Safety at Work Regulations 1999. The HSW Act and regulations cover process fire precautions.

17.16.2 Application – Regulation 3

The regulations apply to employers with respect to every workplace and a person who has, to any extent, control of a workplace (including maintenance, repair or safety). Except for:

> a construction site
> mine or offshore installation
> ship, aircraft, locomotive or rolling stock
> work in fields, woods or other agricultural or forestry land which is not inside a building
> prisons or if they would prevent the armed forces or emergency services from carrying out their duties.

17.16.3 Fire-fighting and fire detection – Regulation 4

Where necessary every workplace shall have:

> Appropriate fire-fighting equipment, fire detectors and alarms
> non-automatic equipment that is easily accessible, simple to use and indicated by signs.

Account should be taken of:

> dimensions and use of the building
> the equipment they contain
> the physical and chemical properties of the substances likely to be present
> the maximum number of people that may be present at any one time.

An employer shall:

> take measures for fire-fighting adapted to the nature of the activities, size and all the persons who may be present;
> nominate employees to implement the measures and ensure that the number of nominees, their training and equipment is sufficient for the risk;
> arrange any contacts with external emergency services.

17.16.4 Emergency routes and exits – Regulation 5

Where necessary to safeguard people and relevant to the risks at the workplace, the following shall be complied with:

> routes to emergency exits and the exits themselves must be clear at all times
> emergency routes and exits to lead as directly as possible to a place of safety
> evacuation must be possible quickly and safely
> the number, type and distribution of emergency routes and exits shall be adequate for the risk and numbers of people
> emergency doors must open in the direction of escape
> sliding or revolving doors not to be used for emergency exits
> emergency doors shall not be locked or fastened so they cannot be easily and immediately opened by any person who may require them in an emergency
> emergency routes and exits must be indicated by signs
> emergency routes and exits requiring illumination must have adequate emergency lighting.

17.16.5 Maintenance – Regulation 6

Where necessary for safety, equipment shall be subject to a suitable system of maintenance and must be kept properly maintained at all times.

17.16.6 Other requirements

Most of the rest of the regulations is about enforcement and penalties. Fire authority officers have similar powers to inspectors under the HSW Act and can issue improvement and prohibition notices. Cases can be tried in a magistrates' court or on indictment for which there is a maximum penalty of two years' imprisonment.

17.16.7 References

Fire Safety, An employers guide. HMSO 1999 HSE Books
ISBN 0-11-341229-0
Fire safety at work: Fire Precautions Act 1971 guide. 1989 HMSO
ISBN 0-11-340905-2

17.17 Health and Safety (First Aid) Regulations 1981

These Regulations set out employers' duties to provide adequate first-aid facilities. They define first aid as:

➤ treatment for the purposes of preserving life and minimizing the consequences of injury and illness until medical help is obtained
➤ treatment of minor injuries which would otherwise receive no treatment or which do not need treatment by a medical practitioner or nurse.

17.17.1 Duty of the employer – Regulation 3

An employer shall provide, or ensure that there are provided:

➤ adequate and appropriate facilities and equipment
➤ qualified first aiders to render first aid
➤ an appointed person, being someone to take charge of situations as well as first-aid equipment and facilities, where medical aid needs to be summoned. An appointed person will suffice where:
  ➤ the nature of the work is such that there are no specific serious hazards (offices, libraries, etc.), the workforce is small, the location makes further provision unnecessary
  ➤ there is temporary (not planned holidays) or exceptional absence of the first aider.

There must always be at least an appointed person in every workplace during working hours.

Employers must make an assessment of the first-aid requirements that are appropriate for each workplace.

17.17.2 Employees information – Regulation 4

Employees must be informed of the arrangements for first-aid, including the location of facilities, equipment and people.

17.17.3 Self-employed – Regulation 5

The self-employed shall provide such first-aid equipment as is appropriate to render first aid to themselves.

17.17.4 References

HSE, L74 First aid at work. Health and Safety (First Aid) Regulations 1981. Approved code of practice and guidance, HSE Books
ISBN 0-7176-1050-0

17.18 Health and Safety (Information for Employees) Regulations 1989

17.18.1 General requirements

These regulations require that the Approved Poster entitled, *Health and safety – what you should know*, is displayed or the Approved Leaflet is distributed.

This information tells employees in general terms about the requirements of health and safety law.

Employers must also inform employees of the local address of the enforcing authority (either the HSE or the local authority) and the Employment Medical Advisory Service. This should be marked on the poster or supplied with the leaflet.

Since 1 July 2000 the latest version of the poster must be displayed or distributed. References to obsolete legal requirements are removed and the revised text focuses on the modern framework of general duties, supplemented by the basics of health and safety management and risk assessment. It includes two additional boxes: one for details of trade union or other safety representatives and one for competent persons appointed to assist with health and safety and their responsibilities.

17.18.2 References

Health and Safety (Information for Employees) Regulations 1989 (SI No 682)
Health and Safety Information for Employees (Modifications and Repeals) Regulations 1995
Poster ISBN 0-7176-2493-5
Leaflet ISBN 0-7176-1702-5

17.19 Ionising Radiations Regulations 1999

17.19.1 Introduction

The Ionising Radiations Regulations 1999 (IRR99) implement the majority of the Basic Safety Standards Directive 96129/Euratom (BSS Directive). From 1 January 2000, they replaced the Ionising Radiations Regulations 1985 (IRR85)(except for regulation 26 (special hazard assessments)).

The main aim of the Regulations and the supporting ACOP is to establish a framework for ensuring that exposure to ionising radiation arising from work activities, whether from man-made or natural radiation and from external radiation (e.g. X-ray set) or internal radiation (e.g. inhalation of a radioactive substance), is kept as low as reasonably practicable and does not exceed dose limits specified for individuals. IRR99 also:

(a) replace the Ionising Radiations (Outside Workers) Regulations 1993 (OWR93), which were made to implement the Outside Workers Directive 90/641/Euratom and

(b) implement a part of the Medical Exposures Directive 97143/Euratom in relation to equipment used in connection with medical exposures.

The guidance which accompanies the Regulations and ACOP gives detailed advice about the scope and duties of the requirements imposed by IRR99. It is aimed at employers with duties under the Regulations but should also be useful to others such as radiation protection advisers, health and safety officers, radiation protection supervisors and safety representatives.

17.19.2 Essentially, work with ionising radiation means

(a) a practice, which involves the production, processing, handling, use, holding, storage, transport or disposal of artificial radioactive substances and some naturally occurring sources, or the use of electrical equipment emitting ionising radiation at more than 5 kV (see definition of practice in Regulation 2(1))

(b) work in places where the radon gas concentration exceeds the values in Regulation 3(1)(b) or

(c) work with radioactive substances containing naturally occurring radionuclides not covered by the definition of a practice.

17.19.3 Radiation employers

Radiation employers are essentially those employers who work with ionising radiation, i.e. they carry out:

(a) a practice (see definition in Regulation 2(1)), or

(b) work in places where the radon gas concentration exceeds the values in regulation 3(1)(b); or

(c) work with radioactive substances containing naturally occurring radionuclides not covered by the definition of a practice.

426

17.19.4 Duties of self-employed people

A self-employed person who works with ionising radiation will simultaneously have certain duties under these Regulations, both as an employer and as an employee. For example, self-employed persons may need to take such steps as:

(a) carrying out assessments under Regulation 7

(b) providing control measures under Regulation 8 to restrict exposure

(c) designating themselves as classified persons under Regulation 20

(d) making suitable arrangements under Regulation 21 with one or more approved dosimetry services (ADS) for assessment and recording of doses they receive

(e) obtaining a radiation passbook and keeping it up to date in accordance with Regulation 21

(f) if they carry out services as an outside worker; making arrangements for their own training as required by Regulation 14

(g) ensuring they use properly any dose meters provided by an ADS as required by regulation.

17.19.5 General requirements

Some of the major considerations are:

➤ notification to HSE of specific work unless specified in schedule 1 to the regulations (Regulation 6)

➤ carrying out prior risk assessment by a radiation employer before commencing a new activity (Regulation 7)

➤ use of personal protective equipment (Regulation 9)

➤ maintenance and examination of engineering controls and PPE (Regulation 10)

➤ dose limitations (Regulation 11)

➤ contingency plans for emergencies (Regulation 12)

➤ radiation protection adviser appointment (Regulation 13)

➤ information instruction and training (Regulation 14).

17.19.6 Prior risk assessment

Where a radiation employer is required to undertake a prior risk assessment, the following matters need to be considered, where they are relevant:

(a) the nature of the sources of ionising radiation to be used, or likely to be present, including accumulation of radon in the working environment

(b) estimated radiation dose rates to which anyone can be exposed

(c) the likelihood of contamination arising and being spread

(d) the results of any previous personal dosimetry or area monitoring relevant to the proposed work

(e) advice from the manufacturer or supplier of equipment about its safe use and maintenance

(f) engineering control measures and design features already in place or planned

(g) any planned systems of work; estimated levels of airborne and surface contamination likely to be encountered

(h) the effectiveness and the suitability of personal protective equipment to be provided

(i) the extent of unrestricted access to working areas where dose rates or contamination levels are likely to be significant; possible accident situations, their likelihood and potential severity

(j) the consequences of possible failures of control measures – such as electrical interlocks, ventilation systems and warning devices – or systems of work; steps to prevent identified accident situations, or limit their consequences.

This prior risk assessment should enable the employer to determine:

(a) what action is needed to ensure that the radiation exposure of all persons is kept as low as reasonably practicable (Regulation 8(1))

(b) what steps are necessary to achieve this control of exposure by the use of engineering controls, design features, safety devices and warning devices (Regulation 8(2)(a)) and, in addition, by the development of systems of work (Regulation 8(2)(b))

(c) whether it is appropriate to provide personal protective equipment and if so what type would be adequate and suitable (Regulation 8(2)(c))

(d) whether it is appropriate to establish any dose constraints for planning or design purposes, and if so what values should be used (Regulation 8(3))

(e) the need to alter the working conditions of any female employee who declares she is pregnant or is breastfeeding (Regulation 8(5))

(f) an appropriate investigation level to check that exposures are being restricted as far as reasonably practicable (Regulation 8(7))

(g) what maintenance and testing schedules are required for the control measures selected (Regulation 10)

(h) what contingency plans are necessary to address reasonably foreseeable accidents (Regulation 12)

(i) the training needs of classified and non-classified employees (Regulation 14)

(j) the need to designate specific areas as controlled or supervised areas and to specify local rules (Regulations 16 and 17)

(k) the actions needed to ensure restriction of access and other specific measures in controlled or supervised areas (Regulation 18)

(l) the need to designate certain employees as classified persons (Regulation 20)

(m) the content of a suitable programme of dose assessment for employees designated as classified persons and for others who enter controlled areas (Regulations 18 and 21)

(n) the responsibilities of managers for ensuring compliance with these Regulations

(o) an appropriate programme of monitoring or auditing of arrangements to check that the requirements of these Regulations are being met.

17.19.7 References

Work with Ionising radiation, Ionising Radiations Regulations, Approved Code of Practice and Guidance. HSC, L121, 2000 HSE Books
> ISBN 0-7176-1746-7

This is a large document and need only be studied by those with a specific need to control IR. The HSE have also produced a number of information sheets which are available free on the Internet at their web site.

17.20 Lifting Operations and Lifting Equipment Regulations 1998

17.20.1 Introduction

This summary gives information about the Lifting Operations and Lifting Equipment Regulations 1998 (LOLER) which came into force on 5 December 1998.

In the main, LOLER replaced existing legal requirements relating to the use of lifting equipment, for example the Construction (Lifting Operations) Regulations 1961, the Docks Regulations 1988 and the Lifting Plant and Equipment (Records of Test and Examination etc.) Regulations 1992.

The Regulations aim to reduce risks to people's health and safety from lifting equipment provided for use at work. In addition to the requirements of LOLER, lifting equipment is also subject to the requirements of the Provision and Use of Work Equipment Regulations 1998 (PUWER).

Generally, the Regulations require that lifting equipment provided for use at work is:

➤ strong and stable enough for the particular use and marked to indicate safe working loads;
➤ positioned and installed to minimize any risks;
➤ used safely, i.e. the work is planned, organized and performed by competent people;
➤ subject to ongoing thorough examination and, where appropriate, inspection by competent people.

17.20.2 Definition

Lifting equipment includes **any equipment used at work for lifting or lowering loads,** including attachments used for anchoring, fixing or supporting it. The Regulations cover a wide range of equipment including cranes, fork lift trucks, lifts, hoists, mobile elevating work platforms, and vehicle inspection platform hoists. The definition also includes lifting accessories such as chains, slings, eyebolts, etc. LOLER **does not** apply to escalators, these are covered by more specific legislation, i.e. the Workplace (Health, Safety and Welfare) Regulations 1992.

If employees are allowed to provide their own lifting equipment, then this too is covered by the Regulations.

17.20.3 Application

The Regulations apply to an employer or self-employed person providing lifting equipment for use at work, or who has control of the use of lifting equipment. They do not apply to equipment to be used primarily by members of the public, for example, lifts in a shopping centre. However, such circumstances are covered by the HSW Act 1974.

LOLER applies to the way lifting equipment is used in industry and commerce. LOLER applies only to work activities, for example:

➤ a crane on hire to a construction site
➤ a contract lift

➤ a passenger lift provided for use of workers in an office block
➤ refuse collecting vehicles lifting on a public road
➤ patient hoist
➤ fork lift truck.

These Regulations add to the requirements of PUWER 98 and should be interpreted with them. For example, when selecting lifting equipment, PUWER Regulation 4, regarding suitability, should be considered in connection with:

➤ ergonomics
➤ the conditions in which the equipment is to be used
➤ safe access and egress
➤ preventing slips, trips and falls
➤ protecting the operator.

While employees do not have duties under LOLER, they do have general duties under the HSW Act and the Management of Health and Safety at Work Regulations (MHSWR), for example, to take reasonable care of themselves and others who may be affected by their actions and to cooperate with others.

The Regulations cover places where the HSW Act applies – these include factories, offshore installations, agricultural premises, offices, shops, hospitals, hotels, places of entertainment, etc.

17.20.4 Strength and stability – Regulation 4

Lifting equipment shall be of adequate strength and stability for each load, having regard in particular to the stress induced at its mounting or fixing point.

Every part of a load and anything attached to it and used in lifting it shall be of adequate strength.

Account must be taken of the combination of forces to which the lifting equipment will be subjected, as well as the weight of any lifting accessories. The equipment should include an appropriate factor of safety against failure.

Stability needs to take into account the nature, load-bearing strength, stability, adjacent excavations and slope of the surface. For mobile equipment, keeping rails free of obstruction and tyres correctly inflated must be considered.

17.20.5 Lifting equipment for lifting persons – Regulation 5

To ensure safety of people being lifted, there are additional requirements for such equipment. The use of equipment not specifically designed for raising and lowering people should only be used in exceptional circumstances.

The regulation applies to all lifting equipment used for raising and lowering people and requires that lifting equipment for lifting persons shall:

➤ prevent a person using it from being crushed, trapped or struck, or falling from the carrier;
➤ prevent, so far as is reasonably practicable, persons using it while carrying out work, from the carrier being crushed, trapped or struck or falling from the carrier;

➤ have suitable devices to prevent the risk of the carrier falling. If a device cannot be fitted, the carrier must have:

    ➤ an enhanced safety coefficient suspension rope or chain;

    ➤ the rope or chain inspected every working day by a competent person;

➤ be such that a person trapped in any carrier is not thereby exposed to any danger and can be freed.

17.20.6 Positioning and installation – Regulation 6

Lifting equipment must be positioned and installed so as to reduce the risks, so far as is reasonably practicable, from:

➤ equipment or a load striking another person

➤ a load drifting, falling freely or being released unintentionally and it is otherwise safe.

Lifting equipment should be positioned and installed to minimize the need to lift loads over people and to prevent crushing in extreme positions. It should be designed to stop safely in the event of a power failure and not release its load. Lifting equipment, which follows a fixed path, should be enclosed with suitable and substantial interlocked gates and any necessary protection in the event of power failure.

17.20.7 Marking of lifting equipment – Regulation 7

Machinery and accessories for lifting loads shall be clearly marked to indicate their safe working loads (SWL), and:

➤ where the SWL depends on the configuration of the lifting equipment:

    ➤ the machinery should be clearly marked to indicate its SWL for each configuration

    ➤ information which clearly indicates its SWL for each configuration should be kept with the machinery

➤ accessories for lifting (e.g. hooks, slings) are also marked in such a way that it is possible to identify the characteristics necessary for their safe use (e.g. if they are part of an assembly)

➤ lifting equipment which is designed for lifting people is appropriately and clearly marked

➤ lifting equipment not designed for lifting people, but which might be used in error, should be clearly marked to show it is not for lifting people.

17.20.8 Organization of lifting operations – Regulation 8

Every lifting operation, that is lifting or lowering of a load, shall be:

➤ properly planned by a competent person

➤ appropriately supervised

➤ carried out in a safe manner.

The person planning the operation should have adequate practical and theoretical knowledge and experience of planning lifting operations. The plan will need to address the risks identified by the risk assessment and identify the resources, the procedures and the responsibilities required so that any lifting operation is carried out safely. For routine simple lifts a plan will normally be left to the people using the lifting equipment. For complex lifting operations, for example, where two cranes are used to lift one load, a written plan may need to be produced each time.

The planning should take account of, avoiding suspending loads over occupied areas, visibility, attaching/detaching and securing loads, the environment, location, overturning, proximity to other objects, lifting of people and pre-use checks of the equipment.

17.20.9 Thorough examination and inspection – Regulation 9

Before using lifting equipment for the first time by an employer, it must be thoroughly examined for any defect unless either:

➤ the lifting equipment has not been used before
➤ an EC declaration of conformity (where one should have been drawn up) has been received or made not more than 12 months before the lifting equipment is put into service
➤ if it is obtained from another undertaking, it is accompanied by physical evidence of an examination.

A copy of this thorough examination report shall be kept for as long as the lifting equipment is used (or, for a lifting accessory, two years after the report is made) (Regulation 11).

Where safety depends on the installation conditions, it shall be thoroughly examined:

➤ after installation and before being put into service
➤ after assembly and before being put into service at a new site or in new location to ensure that it has been installed correctly and is safe to operate.

A copy of the thorough examination report shall be kept for as long as the lifting equipment is used at the place it was installed or assembled (Regulation 11).

Lifting equipment, which is exposed to conditions causing deterioration, which may result in dangerous situations, shall be:

➤ thoroughly examined at least every 6 months (for lifting equipment for lifting persons, or a lifting accessory); at least every 12 months (for other lifting equipment); or in accordance with an examination scheme; and each time that exceptional circumstances, liable to jeopardize the safety of the lifting equipment, have occurred and a copy of the report kept until the next report is made, or for two years (whichever is longer)
➤ inspected, if appropriate, by a competent person at suitable intervals between 'thorough examinations' (and a copy of the record kept until the next record is made).

All lifting equipment shall be accompanied by physical evidence that the last 'thorough examination' has been carried out before it leaves an employer's undertaking (or before it is used after leaving another undertaking).

The user, owner, manufacturer or some other independent party may draw up examination schemes provided they have the necessary competence. Schemes should specify the intervals at which lifting equipment should be thoroughly examined and, where appropriate, those parts that need to be tested. The scheme should take

432

account, for example, of its condition, the environment in which it is used, the number of lifting operations and the loads lifted.

The 'competent person' carrying out a thorough examination should have appropriate practical and theoretical knowledge and experience of the lifting equipment to be examined to enable them to detect defects or weaknesses and to assess their importance in relation to the safety and continued use of the lifting equipment. They should also determine whether a test is necessary and the most appropriate method for carrying it out.

17.20.10 Reports and defects – Regulation 10

The person making a 'thorough examination' of lifting equipment shall:

➤ notify the employer forthwith of any defect which, in their opinion, is or could become, dangerous;

➤ as soon as is practicable (within 28 days) write an authenticated report to:
  ➤ the employer
  ➤ any person who hired or leased the lifting equipment, containing the information specified in Schedule 1.

➤ send a copy (as soon as is practicable) to the relevant enforcing authority where there is, in their opinion, a defect with an existing or imminent risk of serious personal injury (this will always be HSE if the lifting equipment has been hired or leased).

Every employer notified of a defect following a 'thorough examination' of lifting equipment should ensure that it is not used:

➤ before the defect is rectified
➤ after a time specified in the schedule accompanying the report.

The person making an 'inspection' shall also notify the employer when, in his opinion, a defect is, or could become, dangerous and, as soon as is practicable, make a record of the inspection in writing.

17.20.11 Reports – Schedule 1

Schedule 1 lists the information to be contained in a report of a thorough examination, for example, name and address; identity of equipment; date of last thorough examination; SWL; appropriate interval; any dangerous or potentially dangerous defects; repairs required; date of next examination and test; details of the competent person.

17.20.12 References

Safe use of lifting equipment, Lifting Operations and Lifting Equipment Regulations 1998 Approved Code of Practice and Guidance, L113, 1998 HSE Books
 ISBN 0-7176-1628-2
Simple guide to the LOLER, 1999 HSE, INDG290, HSE Books
 ISBN 0-7176-2430-7

LOLER 1998 Lifting Operations and Lifting Equipment Regulations 1998, Open learning guidance. 1999 HSE Books
ISBN 0-7176-2464-1

17.21 Management of Health and Safety at Work Regulations 1999

17.21.1 General

These Regulations give effect to the European Framework Directive on health and safety. They supplement the requirements of the Health and Safety at Work Act etc. 1974 and specify a range of management issues, most of which must be carried out in all workplaces. The aim is to map out the organization of precautionary measures in a systematic way, and to make sure that all staff are familiar with the measures and their own responsibilities.

17.21.2 Risk assessment – Regulation 3

Every employer is required to make a 'suitable and sufficient' assessment of risks to employees, and risks to other people who might be affected by the organization, such as visiting contractors and members of the public. A systematic investigation of risks involved in all areas and operations is required, together with identification of the persons affected, a description of the controls in place and any further action required to reduce risks.

The risk assessments must take into account risks to new and expectant mothers and young people. Fire risk assessments must also be included.

Significant findings from the assessments must be written down (or recorded by other means, such as on a computer) when there are five or more employees. The assessments need to be reviewed regularly and if necessary, when there have been significant changes, they should be modified.

17.21.3 Principles of prevention – Regulation 4

The following principles must be adopted when implementing any preventative and protective measures:

➤ avoiding risks
➤ evaluating the risks which cannot be avoided
➤ combating the risks at source
➤ adapting the work to the individual, especially as regards the design of workplaces, the choice of work equipment and the choice of working and production methods, with a view, in particular, to alleviating monotonous work and work at a predetermined work-rate and to reducing their effect on health
➤ adapting to technical progress
➤ replacing the dangerous by the non-dangerous or the less dangerous

> developing a coherent overall prevention policy which covers technology, organization of work, working conditions, social relationships and the influence of factors relating to the working environment
> giving collective protective measures priority over individual protective measures
> giving appropriate instruction to employees.

17.21.4 Effective arrangements for health and safety – Regulation 5

Formal arrangements must be devised (and recorded) for effective planning, organization, control, monitoring and review of safety measures. This will involve an effective health and safety management system to implement the policy. Where there are five or more employees the arrangements should be recorded.

Planning involves a systematic approach to risk assessment, the selection of appropriate risk controls and establishing priorities with performance standards.

Organization involves consultation and communication with employees; employee involvement in risk assessment; the provision of information and securing competence with suitable instruction and training. Control involves clarifying responsibilities and making sure people adequately fulfil their responsibilities. It involves adequate and appropriate supervision.

Monitoring should include the measurement of how well the policy is being implemented and whether hazards are being controlled properly. It covers inspections of the workplace and management systems: the investigation of incidents and accidents to ascertain the underlying causes and effect a remedy.

Review is essential to look at the whole of the health and safety management system to ensure that it is effective and achieving the correct standard of risk control.

17.21.5 Health surveillance – Regulation 6

In appropriate circumstances health surveillance of staff may be required – the Approved Code of Practice describes more fully when this duty will arise. Health surveillance is considered relevant when there is an identifiable disease or poor health condition; there are techniques to detect the disease; there is a reasonable likelihood that the disease will occur; and surveillance is likely to enhance the protection of the workers concerned. A competent person, who will range from a manager, in some cases, to a fully qualified occupational medical practitioner in others, should assess the extent of the surveillance.

17.21.6 Competent assistance – Regulation 7

Every employer is obliged to appoint one or more 'competent person(s)' to advise and assist in undertaking the necessary measures to comply with the relevant statutory requirements. They may be employees or outside consultants. The purpose is to make sure that all employers have access to health and safety expertise. Preference should be given to an in-house appointee who may be backed up by external expertise.

The competence of the person(s) appointed is to be judged in terms of their training, knowledge and experience of the work involved; it is not necessarily dependent upon particular qualifications. In simple situations, it may involve knowledge of relevant best practice, knowing one's limitations and taking external advice when necessary. In more complex situations or risks, fully qualified and appropriately experienced practitioners will be required.

Appointed competent persons must be provided with adequate information, time and resources to do their job.

17.21.7 Procedures for serious and imminent danger and contact with external services – Regulations 8 and 9

Procedures must be established for dealing with serious and imminent dangers, including fire evacuation plans and arrangements for other emergencies. A sufficient number of competent persons must be appointed to evacuate the premises in the event of an emergency. The procedures should allow for persons at risk to be informed of the hazards and how and when to evacuate to avoid danger. In shared workplaces employers must cooperate. Access to dangerous areas should be restricted to authorized and properly trained staff. Any necessary contact arrangements with external services for first aid, emergency medical care and rescue work must be set up.

17.21.8 Information for employees – Regulation 10

Information must be provided to staff on the risk assessment, risk controls, emergency procedures, the identity of the people appointed to assist on health and safety matters and risks notified by others.

The information provided must take into account the level of training, knowledge and experience of the employees. It must take account of language difficulties and be provided in a form that can be understood by everyone. The use of translations, symbols and diagrams should be considered. Where children under school leaving age are at work, information on the risk assessments and control measures must be provided to the child's parent or guardians of children at work before the child starts work. It can be provided verbally or directly to the parent, guardians or school.

17.21.9 Cooperation and coordination – Regulations 11, 12 and 15

Where two or more employers share a workplace, each must:

➤ cooperate with other employers in health and safety matters;
➤ take reasonable steps to coordinate their safety precautions;
➤ inform the other employers of the risks to their employees, i.e. risks to neighbours' employees;
➤ where people from outside organizations are present to do work they, and their employers, have to be provided with appropriate information on risks, health and the necessary precautions to be taken;
➤ Temporary staff and staff with fixed-term contracts as well as permanent employees must be supplied with health and safety information before starting work (Regulations 12 and 15);

➤ Regulation 11 does not apply to multi-occupied premises or sites where each unit, under the control of an individual tenant employer or self-employed person, is regarded as a separate workplace. In other cases, common areas may be shared workplaces, such as a reception area or canteen or they may be under the control of a person to whom Section 4 of HSW Act applies. Suitable arrangements may need to be put in place for these areas.

17.21.10 Capabilities and training – Regulation 13

When giving tasks to employees, their capabilities with regard to health and safety must be taken into account.

Employees must be provided with adequate health and safety training:

➤ on recruitment
➤ on being exposed to new or increased risks
➤ on the introduction of new procedures, systems or technology. Training must be repeated periodically and take place in working hours (or while being paid).

17.21.11 Duties on employees – Regulation 14

Equipment and materials must be used properly in accordance with instructions and training. Obligations on employees are extended to include certain requirements to report serious and immediate dangers and any shortcomings in the employer's protection arrangements.

17.21.12 New or expectant mothers – Regulations 16, 17 and 18

Where work is of a kind that could present a risk to new or expectant mothers working there or their babies, the risk assessments must include an assessment of such risks. When the risks cannot be avoided the employer must alter a women's working conditions or hours to avoid the risks; offer suitable alternative work; or suspend from work on full pay. The woman must notify the employer in writing of her pregnancy, that she has given birth within the last six months or she is breastfeeding.

17.21.13 Young persons – Regulation 19

Employers must protect young persons at work from risks to their health and safety which are the result of lack of experience, or absence of awareness of existing or potential risks or because they have not yet fully matured. Young persons may not be employed in a variety of situations enumerated in the regulations, which pose a significant risk to their health and safety. The exception to this is young persons over school leaving age:

➤ where the work is necessary for their training
➤ where they will be supervised by a competent person
➤ where the risk will be reduced to the lowest level that is reasonably practicable.

17.21.14 Provisions as to liability – Regulation 21

A new provision has been added to prevent a defence for an employer by reason of any act or default by an employee or a competent person appointed under Regulation 7.

17.21.15 References

The Management of Health and Safety at Work Regulations 1999 HMSO
ISBN 0-11-025051-6
Management of health and safety at work: Approved Code of Practice and Guidance, HSC L21, Second Edition 2000 HSE Books
ISBN 0-7176-2488-9
Five Steps to Risk Assessment, HSE IND (G)163(rev1). 1999 HSE Books

17.22 Manual Handling Operations Regulations 1992

17.22.1 General

The Regulations apply to the manual handling (any transporting or supporting) of loads, i.e. by human effort, as opposed to mechanical handling by fork lift truck, crane, etc. Manual handling includes lifting, putting down, pushing, pulling, carrying or moving. The human effort may be applied directly to the load, or indirectly by pulling on a rope, chain or lever. Introducing mechanical assistance, like a hoist or sack truck, may reduce but not eliminate manual handling, since human effort is still required to move, steady or position the load.

The application of human effort for purposes other than transporting or supporting a load, for example, pulling on a rope to lash down a load or moving a machine control, is not a manual handling operation. A load is a discrete movable object, but it does not include an implement, tool or machine while being used.

Injury in the context of these Regulations means to any part of the body. It should take account of the physical features of the load which might affect grip or cause direct injury, for example, slipperiness, sharp edges, and extremes of temperature. It does not include injury caused by any toxic or corrosive substance which has leaked from a load, is on its surface or is part of the load.

17.22.2 Duties of employers – avoidance of manual handling – Regulation 4(1)(a)

Employers should take steps to avoid the need for employees to carry out MHO which involves a risk of their being injured.

The guidance suggests that a preliminary assessment should be carried out when making a general risk assessment under the Management of Health and Safety at Work Regulations 1999. Employers should consider whether the operation can be eliminated, automated or mechanized.

17.22.3 Duties of employers – assessment of risk – Regulation 4(1)(b)(I)

Where it is not reasonably practicable to avoid MHO, employers must make a suitable and sufficient risk assessment of all such MHO in accordance with the requirements of Schedule 1 to the Regulations (shown later). This duty to assess the risk takes into account the **task**, the **load**, the **working environment** and **individual capability**.

17.22.4 Duties of employers – reducing the risk of injury – Regulation 4(1)(b)(ii)

Where it is not reasonably practicable to avoid the MHO at which there is a risk of injury, employers must take steps to reduce the risk of injury to the lowest level reasonably practicable.

The structured approach (considering the task, the load, the working environment and the individual capability) is recommended in the guidance. The steps taken will involve ergonomics, changing the load, mechanical assistance, task layout, work routines, personal protective equipment, team working and training.

17.22.5 Duties of employers – additional information on the load – Regulation 4(1)(b)(iii)

Employers must take appropriate steps where manual handling cannot be avoided to provide general indications and, where practicable precise information on:

> the weight of each load
> the heaviest side of any load which does not have a central centre of gravity.

The information is probably best marked on the load. Sections 3 and 6 of the HSW Act may place duties on originators of loads, like manufacturers or packers.

17.22.6 Duties of employers – reviewing assessment – Regulation 4(2)

The assessment should be reviewed if there is reason to suspect that it is no longer valid or there have been significant changes in the particular MHO.

17.22.7 Duty of employees

Each employee, while at work, has to make proper use of any system of work provided for their use. This is in addition to other responsibilities under the HSW Act and the Management of Health and Safety at Work Regulations.

The provisions do not include well intentioned improvisation in an emergency, for example, rescuing a casualty or fighting a fire.

Table 17.3 Schedule 1 to the Manual Handling Operations Regulations

Factors to which the employer must have regard and questions they must consider when making an assessment of manual handling operations	
Factors	Questions
1 The tasks	Do they involve: Holding or manipulating loads at distance from trunk? Unsatisfactory bodily movement or posture, especially: Twisting the trunk? Stooping? Reaching upwards? Excessive movement of loads especially: Excessive lifting or lowering distances? Excessive carrying distances? Excessive pulling or pushing of loads? Risk of sudden movement of loads? Frequent or prolonged physical effort? Insufficient rest or recovery periods? A rate of work imposed by a process?
2 The loads	Are they: Heavy? Bulky or unwieldy? Difficult to grasp? Unstable, or with contents likely to shift? Sharp, hot or otherwise potentially damaging?
3 The working environment	Are there: Space constraints preventing good posture? Uneven, slippery or unstable floors? Variations in level of floors or work surfaces? Extremes of temperature or humidity? Conditions causing ventilation problems or gusts of wind? Poor lighting conditions?
4 Individual capability	Does the job: Require unusual strength, height, etc.? Create a hazard to those who might reasonably be considered to be pregnant or have a health problem? Require special information or training for its safe performance?
5 Other factors	Is movement or posture hindered by personal protective equipment or by clothing?

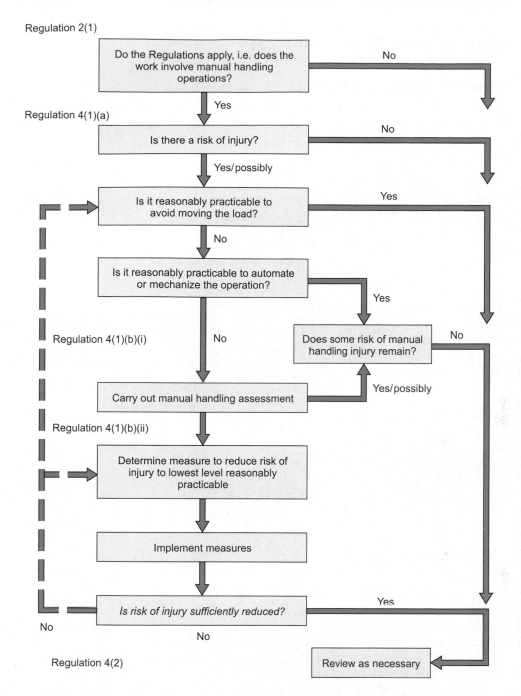

Regulation 2(1)

Regulation 4(1)(a)

Regulation 4(1)(b)(i)

Regulation 4(1)(b)(ii)

Regulation 4(2)

Figure 17.1 Manual Handling Operations Regulations – flow chart. Source HSE. Crown copyright material is reproduced with the permission of the Controller of HMSO and the Queen's Printer for Scotland.

17.22.8 References

Manual Handling, Manual Handling Operations Regulations 1992, Guidance on Regulations Revised, L23, 1998 HSE Books
        ISBN 0-7176-2415-3
Manual Handling, Solutions you can handle, HSG 115, 1994 HSE Books
        ISBN 0-7176-0693-7
Getting to Grips with manual handling. A short guide for employers revised. INDG 143
        ISBN 0-7176-1754-8

17.23 Noise at Work Regulations 1989

17.23.1 Action levels

Requirements under the Regulations to protect people from exposure to harmful noise progressively take effect. The main requirements apply when noise exposure is likely to be at or above any of the three *Action Levels* (AL) as summarized in Table 17.4.

Two of the action levels are expressed as a daily personal noise exposure, shortened to $L_{EP.d}$. These are dependent on the noise level in the workplace and the length of exposure. The first AL is 85 dB(A) and the second AL is 90 dB(A).

Table 17.4 Requirements under Noise at Work Regulations

Action required where $L_{EP.d}$ to be: (see note 1 opposite)	Below 85 dB(A)	85 dB(A) 1st AL	90 dB(A) 2nd AL
Employers' duties General duty to reduce risk Risk of hearing damage to be reduced to the lowest level reasonably practicable (Regulation 6)	✓	✓	(2) ✓
Assessment of noise exposure Noise assessments to be made by a competent person (Regulation 4) Record of assessments to be kept until a new one is made (Regulation 5)		✓ ✓	✓ ✓
Noise reduction Reduce exposure to noise as far as is reasonably practicable by means other than ear protection (Regulation 7)			✓
Provision of information to workers Provide adequate information, instruction and training about risks to hearing, what employees should do to minimize risk, how they can obtain ear protectors if they are exposed to between 85 and 90 dB(A), and their obligations under the Regulations (Regulation 11). Mark ear protection zones with notices, so far as reasonably practicable (Regulation 9)		✓	✓ ✓
Ear protectors Ensure, so far as is reasonably practicable, that protectors are: Provided to employees who ask for them (Regulation 8(1)) Provided to all exposed (Regulation 8(2)) Maintained and repaired (Regulation 10(1)(b)) Used by all exposed (Regulation 10(1)(a)) Ensure, so far as reasonably practicable, that all who go into a marked ear protection zone use ear protectors (Regulation 9(1)(b))		✓ ✓	✓ ✓ ✓ ✓(3)

Action required where $L_{EP.d}$ to be: (see note 1 opposite)	Below 85 dB(A)	85 dB(A) 1st AL	90 dB(A) 2nd AL
Maintenance and use of equipment Ensure, so far as is practicable, that: All equipment provided under the Regulations is used, except for the ear protectors provided between 85 and 90 dB(A) (Regulation 10(1)(a))		✓	✓
Ensure all equipment is maintained (Regulation 10(1)(b))		✓	✓
Employees' duties Use of equipment So far as practicable, Use ear protectors (Regulation 10(2))			✓
Use any other protective equipment (Regulation 10(2))		✓	✓
Report any defects discovered to employer (Regulation 10(2))		✓	✓
Machine makers' and suppliers' duties Provision of information Provide information on noise likely to be generated (Regulation 12)		✓	✓

Notes

(1) The dB(A) action levels are values of daily personal exposure ($L_{EP.d}$)   (Source HSE)

(2) All the action indicated at 90 dB(A) are also required where the peak sound pressure is at or above 200 Pa (140 dB re 20 µPa or for practical purposes 140 dB(C)

(3) This requirement applies to all who enter the zones, even if they do not stay long enough to receive an exposure of 90 dB(A) $L_{EP.d}$

The third AL is a peak pressure to prevent instantaneous hearing damage. It is set at 200 pascals and often expressed as 140 dB re 20 μPa. This peak pressure for most practical purposes is equivalent to a 'C' weighted peak level of 140 dB(C).

The Noise Regulations require employers to take certain basic steps where any employees are likely to be exposed to noise at or above the first AL. Additional action must be taken where any employees are likely to be exposed at or above the second AL.

17.23.2 Self-employed and home workers

Regulation 2(2) requires self-employed people to fulfil the responsibilities of both employer and employee. Employers must also take action to protect people who are employed by them to work at home.

17.23.3 More than one employer involved

Where more than one employer is involved, the Regulations place duties on all the employers and each will have a responsibility.

➤ to their own employees
➤ so far as is reasonably practicable, to anyone else at work who is affected by their activities.

17.23.4 Exemptions – Regulation 13

HSE can give exemptions to certain Regulations where the exposure averaged over a week can be guaranteed to be below 90 dB(A) or where ear protection would be likely to cause risk to the user, compliance is not reasonably practicable.

17.23.5 References

Noise at Work Regulations 1989 (SI No 1790).
Reducing Noise at Work. Guidance on The Noise At Work Regulations 1989, L108, 1998 HSE Books
    ISBN 0-7176-1511-1
Introducing the Noise at Work Regulations. A brief guide to the requirements for controlling noise at work. INDG75, HSE Books, 189
    ISBN 0-7176-0961-8
Ear Protection Employers duties explained. INDG298, 1999 HSE Books

17.24 Personal Protective Equipment at Work Regulations 1992

17.24.1 Introduction

The effect of the Personal Protective Equipment (PPE) at Work regulations is to ensure that certain basic duties governing the provision and use of PPE apply to all situations where PPE is required. The Regulations follow sound principles for the effective and economical use of PPE, which all employers should follow.

PPE, as defined, includes all equipment (including clothing affording protection against the weather), which is intended to be worn or held by a person at work and which protects them against one or more risks to their health and safety. Waterproof, weatherproof or insulated clothing is covered only if its use is necessary to protect against adverse climatic conditions.

Ordinary working clothes and uniforms, which do not specifically protect against risks to health and safety, and protective equipment worn in sports competitions are not covered.

Where there is overlap in the duties in these Regulations and those covering lead, ionizing radiations, asbestos, hazardous substances (COSHH), noise, and construction head protection then the specific legislative requirements should prevail.

17.24.2 Provision of PPE – Regulation 4

Every employer shall ensure that suitable PPE is provided to their employees who may be exposed to risks to their health and safety except where it has been adequately or more effectively controlled by other means. (Management Regulations require PPE to be the last choice in the principles of protection.)

PPE shall not be suitable unless:

➤ it is appropriate for the risks and the conditions of use

➤ it takes account of ergonomic requirements and the state of health of the wearer
➤ it is capable of fitting the wearer correctly, by adjustments if necessary
➤ it is, so far as is practicable, able to combat the risks without increasing overall risks
➤ it complies with UK legislation on design or manufacture, i.e. it has a CE marking.

17.24.3 Compatibility – Regulation 5

Where more than one health and safety risk necessitates the wearing of multiple PPE simultaneously then they shall be compatible and remain effective.

17.24.4 Assessment – Regulation 6

Before choosing any PPE, employers must ensure that an assessment is made to determine whether the PPE is suitable.

The assessment shall include:

➤ assessing risks which have not been avoided by other means
➤ a definition of the characteristics that PPE must have to be effective, taking into account any risks created by the PPE itself
➤ a comparison of available PPE with the required characteristics.

The assessment should be reviewed if it is no longer valid or there have been significant changes. In simple cases it will not be necessary to record the assessment but, in more complex cases, written records should be made and kept readily available for future reference.

17.24.5 Maintenance – Regulation 7

Every employer (and self-employed person) shall ensure that any PPE provided is maintained, including replaced and cleaned, in an efficient state, in efficient working order and in good repair.

The guide emphasizes the need to set up an effective system of maintenance for PPE. This should be proportionate to the risks and appropriate to the particular PPE. It could include, where appropriate, cleaning, disinfection, examination, replacement, repair and testing. For example, mechanical fall arrestor equipment or sub-aqua breathing apparatus will require planned preventative maintenance with examination, testing and overhaul. Records should be kept of the maintenance work. Gloves may only require periodic inspection by the user as necessary, depending on their use.

Spare parts must be compatible and be the proper part suitably CE marked where applicable. Manufacturers' maintenance schedules and instructions should be followed unless alternative schemes are agreed with the manufacturer or agent.

In some cases these requirements can be fulfilled by using disposable PPE which can be discarded after use or when their life has expired. Users should know when to discard and replace disposable PPE.

17.24.6 Accommodation – Regulation 8

When an employer or self-employed person has to provide PPE they must ensure that appropriate accommodation is provided to store it when not in use.

The type of accommodation will vary and may just be suitable hooks for special clothing and small portable cases for goggles. It should be separate from normal outer clothing storage arrangements and protect the PPE from contamination or deterioration.

17.24.7 Information, instruction and training – Regulation 9

Employers shall provide employees with adequate and appropriate information, instruction and training on:

➤ the risks which the PPE will avoid or limit
➤ the purpose for which and the manner in which PPE should be used
➤ any action required of the employee to maintain the PPE.

The guidance suggests the training should include:

➤ an explanation of the risks and why PPE is needed
➤ the operation, performance and limitations of the equipment
➤ instructions on the selection, use and storage of PPE
➤ problems that can affect PPE relating to other equipment, working conditions, defective equipment, hygiene factors and poor fit
➤ the recognition of defects and how to report problems with PPE
➤ practice in putting on wearing and removing PPE
➤ practice in user cleaning and maintenance
➤ how to store safely.

17.24.8 Use and reporting of defects – Regulation 10 and 11

Every employer shall take all reasonable steps to ensure that PPE is properly used. Every employee shall:

➤ use PPE provided in accordance with training and instructions;
➤ return it to the accommodation provided after use;
➤ report any loss or obvious defect.

17.24.9 References

Personal Protective Equipment Regulations 2002. SI 2002 No 1144, Stationery Office
Personal protective equipment at work. Personal Protective Equipment at Work Regulations 1992. Guidance on Regulations. L25, 1992 HSE Books
ISBN 0-7176-0415-2
A short guide to the Personal Protective Equipment at Work Regulations 1992, INDG174 1997 HSE Books
ISBN 0-7176-0889-1

17.25 Provision and Use of Work Equipment Regulations 1998 (except part IV)

17.25.1 Introduction

The Provision and Use of Work Equipment Regulations 1998 (PUWER) are made under the HSW Act and their primary aim is to ensure that work equipment is used without risks to health and safety, regardless of its age, condition or origin. The requirements of PUWER that are relevant to woodworking machinery are set out in the *Safe use of woodworking machinery Approved Code of Practice*. PUWER has specific requirements for risk assessment which are covered under the Health and Safety Management Regulations 1999.

Part IV of PUWER is concerned with power presses and is not part of the Certificate syllabus and is therefore not covered in this summary.

17.25.2 Definitions

Work equipment means any machinery, appliance, apparatus, tool or installation for use at work.

Use in relation to work equipment means any activity involving work equipment and includes starting, stopping, programming, setting, transporting, repairing, modifying, maintaining, servicing and cleaning.

17.25.3 Duty holders – Regulation 3

Under PUWER the following groups of people have duties placed on them:

➤ employers
➤ the self-employed
➤ people who have control of work equipment, for example plant hire companies.

In addition to all places of work the Regulations apply to common parts of shared buildings, industrial estates and business parks; to temporary works sites including construction; to home working (but not to domestic work in a private household); to hotels, hostels and sheltered accommodation.

17.25.4 Suitability of work equipment – Regulation 4

Work equipment:

➤ has to be constructed or adapted so that it is suitable for its purpose
➤ it has to be selected with the conditions of use and the users' health and safety in mind
➤ it may only be used for operations for which, and under conditions for which it is suitable.

This covers all types of use and conditions and must be considered for each particular use or condition. For example: scissors may be safer than knives with unprotected blades and should therefore be used for cutting operations where practicable; risks imposed by wet, hot or cold conditions must be considered.

17.25.5 Maintenance – Regulation 5

The regulation sets out the general requirement to keep work equipment maintained in:

➤ an efficient state
➤ efficient working order
➤ good repair.

Compliance involves all three criteria. In addition, where there are maintenance logs for machinery, they must be kept up to date.

In many cases this will require routine and planned preventive maintenance of work equipment. When checks are made priority must be given to:

➤ safety
➤ operating efficiency and performance
➤ the equipment's general condition.

17.25.6 Inspection – Regulation 6

Where the safety of work equipment depends on the installation conditions, it must be inspected:

➤ after installation and before being put into service for the first time
➤ after assembly at a new site or in a new location

to ensure that it has been installed correctly and is safe to operate.

Where work equipment is exposed to conditions causing deterioration which is liable to result in dangerous situations it must be inspected:

➤ at suitable intervals
➤ when exceptional circumstances occur.

Inspections must be determined and carried out by competent persons. An inspection will vary from a simple visual external inspection to a detailed comprehensive inspection which may include some dismantling and/or testing. However, the level of inspection would normally be less than that required for a thorough examination under, for example, LOLER for certain items of lifting equipment.

Records of inspections must be kept with sufficient information to properly identify the equipment, its normal location, dates, faults found, action taken, to whom faults were reported, who carried out the inspection, when repairs were made, date of the next inspection. When equipment leaves an employer's undertaking it must be accompanied by physical evidence that the last inspection has been carried out.

17.25.7 Specific risks – Regulation 7

Where the use of work equipment involves specific hazards, its use must be restricted to those persons given the specific task of using it and repairs etc. must be restricted to designated persons.

Designated persons must be properly trained to fulfil their designated task.

Hazards must be controlled using a hierarchy of control measures, starting with elimination where this is possible, then considering hardware measures such as physical barriers and, lastly, software measures such as a safe system of work.

17.25.8 Information, instruction and training – Regulations 8 and 9

Persons who use work equipment must have adequate:

➤ health and safety information
➤ where appropriate, written instructions about the use of the equipment
➤ training for health and safety in methods which should be adopted when using the equipment, any hazards and precaution which should be taken to reduce risks.

Any persons who supervise the use of work equipment should also receive information, instruction and training. The training of young persons is especially important with the need for special risk assessments under the Management Regulations.

Health and safety training should take place within working hours.

17.25.9 Conformity with Community requirements – Regulation 10

The intention of this regulation is to require that employers ensure that equipment, provided for use after 31 December 1992, complies with the relevant essential requirements in various European Directives made under Article 100A of the Treaty of Rome. The requirements of PUWER 98 regulations 11 to 19 and 22 to 29 only apply if the essential requirements do not apply to a particular piece of equipment.

However, PUWER regulations 11–19 and 22–29 **will apply** if:

➤ they include requirements which were not included in the relevant product legislation;
➤ the relevant product legislation has not been complied with (e.g. the guards fitted on a machine when supplied were not adequate).

Employers using work equipment need to check that any new equipment has been made to the requirements of the relevant Directive, has a CE marking, suitable instructions and a Certificate of Conformity.

The Machinery Directive was brought into UK law by the Supply of Machinery (Safety) Regulations 1992 as amended, which duplicate PUWER regulations 11–19 and 22–29.

The employer still retains the duty to ensure that the equipment is safe to use.

17.25.10 Dangerous parts of machinery – Regulation 11

Measures have to be taken which:

➤ prevent access to any dangerous part of machinery or to any rotating stock-bar;
➤ stop the movement of any dangerous part of machinery or rotating stock-bar before any part of a person enters a danger zone.

The measures required follow the normal hierarchy and consist of:

➤ the provision of fixed guards enclosing every dangerous part of machinery

➤ the provision of other guards or protection devices
➤ the provision of jigs, holders, push-sticks or similar protection appliances used in conjunction with the machinery
➤ provision of information, instruction, training and supervision.

All guards and protection devices shall:

➤ be suitable for its purpose
➤ be of good construction, sound material and adequate strength
➤ be maintained in an efficient state, in efficient working order and in good repair
➤ not give rise to increased risks to health and safety
➤ not be easily bypassed or disabled
➤ be situated at sufficient distance from the danger zone
➤ not unduly restrict the view of the operating cycle of the machine where this is relevant
➤ be so constructed or adapted that they allow operations necessary to fit or replace parts and for maintenance work, if possible without having to dismantle the guard or protection device.

17.25.11 Protection against specified hazards – Regulation 12

Exposure to health and safety risks from the following hazards must be prevented or adequately controlled:

➤ any article falling or being ejected from work equipment
➤ rupture or disintegration of work equipment
➤ work equipment catching fire or overheating
➤ the unintended or premature discharge of any article, or of any gas, dust, liquid, vapour or other substance which is produced, used or stored in the work equipment
➤ the unintended or premature explosion of the work equipment or any article or substance produced, used or stored in it.

17.25.12 High or very low temperature – Regulation 13

Work equipment and any article or substance produced, used or stored in work equipment which is at a high or very low temperature must have protection to prevent injury by burn, scald or sear.

This does not cover risks such as from radiant heat or glare.

Engineering methods of control such as insulation, doors, temperature control, guards, etc. should be used where practicable, but there are some cases, like cooker hot plates, where this is not possible.

17.25.13 Controls – Regulations 14 to 18

Where work equipment is provided with (Regulation 14):

➤ starting controls (including restarting after a stoppage)
➤ controls which change speed, pressure or other operating condition which would affect health and safety

it should not be possible to perform any operation except by a deliberate action on the control. This does not apply to the normal operating cycle of an automatic device.

Where appropriate, one or more readily accessible **Stop controls** shall be provided to bring the work equipment to a safe condition in a safe manner (Regulation 15). They must:

➤ bring the work equipment to a complete stop where necessary
➤ if necessary switch off all sources of energy after stopping the functioning of the equipment
➤ operate in priority to starting or operating controls.

Where appropriate one or more readily accessible **emergency stop controls** (Regulation 16) must be provided unless it is not necessary:

➤ by the nature of the hazard
➤ by the time taken for the stop controls to bring the equipment to a complete stop.

Emergency stop controls must have priority over stop controls. They should be provided where other safeguards are not adequate to prevent risk when something irregular happens. They should not be used as a substitute for safeguards or the normal method of stopping the equipment.

All **controls** for work equipment shall (Regulation 17):

➤ be clearly visible and identifiable including appropriate marking where necessary
➤ not expose any person to a risk to their health and safety except where necessary.

Where appropriate employers shall ensure that:

➤ controls are located in a safe place
➤ systems of work are effective in preventing any person being in a danger zone when equipment is started
➤ an audible, visible or other suitable warning is given whenever work equipment is about to start.

Persons in a danger zone as a result of starting or stopping equipment must have sufficient time and suitable means to avoid any risks.

Control systems (Regulation 18) must be safe and chosen so as to allow for failures, faults and constraints. They must:

➤ not create any increased risk to health and safety
➤ not result in additional or increased risks when failure occurs
➤ not impede the operation of any stop or emergency stop controls.

17.25.14 Isolation from sources of energy – Regulation 19

Work equipment must be provided with readily accessible and clearly identified means to isolate it from all sources of energy.

Re-connection must not expose any person using the equipment to any risks.

The main purpose is to allow equipment to be made safe under particular circumstances, such as maintenance, when unsafe conditions occur, or when adverse conditions such as electrical equipment in a flammable atmosphere or wet conditions occur.

If isolation may cause a risk in itself, special precautions must be taken, for example, a support for a hydraulic press tool which could fall under gravity if the system is isolated.

17.25.15 Stability – Regulation 20

Work equipment must be stabilized by clamping or otherwise as necessary to ensure health and safety.

Most machines used in a fixed position should be bolted or fastened so that they do not move or rock in use.

17.25.16 Lighting – Regulation 21

Suitable and sufficient lighting, taking account of the operations being carried out, must be provided where people use work equipment.

This will involve general lighting and in many cases local lighting, such as on a sewing machine. If access for maintenance is required regularly, permanent lighting should be provided.

17.25.17 Maintenance operations – Regulation 22

So far as is reasonably practicable work equipment should be constructed or adapted to allow maintenance operations to be:

➤ conducted while they are shut down;
➤ undertaken without exposing people to risk;
➤ carried out after appropriate protection measures have been taken.

17.25.18 Markings and warnings – Regulations 23 and 24

Work equipment should have all appropriate **markings** for reasons of health and safety made in a clearly visible manner. For example, the maximum safe working load, stop and start controls, or the maximum rotation speed of an abrasive wheel.

Work equipment must incorporate **warnings or warning devices** as appropriate, which are unambiguous, easily perceived and easily understood.

They may be incorporated in systems of work, a notice, a flashing light or an audible warning. They are an active instruction or warning to take specific precautions or actions when a hazard exists.

17.25.19 Part III – Mobile work equipment – Regulations 25–30

The main purpose of this section is to require additional precautions relating to work equipment while it is travelling from one location to another or where it does work while moving. If the equipment is designed primarily for travel on public roads the Road

Vehicles (Construction and Use) Regulations 1986 will normally be sufficient to comply with PUWER 98.

Mobile equipment would normally move on wheels, tracks, rollers, skids, etc. Mobile equipment may be self-propelled, towed or remote controlled and may incorporate attachments. Pedestrian controlled work equipment such as lawn mowers is not covered by Part III.

Employees carried on mobile work equipment – Regulation 25
No employee may be carried on mobile work equipment unless:

➤ it is suitable for carrying persons
➤ it incorporates features to reduce risks as low as is reasonably practicable, including risks from wheels and tracks.

Rolling over of mobile work equipment – Regulations 26
Where there is a risk of overturning it must be minimized by:

➤ stabilizing the equipment;
➤ fitting a structure so that it only falls on its side, TOPS (tip over protection structure);
➤ fitting a structure which gives sufficient clearance for anyone being carried if it turns over further, ROPS (roll over protection structure);
➤ a device giving comparable protection;
➤ fitting a suitable restraining system for people if there is a risk of being crushed by rolling over.

This Regulation does not apply:

➤ to a fork truck fitted with TOPS or ROPS
➤ where it would increase the overall risks
➤ where it would not be reasonably practicable to operate equipment
➤ to any equipment provided for use before 5 December 1998.

Overturning of fork lift trucks – Regulation 27
Fork lift trucks, which carry an employee, must be adapted or equipped to reduce the risk to safety from overturning to as low as is reasonably practicable.

Self-propelled work equipment – Regulation 28
Where self-propelled work equipment may involve risks while in motion they shall have:

➤ facilities to prevent unauthorized starting
➤ (with multiple rail-mounted equipment) facilities to minimize the consequences of collision
➤ a device for braking and stopping
➤ (where safety constraints so require) emergency facilities for braking and stopping, in the event of failure of the main facility, which have readily accessible or automatic controls
➤ (where the driver's vision is inadequate) devices fitted to improve vision
➤ (if used at night or in dark places) appropriate lighting fitted or otherwise it shall be made sufficiently safe for its use
➤ if there is anything carried or towed that constitutes a fire hazard liable to endanger employees (particularly if escape is difficult such as from a tower crane) appropriate fire-fighting equipment carried, unless it is sufficiently close by.

Remote-controlled self-propelled work equipment – Regulation 29

Where remote-controlled self-propelled work equipment involves a risk while in motion it shall:

> stop automatically once it leaves its control range
> have features or devices to guard against the risk of crushing or impact.

Drive shafts – Regulation 30

Where seizure of the drive shaft between mobile work equipment and its accessories or anything towed is likely to involve a risk to safety:

> the equipment must have means to prevent a seizure
> where it cannot be avoided, every possible measure should be taken to avoid risks
> the shaft should be safeguarded from contacting the ground and becoming soiled or damaged.

17.25.20 Part IV: Power presses

Regulations 31–35 relate to power presses and are not included here as they are excluded from the NEBOSH Certificate syllabus. Details can be found in the Power Press ACOP.

17.25.21 References

Safe use of work equipment, Provision and Use of Work Equipment Regulations 1998, Approved Code of Practice and Guidance, HSC L22 1998 HSE Books
ISBN 0-7176-1626-6
PUWER 1998, Provision and Use of Work Equipment Regulations 1998: Open learning guidance, 1999 HSE Books
ISBN 0-7176-2459-5
Simple guide to the Provision and Use of Work Equipment Regulations 1998, INDG291, 1999 HSE Books
ISBN 0-7176-2429-3
Hiring and leasing out of plant: application of PUWER 98, Regulations 26 and 27, HSE MISC156, 1998 HSE Books

17.26 The Reporting of Injuries, Diseases and Dangerous Occurrences Regulations 1995

17.26.1 Introduction

These Regulations require the reporting of specified accidents, ill-health and dangerous occurrences to the enforcing authorities. The events all arise out of or in connection with work activities covered by the HSW Act. They include death, major injury and more than three-day lost-time accidents. Schedules to the Regulations specify the details of cases of ill-health and dangerous occurrences.

For most businesses reportable events will be quite rare and so there is little for them to do under these Regulations apart from keeping the guidance and forms available and being aware of the general requirements.

17.26.2 Definitions – Regulation 2

Accident includes an act of non-consensual physical violence done to a person at work. This means that injuries, through physical violence to people not at work are not reportable. Neither is any injury, which occurs between workers over a personal matter or carried out by a visiting relative of a person at work to that person. However, if a member of the public caused injury to a person at work through physical violence, that is reportable.

Incidents involving acts of violence may well be reportable to the police which is outside the requirements of these Regulations.

17.26.3 Notification and reporting of major injuries and dangerous occurrences – Regulation 3(1), 4 and Schedules 1 and 2

The HSE or Local Authority shall be notified immediately by the quickest practicable means and sent a report form F2508 (or other approved means) within 10 days following:

(a) Death of a person as a result of an accident arising out of or in connection with work. Also if death occurs within one year of an accident, the Authorities must be informed

(b) Major Injury to a person as a result of an accident arising out of or in connection with work

(c) where a person not at work suffers an injury as a result of or in connection with work, and that person is taken from the site to hospital for treatment

(d) where a person not at work suffers a major injury as a result of an accident arising out of or in connection with work at a hospital

e) where there is a dangerous occurrence.

17.26.4 Reporting of three-day plus accidents – Regulation 3(2)

Where a person at work is incapacitated for work of a kind which they might reasonably be expected to do, either under their own contract of employment or in the normal course of employment, for more than three consecutive days (excluding the day of the accident, but including any days which would not have been working days) because of an injury resulting from an accident at work, the responsible person shall within 10 days send a report on form F2508 or other approved form, unless it has been reported under Regulation 3(1) as a Major Injury etc.

17.26.5 Reporting of cases of disease – Regulation 5

Where a medical practitioner notifies the employer's responsible person that an employee suffers from a reportable work-related disease, a completed disease report form (F2508A) should be sent to the enforcing authority. The full list is contained in schedule 3 to the regulations which is summarized in this guide.

17.26.6 Which enforcing authority?

Local Authorities are responsible for retailing, some warehouses, most offices, hotels and catering, sports, leisure, consumer services, and places of worship.

The HSE are responsible for all other places of work.

17.26.7 Records

A record of each incident reported must be kept at the place of business for at least 3 years.

Major injuries schedule 1

1 Any fracture, other than to the fingers, thumbs or toes
2 Any amputation
3 Dislocation of the shoulder, hip, knee or spine
4 Loss of sight (whether temporary or permanent)
5 A chemical or hot metal burn to the eye or any penetrating injury to the eye
6 Any injury resulting from an electric shock or electrical burn (including any electrical burn caused by arcing or arcing products) leading to unconsciousness or requiring resuscitation or admittance to hospital for more than 24 hours
7 Any other injury
  (a) leading to hypothermia, heat-induced illness or to unconsciousness
  (b) requiring resuscitation
  (c) requiring admittance to hospital for more than 24 hours
8 Loss of consciousness caused by asphyxia or by exposure to a harmful substance or biological agent
9 Either of the following conditions which result from the absorption of any substance by inhalation, ingestion or through the skin
  (a) acute illness requiring medical treatment
  (b) loss of consciousness
10 Acute illness which requires medical treatment where there is reason to believe that this resulted from exposure to a biological agent or its toxins or infected material.

17.26.9 Dangerous occurrences – Schedule 2 summary

Part I General

1 Lifting machinery, etc.
The collapse, overturning, or the failure of any load-bearing part of lifts and lifting equipment.
2 Pressure systems
The failure of any closed vessel or of any associated pipe-work, in which the internal pressure was above or below atmospheric pressure.

3 Freight containers

The failure of any freight container in any of its load-bearing parts.

4 Overhead electric lines

Any unintentional incident in which plant or equipment comes into contact with overhead power lines.

5 Electrical short circuit

Electrical short circuit or overload attended by fire or explosion.

6 Explosives

The unintentional explosion or ignition of explosives, misfire, the failure of the shots in any demolition operation to cause the intended extent of collapse, the projection of material beyond the boundary of the site, any injury to a person resulting from the explosion or discharge of any explosives or detonator.

7 Biological agents

Any accident or incident which resulted or could have resulted in the release or escape of a biological agent likely to cause severe human infection or illness.

8 Malfunction of radiation generators etc.

Any incident in which the malfunction of a radiation generator or its ancillary equipment used in fixed or mobile industrial radiography, the irradiation of food or the processing of products by irradiation, causes it to fail to de-energize at the end of the intended exposure period; or to fail to return to its safe position at the end of the intended exposure period.

9 Breathing apparatus

Any incident in which breathing apparatus malfunctions while in use, or during testing immediately prior to use.

10 Diving operations

In relation to a diving project the failure or the endangering of diving equipment, the trapping of a diver, any explosion in the vicinity of a diver, any uncontrolled ascent or any omitted decompression.

11 Collapse of scaffolding

The complete or partial collapse of any scaffold which is more than 5 m in height; erected over or adjacent to water, in circumstances such that there would be a risk of drowning to a person falling from the scaffold into the water; or the suspension arrangements of any slung or suspended scaffold which causes a working platform or cradle to fall.

12 Train collisions

Any unintended collision of a train with any other train or vehicle.

13 Wells

Any dangerous occurrence at a well other than a well sunk for the purpose of the abstraction of water.

14 Pipelines or pipeline works

A dangerous occurrence in respect of a pipeline or pipeline works.

15 Fairground equipment

The failure of any load-bearing part or the derailment or the unintended collision of cars or trains.

16 Carriage of dangerous substances by road

Any incident involving a road tanker or tank container used for the carriage of dangerous goods in which the tanker overturns, is seriously damaged, there is an uncontrolled release or there is a fire.

Dangerous occurrences which are reportable except in relation to offshore work-places

17 Collapse of building or structure

Any unintended collapse or partial collapse of any building or structure: under construction; reconstruction; alteration or demolition which involves a fall of more than 5 tonnes of material; any floor or wall of any building; or any false-work.

18 Explosion or fire

An explosion or fire occurring in any plant or premises which results in the suspension of normal work for more than 24 hours.

19 Escape of flammable substances

The sudden, uncontrolled release inside a building:

    (i) of 100 kg or more of a flammable liquid
    (ii) of 10 kg or more of a flammable liquid above its boiling point
    (iii) of 10 kg or more of a flammable gas

500 kg or more of any of the substances, if released in the open air.

20 Escape of substances

The accidental release or escape of any substance in a quantity sufficient to cause the death, major injury or any other damage to the health of any person.

17.26.10 Reportable diseases – schedule 3 brief summary

These include:

➤ certain poisonings
➤ some skin diseases such as occupational dermatitis, skin cancer, chrome ulcer, oil folliculitis/acne
➤ lung diseases, such as occupational asthma, farmer's lung, pneumoconiosis, asbestosis, mesothelioma
➤ infections such as leptospirosis, heptatitis, tuberculosis, anthrax, legionellosis and tetanus
➤ other conditions, such as occupational cancer, certain musculoskeletal disorders, decompression illness and hand-arm vibration syndrome.

The details can be checked by consulting the guide, looking at the pad of report forms, checking the HSE's web site, ringing the HSE's Infoline or the Incident Control Centre (ICC).

17.26.11 References

A guide to the Reporting of Injuries, Diseases and Dangerous Occurrences Regulations 1995, L73, 1999 HSE Books
    ISBN 0-7176-2431-5
Everyone's guide to RIDDOR, HSE 31(rev1) HSE Books
    ISBN 0-7176-2441-2
RIDDOR Information for doctors HSE 32. 1996 HSE Books
RIDDOR reporting: Information about the new incident centre, MISC310, HSE

17.27 Safety Representatives and Safety Committees Regulations 1977

These Regulations, made under the HSW Act Section 2(4), prescribe the cases in which recognized trade unions may appoint safety representatives, specify the functions of such representatives, and set out the obligations of employers towards them.

Employers' obligations to consult non-union employees are contained in the Health and Safety (Consultation with Employees) Regulations 1996.

17.27.1 Appointment – Regulation 3

A recognized trade union may appoint safety representatives from among employees in all cases where one or more employees are employed. When the employer is notified in writing the safety representatives have the functions set out in Regulation 4.

A person ceases to be a safety representative when:

➤ the appointment is terminated by the trade union
➤ they resign
➤ employment ceases.

A safety representative should have been with the employer for two years or have worked in similar employment for at least two years.

17.27.2 Functions – Regulation 4

These are functions and not duties. They include:

➤ representing employees in consultation with the employer
➤ investigating potential hazards and dangerous occurrences
➤ investigating the causes of accidents
➤ investigating employee complaints relating to health, safety and welfare
➤ making representations to the employer on health safety and welfare matters
➤ carrying out inspections
➤ representing employees at the workplace in consultation with enforcing inspectors
➤ receiving information
➤ attending safety committee meetings.

Safety representatives must be afforded time off with pay to fulfil these functions and to undergo training.

17.27.3 Employers' duties – Regulation 4a

Every employer shall consult safety representatives in good time regarding:

➤ the introduction of any measure which may affect health and safety
➤ the arrangements for appointing or nominating competent person(s) under the Management Regulations
➤ any health and safety information required for employees
➤ the planning and organizing of any health and safety training for employees

> the health and safety consequences of introducing new technology.

Employers must provide such facilities and assistance as safety representatives may reasonably require to carry out their functions.

17.27.4 Inspections – Regulations 5 and 6

Following reasonable notice in writing, safety representatives may inspect the workplace every quarter (or more frequently by agreement with the employer). They may inspect the workplace at any time, after consultation, when there have been substantial changes in the conditions of work or there is new information on workplace hazards published by the HSC or HSE.

Following an injury, disease or dangerous occurrence subject to Reporting of Injuries Diseases and Dangerous Occurrences Regulations, and after notifying the employer, where it is reasonably practicable to do so, safety representatives may inspect the workplace if it is safe.

Employers must provide reasonable assistance and facilities, including provision for independent investigation and private discussion with employees. The employer may be present in the workplace during inspections.

17.27.5 Information – Regulation 7

Having given reasonable notice to the employer, safety representatives are entitled to inspect and take copies of any relevant statutory documents (except any health record of an identified person).

An exempt document is one:

> which could endanger national security
> which could cause substantial commercial injury on the employer
> contravenes a prohibition
> that relates to an individual without their consent
> which has been obtained specifically for legal proceedings.

17.27.6 Safety committees – Regulation 9

When at least two safety representatives have requested in writing that a safety committee is set up, the employer has three months to comply. The employer must consult with the safety representatives and post a notice stating its composition and the workplaces covered by it, in a place where it can be easily read by employees. The guidance gives details on the composition and running of safety committees.

17.27.7 Complaints – Regulation 11

If the employer fails to permit safety representatives time off or fails to pay for time off, complaints can be made to an industrial tribunal within 3 months of the incident.

17.27.8 References

HSC, Safety Representatives and safety committees. (The Brown Book) Approved Code of Practice and Guidance on the Regulations, third edition, L87, 1996 HSE Books

ISBN 0-7176-1220-1

17.28 Health and Safety (Safety Signs and Signals) Regulations 1996

17.28.1 Introduction

These regulations came into force in April 1996, but if existing signs comply with BS 5378 and for fire safety signs BS 5499 no changes are required. The results of the risk assessment made under the Management of Health Safety At Work Regulations will have identified situations where there may be a residual risk when warnings or further information are necessary. If there is no significant risk there is no need to provide a sign.

17.28.2 Definitions – Regulation 2

➤ 'Safety sign' means a sign referring to a specific object, activity or situation and providing information or instruction about health and safety at work by means of a signboard, a safety colour, an illuminated sign, an acoustic sign, a verbal communication or a hand signal.

➤ 'Signboard' means a sign which provides information or instructions by a combination of geometric shape, colour and a symbol or pictogram and which is rendered visible by lighting of sufficient intensity.

➤ 'Hand signal' means a movement or position of the arms or hands or a combination, in coded form, for guiding persons who are carrying out manoeuvres which create a risk to the health and safety of people at work.

➤ 'Acoustic signal' means a coded sound signal which is released and transmitted by a device designed for that purpose, without the use of a human or artificial voice.

➤ 'Verbal communication' means a predetermined spoken message communicated by a human or artificial voice.

17.28.3 Provision and maintenance of safety signs – Regulation 4

The Regulations require employers to use and maintain a safety sign where there is a significant risk to health and safety that has not been avoided or controlled by other means, like engineering controls or safe systems of work, and where the use of a sign can help reduce the risk.

They apply to all workplaces and to all activities where people are employed, but exclude signs used in connection with transport or the supply and marketing of dangerous substances, products and equipment.

The Regulations require, where necessary, the use of road traffic signs in work-places to regulate road traffic.

17.28.4 Information, instruction and training – Regulation 5

Every employer shall ensure that:

➤ comprehensible and relevant information on the measures to be taken in connection with safety signs is provided to each employee
➤ each employee receives suitable and sufficient instruction and training in the meaning of safety signs.

17.28.5 Functions of Colours, Shapes and Symbols in Safety Signs

Safety colours

(a) Red
Red is a safety colour and must be used for any:

➤ prohibition sign concerning dangerous behaviour (e.g. the safety colour on a 'No Smoking' sign). Prohibition signs must be round, with a black pictogram on a white background with red edging and a red diagonal line (top left, bottom right). The red part must take up at least 35% of the area of the sign.
➤ danger alarm concerning stop, shut-down, emergency cut out devices, evacuate (e.g. the safety colour of an emergency stop button on equipment)
➤ fire-fighting equipment.

No forklift trucks

No smoking

Red and white alternating stripes may be used for marking surface areas to show obstacles or dangerous locations.

Yellow
Yellow (or amber) is a safety colour and must be used for any warning sign concerning the need to be careful, take precautions, examine or the like (e.g. the safety colour on hazard signs, such as for flammable material, electrical danger, etc.). Warning signs must be triangular, with a black pictogram on a yellow (or amber) background with black edging. The yellow (or amber) part must take up at least 50% of the area of the sign.

General danger

Explosive

Yellow and black alternating stripes may be used for marking surface areas to show obstacles or dangerous locations.

Yellow may be used in continuous lines showing traffic routes.

Blue

Blue is a safety colour and must be used for any mandatory sign requiring specific behaviour or action (e.g. the safety colour on a 'Safety Helmet Must Be Worn' sign or a 'Pedestrians Must Use This Route' sign). Mandatory signs must be round, with a white pictogram on a blue background. The blue part must take up at least 50% of the area of the sign.

Ear protection
must be worn

Eye protection
must be worn

Green

Green is a safety colour and must be used for: emergency escape signs (e.g. showing emergency doors, exits and routes) and first aid signs (e.g. showing location of first aid equipment and facilities). Escape and first-aid signs must be rectangular or square, with a white pictogram on a green background. The green part must take up at least 50% of the area of the sign. So long as the green takes up at least 50% of the area, it is sometimes permitted to use a green pictogram on a white background, for example where there is a green wall and the reversal provides a more effective sign than one with a green background and white border; no danger (e.g. for 'return to normal').

Means of Escape

First Aid

Other colours

White

White is *not* a safety colour but is used: for pictograms or other symbols on blue and green signs; in alternating red and white stripes to show obstacles or dangerous locations; in continuous lines showing traffic routes.

Black

Black is *not* a safety colour but is used: for pictograms or other symbols on yellow (or amber) signs and, except for fire signs, red signs; in alternating yellow and black stripes to show obstacles or dangerous locations.

Shapes

➤ Round signs must be used for any prohibition (red) sign; mandatory (blue) sign.
➤ Triangular signs must be used for any warning (yellow or amber) sign.
➤ Square or rectangular signs must be used for any emergency escape sign and any first aid sign.

Pictograms and other symbols

The meaning of a sign (other than verbal communication) must not rely on words. However, a sign may be supplemented with words to reinforce the message provided the words do not in fact distract from the message or create a danger.

A sign (other than verbal communication, acoustic signals or hand signals) should use a simple pictogram and/or other symbol (such as directional arrows, exclamation mark, etc.) to effectively communicate its message and so overcome language barriers.

Pictograms and symbols are included in the regulations. Employee training is needed to understand the meaning of these since many are not inherently clear, some are meaningless to anyone who has not had their meaning explained and some can even be interpreted with their opposite meaning.

Pictograms and symbols included in the regulations do not cover all situations for which graphic representation of a hazard or other detail may be needed. Any sign used for a situation not covered in the Regulations, should include:

➤ either the international symbol for general danger (exclamation mark!). If the sign is a warning sign and tests show that the sign is effective
➤ in any other case a pictogram or symbol which has been tested and shown to be effective.

The text of any words used to supplement a sign must convey the same meaning. For example, a round blue sign with a pictogram showing the white outline of a face with a solid white helmet on the head means 'Safety Helmet Must Be Worn' and so any text used must maintain the obligatory nature of the message.

17.28.6 References

The Health and Safety (Safety Signs and Signals) Regulations 1996 SI 1996/341 Stationary Office
Safety Signs and Signals Guidance on the Regulations L64, HSE Books
    ISBN 0-7176-0870-0
Signpost to the Health and Safety (Safety Signs and Signals) Regulations 1996 INDG 184L HSE Books

17.29 Supply of Machinery (Safety) Regulations 1992

17.29.1 Introduction

The Supply of Machinery (Safety) Regulations 1992 entered into force on 1 January 1993, although there was a transitional period to 31 December 1994 during which the manufacturer or importer into the European Community was able to choose between

either the Community regime described below or complying with the health and safety legislation in force on 31 December 1992.

The Supply of Machinery (Safety) (Amendment) Regulations 1994 made a number of changes to the 1992 Regulations, in particular, to widen the scope to include machinery for lifting persons and safety components for machinery. The main provisions of the amending Regulations entered into force on 1 January 1995.

Therefore from **1 January 1995:**

➤ most machinery supplied in the UK, including imports, must:
  ➤ satisfy wide-ranging health and safety requirements, for example, on construction, moving parts and stability
  ➤ in some cases, have been subjected to type-examination by an approved body
  ➤ carry CE marking and other information
➤ the manufacturer or the importer will generally have to be able to assemble a file containing technical information relating to the machine.

17.29.2 Other relevant legislation to the supply of machinery

The two sets of Regulations that will often apply are the:

➤ Electrical Equipment (Safety) Regulations 1994, which apply to most electrically powered machinery used in workplaces
➤ Electromagnetic Compatibility Regulations 1992 which cover equipment likely to cause electromagnetic disturbance, or whose performance is likely to be affected by electromagnetic disturbance.

In some cases, other laws may apply such as the Simple Pressure Vessels (Safety) Regulations 1991 or the Gas Appliances (Safety) Regulations 1995. All these Regulations implement European Directives and contain various requirements. The existence of CE marking on machinery should indicate that the manufacturer has met **all** of the requirements that are relevant.

Special transitional arrangements remain for products covered by existing Directives on roll-over and falling-object protective structures and industrial trucks and for safety components and machinery for lifting persons.

Failure to comply with these requirements:

➤ will mean that the machinery cannot legally be supplied in the UK;
➤ could result in prosecution and penalties, on conviction, of a fine up to £5000 or, in some cases, of imprisonment for up to three months, or of both.

The same rules apply everywhere in the European Economic Area (EEA), so machinery complying with the Community regime may be supplied in any EEA State.

17.29.3 Coverage

Machinery, described as:

➤ an assembly of linked parts or components, at least one of which moves, with the appropriate actuators, control and power circuits, joined together for a specific application, in particular for the processing, treatment, moving or packaging of a material

> an assembly of machines which, in order to achieve the same end, are arranged and controlled so that they function as an integral whole
> interchangeable equipment modifying the function of a machine which is supplied for the purpose of being assembled with a machine (or a series of different machines or with a tractor) by the operator himself in so far as this equipment is not a spare part or a tool.

Safety components for machinery, described as:

> components which are supplied separately to fulfil a safety function when in use and the failure or malfunctioning of which endangers the safety or health of exposed persons.

17.29.4 Exceptions

The Regulations do not apply to machinery or safety components:

> listed in Annex A
> previously used in the European Community or, since 1 January 1994 the European Economic Area (e.g. second-hand)
> for use outside the European Economic Area which does not carry CE marking
> where the hazards are mainly of electrical origin (such machinery is covered by the Electrical Equipment (Safety) Regulations 19941)
> to the extent that the hazards are wholly or partly covered by other Directives, from the date those other Directives are implemented into UK law
> machinery first supplied in the European Community before 1 January 1993
> safety components or machinery for lifting persons first supplied in the European Economic Area before 1 January 1995.

The Regulations do not apply to safety components and machinery for lifting persons first supplied in the European Economic Area before 1 January 1995. Such products first supplied on or after 1 January 1995 must comply **either** with the Supply of Machinery (Safety) Regulations or the UK health and safety legislation in force relating to these items on 14 June 1993. All such products first supplied after 1 January 1997 must comply with the Supply of Machinery (Safety) Regulations.

17.29.5 Machinery excluded – Annex A to the Regulations

> Machinery whose only power source is directly applied manual effort, unless it is a machine used for lifting or lowering loads.
> Machinery for medical use used in direct contact with patients.
> Special equipment for use in fairgrounds and/or amusement parks.
> Steam boilers, tanks and pressure vessels.
> Machinery specially designed or put into service for nuclear purposes, which, in the event of failure, may result in an emission of radioactivity.
> Radioactive sources forming part of a machine.
> Firearms.
> Storage tanks and pipelines for petrol, diesel fuel, highly flammable liquids and dangerous substances.

- Means of transport, i.e. vehicles and their trailers intended solely for transporting passengers by air or on road, rail or water networks. Also, transport, which is designed for transporting goods by air, on public road or rail networks or on water. Vehicles used in the mineral extraction industry are not excluded.
- Sea-going vessels and mobile offshore units together with equipment on board, such as vessels or units.
- Cableways, including funicular railways, for the public or private transportation of people.
- Agriculture and forestry tractors, as defined by certain European Directives.
- Machines specially designed and constructed for military or police purposes.
- Certain goods and passenger lifts.
- Means of transport of people using rack and pinion rail mounted vehicles.
- Mine winding gear.
- Theatre elevators.
- Construction site hoists intended for lifting individuals or people and goods.

17.29.6 General requirements

Subject to the exceptions and transitional arrangements described above, the Regulations make it an offence for a 'responsible person' to supply machinery or a safety component unless:

- it satisfies the essential health and safety requirements
- the appropriate conformity assessment procedure has been carried out
- an EC declaration of conformity or declaration of incorporation has been issued
- CE marking has been properly affixed (unless a declaration of incorporation has been issued)
- it is, in fact, safe.

The Regulations also make it an offence for any supplier to supply machinery or a safety component **unless it is safe.**

17.29.7 Essential health and safety requirements

To comply with the Regulations, machinery and safety components must satisfy the essential health and safety requirements (set out in Annex B to the Regulations), which apply to it. The requirements are wide-ranging, and take into account potential dangers to operators and other exposed persons within a 'danger zone'. Aspects covered in Part 1 include: the materials used in the construction of the machinery; lighting; controls; stability; fire; noise; vibration; radiation; emission of dust, gases, etc.; and maintenance. Part 2 has additional requirements for agri-foodstuffs machinery, portable hand-held machinery, and machinery for working wood and analogous materials. Part 3 deals with particular hazards associated with mobility, Part 4 with those associated with lifting, Part 5 those with underground working and Part 6 those with the lifting or moving of persons. The requirements also comment on instructions (including translation requirements) and marking.

When applying the essential health and safety requirements, technical and economic limitations at the time of construction may be taken into account.

17.29.8 Standards

Machinery and safety components manufactured in conformity with specified, published European standards which have also been published as identically worded national standards ('transposed harmonized standards'), will be presumed to comply with the essential health and safety requirements covered by those standards.

The European Committee for Standardization (CEN) is working to produce a complex of European standards at three levels in support of the Machinery Directive. The first (A) level comprises general principles for the design of machinery. The second (B) level covers specific safety devices and ergonomic aspects. The third (C) level deals with specific classes of machinery by calling up the appropriate standards from the first two levels and addressing requirements specific to the class.

17.29.9 Step 1 – Conformity assessment ('attestation')

The responsibility for demonstrating that the machinery or safety component satisfies the essential health and safety requirements rests on the 'responsible person'.

For most machinery or safety components (other than those listed in Annex D): the 'responsible person', must be able to assemble the technical file described in Annex E.

For machinery or safety components listed in Annex D: the 'responsible person' must follow the special procedures described below.

17.29.10 Step 2 – Declaration procedure

Declaration of conformity: the 'responsible person' must then draw up an EC declaration of conformity, described in Annex G, for each machine or safety component supplied. This declaration is intended to be issued with the machine or safety component and declares that it complies with the relevant essential health and safety requirements or with the example that underwent type-examination.

Declaration of incorporation: alternatively, where the machinery is intended for incorporation into other machinery or assembly with other machinery to constitute machinery covered by the Regulations, the 'responsible person' may draw up a declaration of incorporation, described in Annex H, for each machine.

17.29.11 Step 3 – Marking

Once a declaration of conformity has been issued, the 'responsible person' must affix the CE marking to the machinery.

CE markings must be affixed in a distinct, visible, legible and indelible manner.

The CE marking should not be affixed to safety components or for machinery for which a declaration of incorporation has been issued.

The Regulations make it an offence to affix a mark to machinery, which may be confused with CE marking.

17.29.12 Enforcement

In Great Britain, the Health and Safety Executive is responsible for enforcing the Regulations in relation to machinery and safety components for use at work. Local authority Trading Standards Officers are responsible for enforcement in relation to machinery and safety components for private use.

In Northern Ireland, the Department of Economic Development and the Department of Agriculture are responsible for enforcing the Regulations in relation to machinery and safety components for use at work. District councils are responsible for enforcement in relation to machinery and safety components for private use.

The enforcement authorities have available to them various powers under the Health and Safety at Work etc. Act 1974, the Health and Safety at Work (Northern Ireland) Order 1978 and the Consumer Protection Act 1987, for example, relating to suspension, prohibition and prosecution.

Where machinery bearing the CE marking is safe, but there are breaches of other obligations, the 'responsible person' will be given the opportunity to correct the breach before further enforcement action is taken.

The Machinery Directive, as amended, requires Member States to inform the European Commission of any specific enforcement action taken. The Commission will consider whether the action is justified and advise the parties concerned accordingly.

17.29.13 Penalties

The maximum penalty for contravening the prohibition on supply of non-compliant machinery and safety components is imprisonment for up to three months or a fine of up to £5000 or both. The penalty for other contraventions of the Regulations is a fine up to the same amounts. It is for the Courts to decide the penalty in any given case, taking into account the severity of the offence.

The Regulations provide a defence of due diligence. They also provide for proceedings to be taken against a person other than the principal offender, if it is the other person's fault, and against officers of a company or other body corporate.

17.29.14 References

Supplying new machinery advice to suppliers, INDG270, 1998 HSE Books
        ISBN 0-7176-1560-X
Buying new machinery A short guide to the law, INDG271, 1998 HSE Books
        ISBN 0-7176-1559-6

17.30 Workplace (Health, Safety and Welfare) Regulations 1992

17.30.1 General

These regulations were made to implement the European Directive on the minimum safety and health requirements for the workplace. A workplace for these purposes is defined very widely to include any part of non-domestic premises to which people have access while at work and any room, lobby, corridor, staircase or other means of access to or exit from them. The main exceptions to these rules are construction sites, means of transport, mines and quarries and other mineral extraction sites.

The main requirements are summarized below. They are expressed in very general terms, and in each case it will be necessary to turn to the Approved Code or Practice associated with these Regulations for clarification of what is necessary to meet the objectives set.

17.30.2 Health – The Working Environment Regulations 6, 7, 8, 9, 10

Ventilation
Ventilation must be effective in enclosed areas, and any plant used for this purpose must incorporate warning devices to signal breakdowns which might endanger health or safety.

A reasonable temperature
A reasonable temperature must be maintained during working hours and sufficient thermometers must be provided to enable people at work to determine the temperature in any workroom. The temperature should be comfortable without the need for special clothing. Special guidance is available for areas like food processing where it could be very hot or very cold.

Temperature should be at least 16°C, or where strenuous effort is involved 13°C.

Lighting
Lighting must be suitable and sufficient. This should be natural lighting so far as is reasonably practicable.

Emergency lighting
Emergency lighting shall be provided where persons are especially exposed to danger if artificial light fails. Lights should avoid glare and dazzle and should not themselves cause a hazard. They should not be obscured and be properly maintained.

Workplaces, furniture and fittings
Workplaces, furniture and fittings should be kept sufficiently clean. Surfaces inside buildings shall be capable of being kept sufficiently clean.

Floors
Floors should not be slippery and wall surfaces should not increase fire risks.

Wastes
Wastes should not be allowed to accumulate, except in suitable receptacles and should be kept free from offensive waste products and discharges.

Room dimensions
Room dimensions have to allow adequate unoccupied space to work in and to move freely. 11m^3 minimum per person is required, excluding anything over 3 m high and furniture etc.

Workstations

Workstations shall be suitable for any person in the workplace who is likely to work at that workstation and for any work that is likely to be done there.

Outside workstations

Outside workstations shall provide, so far as is reasonably practicable, protection from adverse weather; adequate means of escape in emergencies; and ensure that no person is likely to slip or fall.

Seating

Seating shall be provided where work can or must be done sitting and shall be suitable for the person as well as the work. A footrest shall be provided where necessary.

17.30.3 Safety – Accident Prevention – Regulations 5, 12, 13, 14, 15, 16, 17, 18 and 19

Maintenance

The workplace and equipment, devices and systems shall be maintained (including cleaned as appropriate) in an efficient state, efficient working order and in good repair, and where appropriate subject to a system of maintenance. This generally means planned rather than breakdown maintenance. Systems include ventilation, emergency lighting, safety fences, window cleaning devices, moving walkways are all given as examples.

Floors and traffic routes

Floors and the surface of traffic routes shall be suitably constructed for their intended purpose and free of slope and holes (unless fenced). They should not be uneven or slippery. The traffic routes should be of adequate width and height to allow people and vehicles to circulate safely with ease and they should be kept free of obstructions.

Additional precautions are necessary where pedestrians have to cross or share vehicle routes. Open sides of staircases should be fenced with an upper rail 900 mm or higher and a lower rail. Loading bays should have exits or refuges to avoid people getting crushed by vehicles.

Falls and falling objects

Measures shall be taken to prevent or control risks, so far as is reasonably practicable, to people, arising out of falling a distance likely to cause injury; and/or being struck by a falling object likely to cause injury. Any areas where such risks exist shall be clearly marked.

Tanks, pits

Where there is a risk of falling into a tank, pit or structure containing a dangerous substance that is likely to:

➤ scald or burn
➤ be poisonous or corrosive
➤ have an asphyxiating gas fume or vapour
➤ have any granular or free flowing substance likely to cause harm

measures must be taken to securely fence or cover the tank, pit or structure.

Ladders and roofs

Attention is drawn to precautions for persons working on roofs, or climbing on vehicles. Advice is given on the use of fixed ladders, stacking, racking and the temporary removal of fencing.

Glazing

Windows and transparent doors and partitions must be appropriately marked and protected against breakage.

Windows

Windows and skylights must open and close safely, and be arranged so that people may not fall out of them. They must be capable of being cleaned safely.

Traffic routes

Pedestrians and vehicles must be able to circulate safely. Separation should be provided between vehicle and people at doors, gateways and common routes. Workplaces should have protection from vehicles.

Doors and gates

Doors, gates and moving walkways have to be of sound construction and fitted with appropriate safety devices.

17.30.4 Welfare – provision of facilities – Regulations 20, 21, 22, 23, 24 and 25

Sanitary conveniences and washing facilities

Suitable and sufficient sanitary conveniences and washing facilities should be provided at readily accessible places. The facilities must be kept clean, adequately ventilated and lit. Washing facilities should have running hot and cold or warm water, soap and clean towels or other method of cleaning or drying. If necessary, showers should be provided. Men and women should have separate facilities unless each facility is in a separate room with a lockable door and is for use by only one person at a time.

Drinking water

An adequate supply of wholesome drinking water, with an upward drinking jet or suitable cups, should be provided. Water should only be supplied in refillable enclosed containers where it cannot be obtained directly from a mains supply.

Accommodation for clothing and facilities for changing

Adequate, suitable and secure space should be provided to store workers' own clothing and special clothing. The facilities should allow for drying clothing. Changing facilities should also be provided for workers who change into special work clothing.

Facilities to rest and to eat meals

Suitable and sufficient, readily accessible, rest facilities should be provided. Arrangements should include suitable facilities to eat meals; adequate seats with backrests and tables; means of heating food (unless hot food is available nearby) and making hot drinks.

Canteens or restaurants

Canteens and restaurants may be used as rest facilities provided there is no obligation to purchase food.

Suitable rest facilities

Suitable rest facilities should be provided for pregnant women and nursing mothers. They should be near sanitary facilities and where necessary, include the facility to lie down.

Non-smokers

Non-smokers should be protected from discomfort caused by tobacco smoke in rest areas and rest rooms.

17.30.5 References

The Workplace (Health, Safety and Welfare) Regulations 1992
ISBN 0-11-025804-5
Workplace Health, Safety and Welfare. Workplace (Health, Safety and Welfare) Regulations 1992 Approved Code of Practice and Guidance. L24, 1992 HSE Books
ISBN 0-7176-0413-6
Workplace health, safety and welfare. A short guide, INDG244, 1997 HSE Books
ISBN 0-7176-1328-3

17.31 Other relevant regulations in brief

There are a number of other Regulations which do not form part of the NEBOSH General Certificate syllabus. Nevertheless, they are important to a wider understanding of health and safety legislation. Very brief summaries are covered here.

17.31.1 Control of Asbestos at Work Regulations 2002

These Regulations came into effect on 21 November 2002 (Regulation 4, 21 May 2004 and Regulation 20, 21 November 2004). These Regulations require employers or the self-employed to prevent exposure at work to asbestos or, where this is not reasonably practicable, to ensure that exposure is kept as low as is reasonably practicable, and in any event below the control limit for asbestos. The Regulations give control limits for the different types of asbestos. Controls must be achieved by measures other than Respiratory Protective Equipment (RPE) and where this cannot be achieved RPE must be worn.

The use of asbestos in new products (with a few limited specialized uses for white asbestos up to 2005) and the sale of second-hand materials containing asbestos, is now banned under the Asbestos (prohibitions) Regulations 1992 as amended in 1999.

The Control of Asbestos at Work Regulations 2002 add a new duty (Regulation 4) to manage the risk from asbestos by requiring duty holders in control of non-domestic premises to:

➤ find out if there are any asbestos containing materials (ACM) in their premises, how much there is and its condition;

➤ to make a presumption that materials contain asbestos unless there is strong evidence to the contrary (e.g. the building was constructed after about 1985);

➤ to produce and maintain up to date a record of the location of ACMs or presumed ACMs;

➤ assess the risk from ACMs;

➤ prepare a plan to show how the risk will be managed;

➤ put the plan into action;

➤ review and monitor the plan as necessary;

➤ provide information on the location of ACM and its condition to anyone who is likely to work on it or disturb it.

The HSE have produced several publications setting out what the new duty will involve. (See INDG223(Rev2) and HSG227, there is also an ACoP L127 from HSE Books.)

17.31.2 Asbestos (Licensing) Regulations 1983 as amended in 1998

The latest amendment came into force on 1 February 1999. The regulations generally ban work with asbestos insulation, asbestos coating or asbestos insulation board, unless it is carried out by a contractor licensed by the HSE.

The purpose of the regulations is to allow the enforcing authorities to monitor closely any significant work with asbestos materials which poses a major threat to people's health.

Exemptions are:

➤ work of short duration

➤ work carried out by employers and the self-employed by themselves or using their own employees in their own premises

➤ air monitoring or collecting samples to identify asbestos.

All work with asbestos **must** comply with the Control of Asbestos at Work Regulations 2002.

17.31.3 Chemicals (Hazard Information and Packaging for Supply) Regulations 2002 as amended in 96, 97, 98, 99 and further expected

This series of Regulations implements a number of EU Directives relating to the supply of dangerous chemicals and has become known as CHIP 3. There are separate regulations covering the carriage of dangerous chemicals.

CHIP 3 suppliers are required to:

➤ identify the hazards (or dangers) of the chemicals they supply which is called classification;

➤ give information about the hazards to the people they supply chemicals to;

➤ package the chemicals safely.

These requirements are called 'supply requirements'.

17.31.4 Dangerous Substances and Explosive Atmospheres Regulations (DSEAR) 2002

These regulations came into force on 9th December 2002. They are designed to implement the safety requirements of the Chemical Agents and Explosive Atmospheres Directives. DSEAR deals with fires, explosions and similar energy-releasing events arising from dangerous substances. Following the introduction of DSEAR, the opportunity will be taken to modernize all the existing laws on the use and storage of petrol.

The regulations follow the modern risk assessment based approach. Technical and organizational measures are required to eliminate or reduce risks as far as is reasonably practicable. There is a requirement to provide equipment and procedures to deal with accidents and emergencies and also provide information and training to employees.

17.31.5 Electrical (Safety) Regulations 1994

These Regulations came into force in January 1995 and relate to the supply of electrical equipment with a working voltage between 50 and 1000 volts and are made under the Consumer Protection Act 1987. They apply to suppliers, which include both landlords and letting agents.

The Regulations apply to all mains voltage household electrical goods and require them to be safe so that there is no risk of injury or death to humans or pets, or risk of damage to property. They do not apply to fixed electrical wiring and built-in appliances like central heating systems. The Regulations also require that instructions be provided where safety depends on the user being aware of certain issues and equipment should be labelled with the CE marking.

There are other electrical consumer regulations, like The Plugs and Sockets Regulations 1994 and the Low Voltage Electrical Equipment Regulations 1989.

17.31.6 Gas appliances (Safety) Regulations 1992

The Gas Appliance Regulations cover the safety standards on new gas appliances which have to:

➤ satisfy safety and efficiency standards
➤ undergo type examination and supervision during manufacture
➤ carry the CE mark and specified information
➤ be accompanied by instructions and warnings in the language of destination.

17.31.7 Gas Safety (Installation and Use) Regulations 1998

The Installation and Use Regulations place duties on gas consumers, installers, suppliers and landlords to ensure that:

➤ only competent people work on gas installations (Registered with CORGI the Council for Registered Gas Installers)

> no one is permitted to use suspect gas appliances
> landlords are responsible, in certain cases, to make sure that fittings and flues are maintained
> with the exception of room sealed appliances there are restrictions on gas appliances in sleeping accommodation
> instantaneous gas water heaters must be room sealed or fitted with appropriate safety devices.

17.31.8 Control of Lead at Work Regulations 2002

These Regulations came into force in November 2002 and impose requirements for the protection of employees who might be exposed to lead at work and others who might be affected by the work. The regulations:

> require occupational exposure levels for lead and lead alkyls
> require blood-lead action and suspension levels for women of reproductive capacity and others
> re-impose a prohibition for women of reproductive capacity and young persons in specified activities
> require an employer to carry out a risk assessment
> require employers to restrict areas where exposures are likely to be significant if there is a failure of controls measures
> impose requirements for the examination and testing of engineering controls and RPE and the keeping of PPE
> impose sampling procedures for air monitoring
> impose requirements in relation to medical surveillance and records of air monitoring
> require information to be given to employees, including emergency warnings
> the keeping of records and identification of containers and pipes.

17.31.9 Occupiers Liability Acts 1957 and 1984 – Civil Law

The 1957 Act concerns the duty that the occupier of premises has towards visitors in relation to the condition of the premises and to things which have or have not been done to them. The Act imposes:

> a duty to take reasonable care to see that a visitor is reasonably safe in using the premises for the purpose for which they were invited or permitted by the occupier to be there
> the common duty of care will differ depending on the visitor, so a greater duty is owed to children
> an occupier can expect that a person in the exercise of his trade or profession will appreciate and guard against normal risks, for example, a window cleaner
> no duty of care is owed to some one exercising a public right of way.

The 1984 Act extends the duty of care to persons other than visitors, i.e. trespassers. The occupier has to take reasonable care in all the circumstances to see that non-visitors do not get hurt on the premises because of its condition or the things done or not done to it. The occupier must cover perceived dangers and must have reasonable grounds to know that the trespassers may be in the vicinity.

17.31.10 Control of Pesticides Regulations 1986

These Regulations made under the Food and Environment Protection Act 1985 all came into force by 1 January 1988. The Regulations apply to any pesticide or any substance used generally for plant control and protection against pests of all types, including antifouling paint used on boats. They do not apply to substances covered by other Acts like the Medicines Act 1968, The Food Safety Act 1990, those used in laboratories and a number of other specific applications.

No person may advertise, sell, supply, store, use a pesticide unless it has received ministerial approval and the conditions of the approval have been complied with. The approval may be experimental, provisional or full, and the Minister has powers to impose conditions. The Regulations also cover the need for users to be competent and have received adequate information and training. Certificates of competence (or working under the supervision of a person with a certificate) are required where pesticides, approved for agricultural use are used commercially.

17.31.11 Pressure Systems Safety Regulations 2000

These Regulations came into effect in February 2000 and replace the Pressure Systems and Transportable Gas Containers Regulations 1989. Transportable gas containers are covered by the Use of Transportable Pressure Receptacles Regulations 1996 (SI 1996 No 2092).

The aim of PSSR is to prevent serious injury from the hazard associated with stored energy as a result of a pressure system or one of its parts failing. The Regulations cover:

➤ steam at any pressure
➤ gases which exert a pressure in excess of 0.5 bar above atmospheric pressure
➤ fluids which may be mixtures of liquids, gases and vapours where the gas or vapour phase may exert a pressure in excess of 0.5 bar above atmospheric pressure.

With the exception of scalding from steam, the Regulations do not consider the effects of the hazardous contents being released following failure. The stored contents are of concern where they can accelerate wear and cause more rapid deterioration and an increased risk of failure.

17.31.12 Working Time Regulations 1998 as amended by 2001 Regulations

These Regulations came into force in October 1998 and, for specified workers, restrict the working week to 48 hours per 7-day period. Individuals can voluntarily agree to disapply the weekly working hours limit. Employers must keep a copy of all such individual agreements. Workers whose working time is not measured or predetermined, or who can themselves determine the duration of their working day are excepted from weekly working time, night work, rest periods and breaks.

The Regulations also cover:

> night workers average normal daily hours limited to eight
> health assessment for night workers
> minimum daily and weekly rest periods
> rest breaks at work
> paid annual leave.

17.31.13 Draft: The Work at Height Regulations

Expected to become law in late 2003 or early 2004 the draft regulations have been published. The regulations are being introduced to comply with the EU Temporary Work at Height Directive which must be implemented in Member States by July 2004.

The draft regulations indicate that they are intended to apply in all sectors of industry including construction. They follow an underpinning hierarchy. The aim of the hierarchy is to:

1. avoid work at height if possible;
2. prevent fall from height by the selection of the most suitable equipment; and
3. minimize the height and consequences of a fall should one occur.

The hierarchy recognizes that work at height can be performed safely in a number of ways depending on the circumstances.
The regulations cover:

> Avoidance of risk from work at height
> Selection of work equipment including access and egress
> Special requirements for:
>> Working platforms
>> Personal suspension equipment
>> Fall arrest equipment
>> ladders
> Rope access and positioning equipment
> Work on fragile materials
> Falling objects
> Organization of work
> Inspection of work platforms and personal suspension equipment
> Cessation of work in adverse weather conditions

Index